# NOTICE

THIS MANUAL IS SOLD FOR HISTORIC RESEARCH PURPOSES ONLY, AS AN ENTERTAINMENT. IT IS NOT INTENDED TO BE USED AS PART OF AN ACTUAL FLIGHT TRAINING PROGRAM. NO BOOK CAN SUBSTITUTE FOR FLIGHT TRAINING BY AN AUTHORIZED INSTRUCTOR. THE LICENSING OF PILOTS IS OVERSEEN BY ORGANIZATIONS AND AUTHORITIES SUCH AS THE FAA AND CAA. OPERATING AN AIRCRAFT WITHOUT THE PROPER LICENSE IS A FEDERAL CRIME.

USAF SERIES

T.O. 1T-38A-1

# T-38A/B FLIGHT MANUAL

AIRCRAFT

AF33(600)38940, F41608-78-D-A005

(AF59-1603 AND LATER AIRCRAFT)

THIS PUBLICATION REPLACES T.O. 1T-38A-1 DATED 1 JANUARY 1972, INCORPORATES T.O. 1T-38A-1-3, DATED 2 MAY 1977 AND SUPERSEDES OPERATIONAL SUPPLEMENTS T.O. 1T-38A-1S-96, -97, -98, -99, -102, -103, -104 AND -105. SEE TECHNICAL ORDER INDEX T.O. 0-1-1-5 AND SUPPLEMENTS THERETO FOR CURRENT STATUS OF TRAINER AIRCRAFT FLIGHT MANUALS, SAFETY AND OPERATIONAL SUPPLEMENTS, AND FLIGHT CREW CHECKLIST.

COMMANDERS ARE RESPONSIBLE FOR BRINGING THIS PUBLICATION TO THE ATTENTION OF ALL AFFECTED PERSONNEL.

PUBLISHED UNDER AUTHORITY OF THE SECRETARY OF THE AIR FORCE

T-38A 1-1 Q

1 JULY 1978

ISBN# 978-1-935327-64-6 1-935327-64-X



AIR FORCE 12 September 1978 — 15,000

# LIST OF EFFECTIVE PAGES

INSERT LATEST CHANGED PAGES. DESTROY SUPERSEDED PAGES.

NOTE: The portion of the text affected by the changes is indicated by a vertical line in the outer margins of the page. Changes to illustrations are indicated by miniature pointing hands. Changes to wiring diagrams are indicated by shaded areas.

Dates of issue for original and changed pages are:

Original . . . . . 0 . . . . . . 1 Jul 78

TOTAL NUMBER OF PAGES IN THIS PUBLICATION IS 248 CONSISTING OF THE FOLLOWING:

| Page No. | *Change No. |
|---|---|
| Title | 0 |
| A | 0 |
| i - iv | 0 |
| 1-1 - 1-40 | 0 |
| 2-1 - 2-15 | 0 |
| 2-16 Blank | 0 |
| 3-1 - 3-27 | 0 |
| 3-28 Blank | 0 |
| 4-1 - 4-24 | 0 |
| 5-1 - 5-7 | 0 |
| 5-8 Blank | 0 |
| 6-1 - 6-11 | 0 |
| 6-12 Blank | 0 |
| 7-1 - 7-4 | 0 |
| 8-1 | 0 |
| 8-2 Blank | 0 |
| 9-1 - 9-11 | 0 |
| 9-12 Blank | 0 |
| A-1 - A-2 | 0 |
| A1-1 - A1-9 | 0 |
| A1-10 Blank | 0 |
| A2-1 - A2-14 | 0 |
| A3-1 - A3-14 | 0 |
| A4-1 - A4-24 | 0 |
| A5-1 - A5-4 | 0 |
| A6-1 - A6-4 | 0 |
| A7-1 - A7-11 | 0 |
| A7-12 Blank | 0 |
| A8-1 - A8-4 | 0 |
| Index 1 - Index 8 | 0 |

**CURRENT FLIGHT CREW CHECKLIST**

T.O. 1T-38A-1CL-1 1 JULY 1978

PRINT CODE
**UN-10-B,**
UN-11-B, C-10-M

* Zero in this column indicates an original page.

# TABLE OF CONTENTS



T-38A 1-2C

**COMMENTS AND QUESTIONS.** Comments, questions, or recommended changes regarding any phase of the Flight Manual program are invited and should be forwarded in accordance with AFR 60-9 thru your Command Headquarters to San Antonio ALC/MMSRE, Kelly AFB, Texas 78241.

**SCOPE.** This manual contains all the information necessary for safe and efficient operation of the T-38A aircraft. These instructions do not teach basic flight principles, but are designed to provide you with a general knowledge of the aircraft, its flight characteristics, and specific normal and emergency operating procedures. Your flying experience is recognized, and elementary instructions have been avoided.

**PERMISSIBLE OPERATIONS.** The flight manual takes a "positive approach" and normally states only what you can do. Unusual operations or configurations which exceed the limitations as specified in this manual are prohibited unless specifically covered herein. Clearance must be obtained from San Antonio ALC/MMSRE, Kelly AFB TX 78241, through the respective major command before any questionable operation is attempted which is not specifically permitted in this manual.

**HOW TO BE ASSURED OF HAVING LATEST DATA.** Refer to T.O. 0-1-1-5 for a listing of all current flight manuals, safety supplements, operational supplements, and checklist. Also, check the flight manual cover page, the title block of each safety and operational supplement, and all status pages contained in the flight manual or attached to formal safety and operational supplements.

**SAFETY AND OPERATIONAL SUPPLEMENTS.** Safety and Operational Supplements are used to get information to you in a hurry. Safety supplements concern safety of flight items. Operational supplements are issued as an expeditious means of reflecting information when mission essential operational procedures are involved. Supplements are issued by teletype (interim) or by printed copy (formal) depending upon the urgency. Supplements are numbered consecutively regardless of whether it is safety or operational. File supplements in reverse numerical order in the front of the Flight Manual.

**CHECKLISTS.** The flight manual contains itemized procedures with necessary amplifications. The checklist contains itemized procedures without the amplifications. Whenever a supplement affects the abbreviated checklist, write in the applicable change on the affected checklist page or if a checklist page is included, cut it out and insert it in your checklist.

**HOW TO GET PERSONAL COPIES.** Each flight crew member is entitled to personal copies of the flight manual, supplements, and checklist. Your publication distribution officer should be contacted to fill your technical order request. T.O. 00-5-1 and T.O. 00-5-2 give detailed information for ordering publications.

**YOUR RESPONSIBILITY—TO LET US KNOW.** Every effort is made to keep the Flight Manual current. However, we cannot correct an error unless we know of its existence. It is essential that you do your part. Any comments, questions, or recommendations should be forwarded, using AF Form 847 in accordance with T.O. 00-5-1, through your Command Headquarters, to: San Antonio ALC/MMSRE, Kelly AFB, Texas 78241.

**CHANGE SYMBOLS.** Changed text is indicated by a black vertical line in either margin of the page. The change symbol shows what part has been changed in the current change.

**WARNINGS, CAUTIONS, AND NOTES.** For your information, the following definitions apply to the "Warnings," "Cautions," and "Notes" found throughout the manual:

**WARNING**

Operating procedures, practices, etc, which will result in personal injury or loss of life if not carefully followed.

**CAUTION**

Operating procedures, practices, etc, which if not strictly observed will result in damage to equipment.

**NOTE**

An operating procedure, condition, etc, which it is essential to emphasize.

## GROUP CODING.

Aircraft having different or additional systems and equipment have been block coded to avoid listing aircraft serial numbers. The Air Force serial numbers of the aircraft included in each block are as follows:

| Block | Air Force Serial Numbers |
|---|---|
| 20 | AF59-1603 thru AF59-1606 |
| 25 | AF60-547 thru AF60-553 |
| 30 | AF60-554 thru AF60-561 |
| 35 | AF60-562 thru AF60-596 |
| 40 | AF61-804 thru AF61-947 |
| 45 | AF62-3609 thru AF62-3752 |
| 50 | AF63-8111 thru AF63-8247 |
| 55 | AF64-13166 thru AF64-13305 |
| 60 | AF65-10316 thru AF65-10475 |
| 65 | AF66-4320 thru AF66-4389 and AF66-8349 thru AF66-8404 |
| 70 | AF67-14825 thru AF67-14859 and AF67-14915 thru AF67-14958 |
| 75 | AF68-8095 thru AF68-8217 |
| 80 | AF69-7073 thru AF69-7088 |
| 85 | AF70-1549 thru AF70-1591 and AF70-1949 thru AF70-1956 |

**US NAVY AIRCRAFT.**

The following aircraft operated by the US Navy are referred to in the manual by block numbers or by their Air Force serial numbers. These USAF serial numbers correspond to the following USN Bureau numbers:

| USAF Serial Number | USN Bureau Number |
|---|---|
| 65-0327 | 10327 |
| 68-8209 | 158198 |
| 68-8212 | 158199 |
| 68-8214 | 158200 |
| 68-8216 | 158201 |

## CODE SYSTEM

(A) T-38A
(B) T-38B Lead in Fighter (LIF)

**NOTE**

- Text, illustrations, and charts applicable to all aircraft are not coded.
- When complete paragraphs are affected, the appropriate code will appear opposite the heading.
- Notes, cautions, and warnings are treated as individual paragraphs with regard to coding.
- Steps of a procedure that have the code preceding the action item apply only to the individual model aircraft.

# THE AIRCRAFT

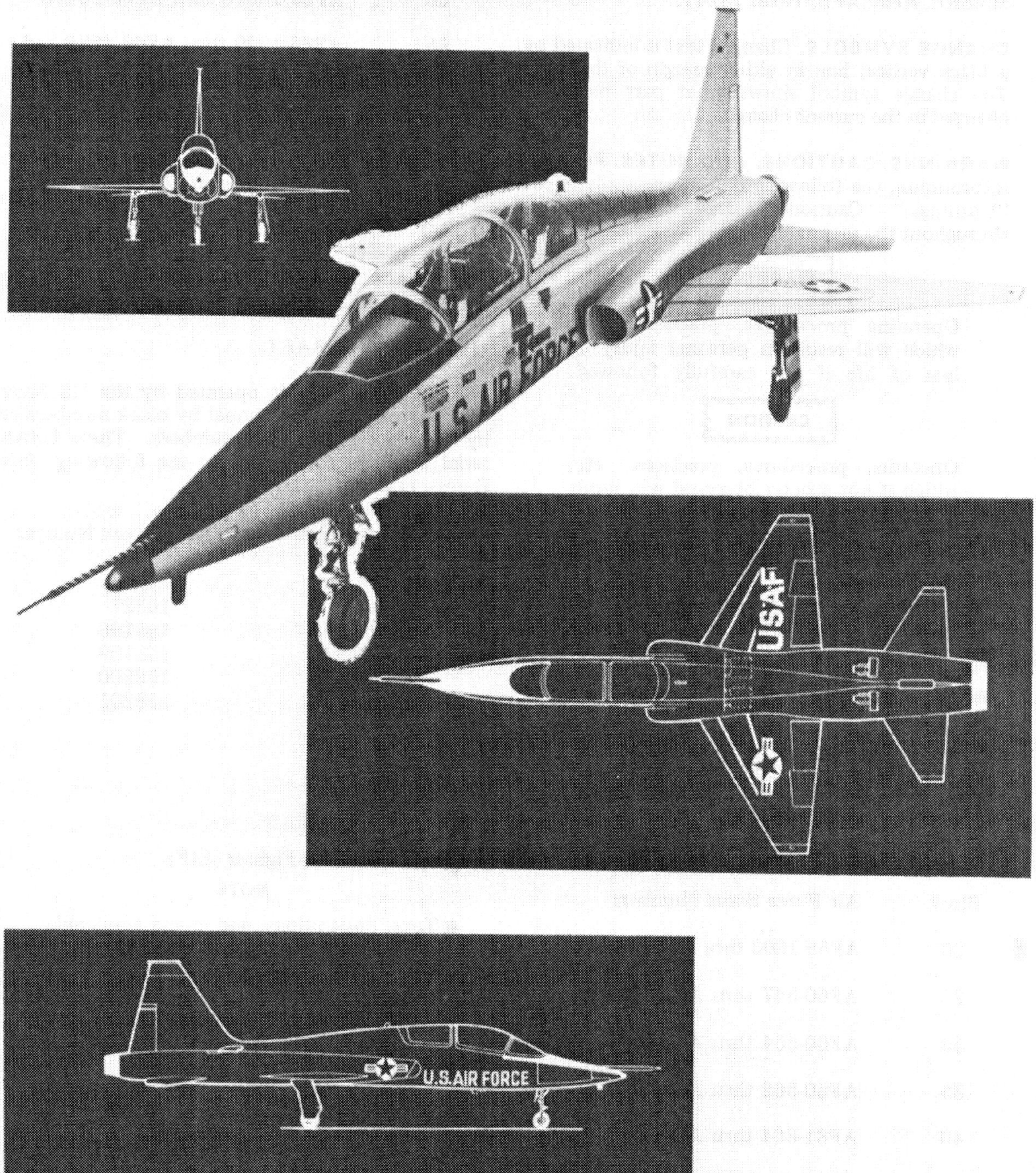


T-38A 1-5 J

# DESCRIPTION

## SECTION I

T-38A 1-100

TABLE OF CONTENTS



### THE AIRCRAFT.

The T-38A aircraft, produced by Northrop Corporation, Aircraft Division, is a two-place, twin-turbojet supersonic trainer. Each cockpit contains an individual jettisonable canopy and ejection seat. A cabin air-conditioning and pressurization system conditions and pressurizes the air in both cockpits. The fuselage is an area-rule (coke bottle) shape, with moderately swept-back wings and empennage. The aircraft is equipped with an all-movable horizontal tail. A speed brake is located on the lower surface of the fuselage center section. The tricycle landing gear has a steerable nosewheel. All flight control surfaces are fully powered by two independent hydraulic systems.

The T-38B is modified by the addition of a centerline pylon and a noncomputing gunsight. Some aircraft have the MXU-553 flight loads recorder installed. Cockpits contain additional controls for the armament system.

### AIRCRAFT DIMENSIONS.

The overall dimensions of the aircraft with normal tire and strut inflation are:

| | |
|---|---|
| Length | 46 ft 4 in. |
| Wingspan | 25 ft 3 in. |
| Height | 12 ft 11 in. |
| Tread | 10 ft 9 in. |
| Wheelbase | 19 ft 5 in. |

# GENERAL ARRANGEMENT DIAGRAM
(TYPICAL)

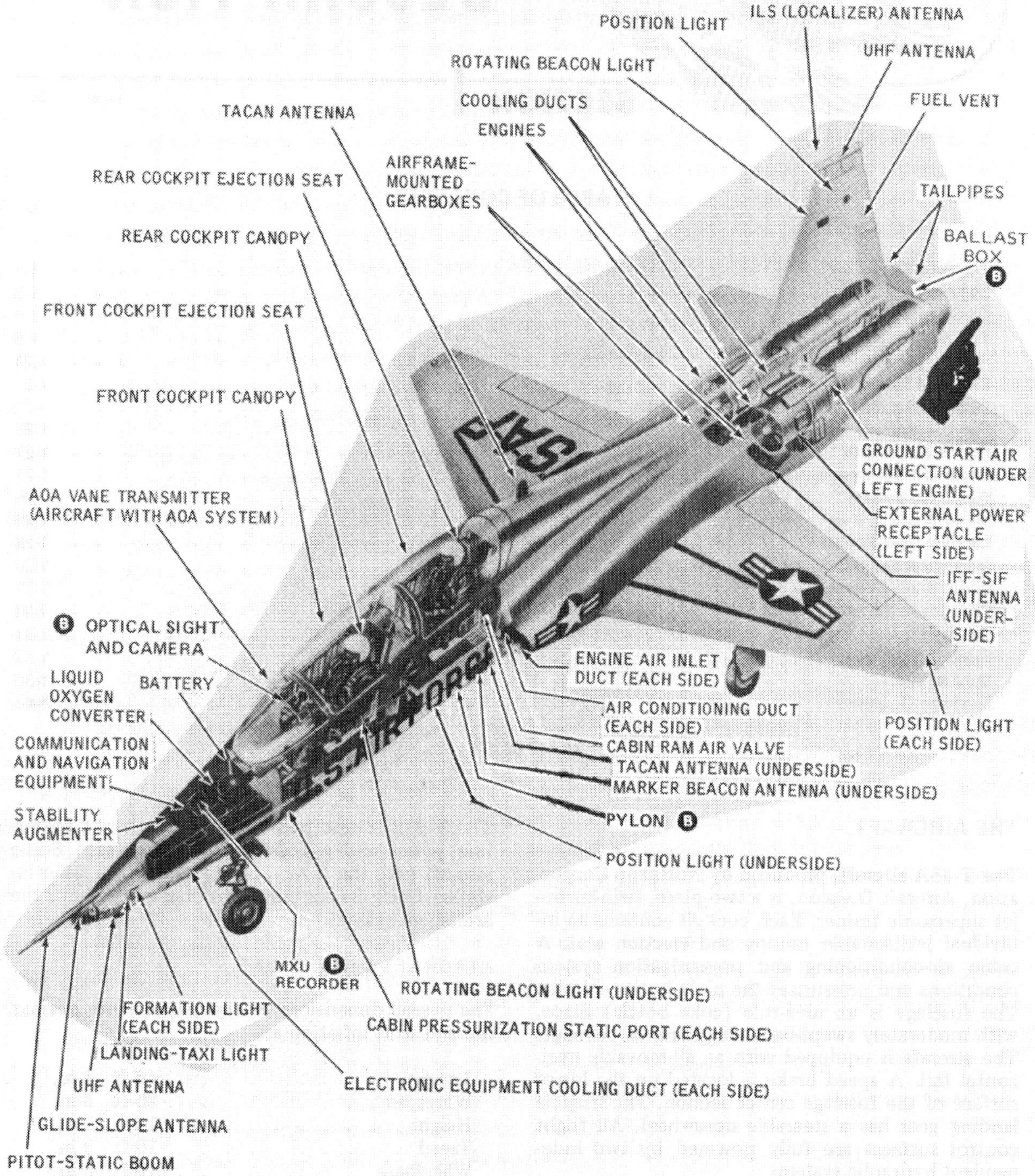


Figure 1-1.

# J85-GE-5 SERIES ENGINE

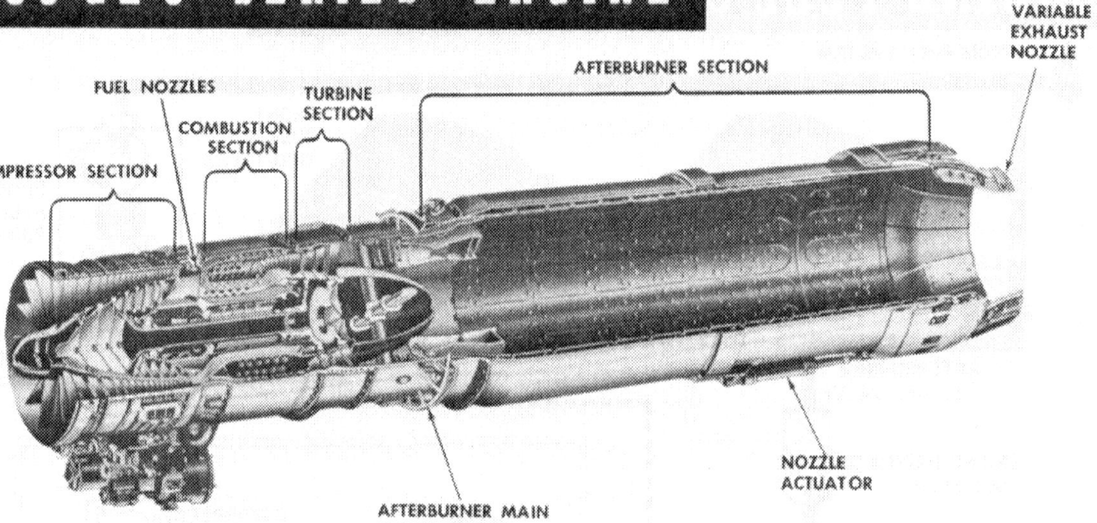


Figure 1-2.

## AIRCRAFT GROSS WEIGHT.

The gross weight of the aircraft fully fueled and including two aircrew is approximately 12,500 pounds. The average gross weight of the B with the AF-B-37K-1 and four practice bombs is 12,700 pounds. The weight with the SUU-20 with six practice bombs and four 2.75 " rockets is 13,060 pounds. The weight with a fully loaded SUU-11 is 12,850 pounds. These weights shall not be used for computing aircraft performance or for any type operation.

## ENGINES.

The aircraft is powered by two J85-GE-5 series, eight-stage, axial-flow, turbojet engines (figure 1-2). Sea level, standard day, static thrust for an installed engine is approximately 2050 pounds at MIL power and approximately 2900 pounds at full MAX power. Air enters thru the variable inlet guide vanes, which direct the flow of air into the compressor. The automatic positioning of the inlet guide vanes and air bleed valves assists in regulating compressor airflow to maintain compressor stall-free operation. Two turbine wheels and the compressor rotor stages are mounted on the same shaft. The exhaust gases are discharged thru a variable area exhaust nozzle. An exhaust gas temperature ($T_5$) sensing system varies the nozzle area to maintain exhaust gas temperature within limits at both MIL and MAX range throttle positions.

## ENGINE FUEL CONTROL SYSTEM.

Each engine has a main fuel control system and an afterburner fuel control system (figure 1-3). The main fuel control system consists primarily of a two-stage engine-driven pump, a main fuel control, and an over-speed governor.

## MAIN FUEL CONTROL.

The main fuel control selects engine power by metering fuel to the main engine combustor as a function of throttle position, engine inlet air temperature, compressor discharge pressure, and engine speed. The control performs the following functions automatically:

a. Regulates engine speed at the selected throttle position, limit engine minimum speed at IDLE and engine maximum speed at MIL and MAX range power.

b. Limits main engine fuel flow to safe levels during starts and during rapid throttle changes, providing protection from overtemperature, stalls, and flameouts.

c. Limits main engine fuel flow to a preset minimum by holding combustor fuel-air ratio at or above the proper level for low power settings and for engine restart during flight.

# ENGINE FUEL CONTROL SYSTEM

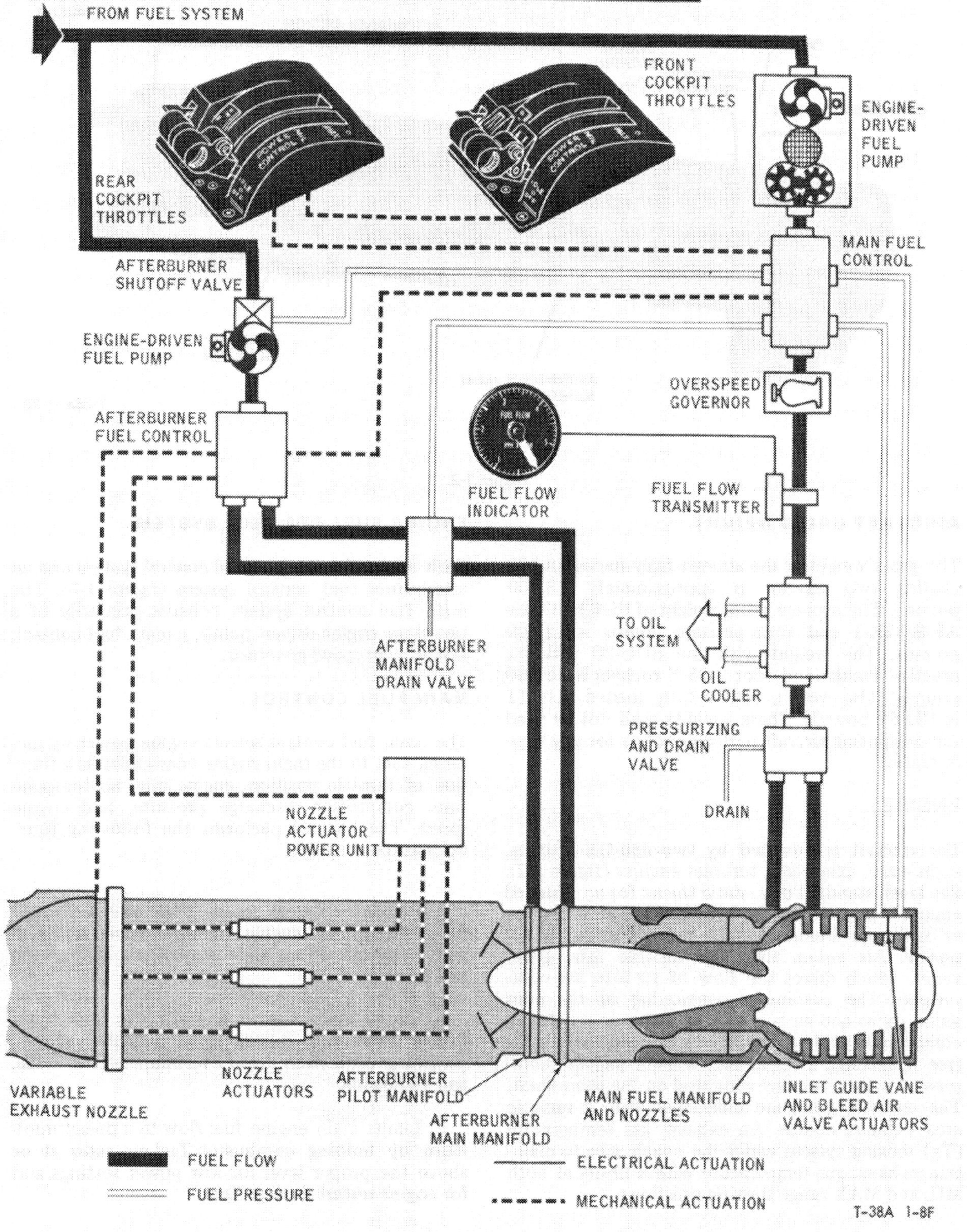


Figure 1-3.

d. Correctly positions the compressor inlet guide vanes and air bleed valves.

AFTERBURNER SYSTEM.

Each afterburner system contains an igniter plug, afterburner pilot manifold, afterburner main manifold, and afterburner fuel pump and control. Afterburner operation is initiated by advancing throttle from the MIL detent into the MAX range. Thrust is variable within MAX range. The total rate of fuel flow at full MAX position for each engine at sea level on a standard day is approximately 7300 pounds per hour with the aircraft at rest and 11,400 pounds per hour at mach 1.

AFTERBURNER FUEL CONTROL.

The primary function of the afterburner fuel control is to initiate and schedule fuel flow to the afterburner main and pilot spraybars. Fuel flow is metered as a function of throttle position and compressor discharge pressure. The control also senses and regulates variable area nozzle position and automatically limits fuel flow to prevent overtemperature in case of a nozzle actuating system malfunction or during rapid throttle advances into MAX range.

THROTTLES.

The throttles (figure 1-4) are provided with a roller ramp-type force gradient, which must be overcome to move the throttles from MIL into MAX range or from IDLE to OFF. The throttles in the front cockpit are equipped with fingerlifts which must be raised before the throttles in either cockpit can be retarded past the IDLE roller ramp to OFF. Friction is ground adjustable only. The throttles, when placed at OFF, mechanically shut off fuel to the engine at the main fuel control and electrically shut off fuel to the engine at the fuel shutoff valves.

NOTE

Throttle movement should be conservative to help minimize blade failures. Abrupt or rapid throttle movements should be avoided. Throttle bursts (throttle movement in one second or less) from idle RPM to MIL should be avoided if possible. These procedures will allow the variable exhaust nozzle to keep pace and match the fuel flow and help to minimize the possibility of compressor blade failures.

PNEUMATIC SYSTEM.

Air taken from the eighth stage compressor is used for hydraulic reservoir and cabin pressurization, air conditioning systems, canopy defogging, engine anti-icing, canopy seal inflation, and for the anti-G system.

ENGINE START AND IGNITION SYSTEM.

Engine starts require compressor motoring (low pressure air supply), DC power to energize the ignition holding relay and AC power for ignitor firing. Two engine start pushbuttons (figures 1-9, 1-10) are located in the left subpanel of each cockpit. For ground starts only, a diverter valve is automatically positioned to direct air to the selected engine. Momentarily pushing a start button positions the diverter valve and arms the ignition circuit for approximately 30 seconds. Moving the throttle to IDLE energizes the ignition exciter, firing main and afterburner igniters and starting fuel flow to the engine. If the left start button is pressed before the right engine 30-second cycle is completed, the diverter valve will move to the neutral position and will remain at neutral until the right engine 30-second start cycle is completed. The valve will then divert air to the left engine, but only for the time that is remaining on the left engine start cycle,

THROTTLE QUADRANT

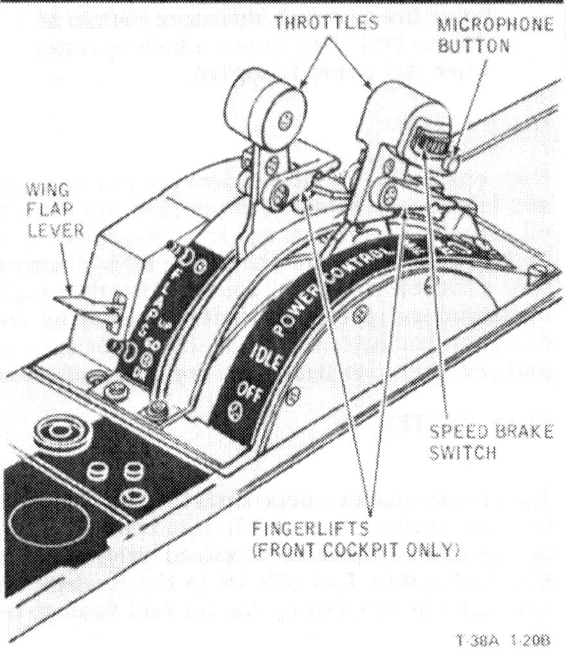


Figure 1-4.

which started when the left engine start button was pushed. The resulting air supply loss may cause an overtemperature condition. Moving the throttles to MAX range energizes the main and afterburner igniters for 30 seconds. The throttles must be retarded to MIL and returned to MAX range to recycle the starting timer. With the throttles at MAX range, the igniters may be energized for longer than 30 seconds by pushing and holding the engine start buttons. AC power from a battery-operated static inverter (figure 1-15) may be used for ground start (one engine) or air starts (either engine). For battery start, the right engine should be started first, as the static inverter supplies ac power for the right engine instruments during the start cycle.

## ENGINE INSTRUMENTS.

A full complement of engine instruments is provided in each cockpit. The front cockpit indicators are primary, since the indicators in the rear cockpit merely repeat the pointer positions of those in the front cockpit. The tachometers are powered independently of the aircraft electrical system; the nozzle position indicators require DC power only. All other engine and quantity indicating instruments require AC power from their respective busses. The right engine instruments may also receive AC power from a battery powered static inverter which is activated upon initiation of the engine start sequence when normal AC power is not available.

**NOTE**

Some front cockpit indicators contain an ON or OFF flag system which operates when AC power is applied.

## OIL SYSTEM.

Each engine has an independent integral oil supply and lubrication system. The reservoir has a normal oil capacity of 4 quarts and an air expansion space of 1 quart. Heat from the engine oil is dissipated thru a fuel-oil cooler. Oil consumption thru engine operation and overboard venting caused by condensation and aerobatic flight should not exceed 1 pint per hour. See figure 1-25 for oil specification.

## FUEL SYSTEM.

The aircraft has an independent fuel supply system for each engine (figure 1-13), interconnected by a dc electrically operated crossfeed valve. The left and right system fuel cells are in the fuselage. The left engine is supplied by the forward fuselage cell and the forward and aft dorsal cells; the right engine, by the center and aft fuselage cells. A single ac electrically driven fuel boost pump in each system supplies fuel under pressure to the engine-driven fuel pump during normal operation. The left system boost pump is in the inverted flight compartment of the forward fuselage cell, and the right system boost pump is in the inverted flight compartment of the aft fuselage cell. Without the aid of the boost pump, each engine can be supplied with fuel by gravity flow from its respective system. Normally, sufficient fuel will flow by gravity to maintain MAX power from sea level up to approximately 25,000 feet; however, by specifications, gravity flow is guaranteed only to 6,000 feet, and flameouts have occurred as low as 15,000 feet. Thru crossfeed operation, both systems may supply fuel to either engine with or without boost pump pressure (one engine off, crossfeed on, boost pumps functioning or failed). Also, one system under boost pump pressure will supply fuel to both engines. (Both engines operating, crossfeed on, one boost pump OFF.) Caution lights indicate fuel low level and low pressure conditions. See figure 1-25 for fuel specification and for fuel quantity data. Refer to fuel management, Section VII, for proper cross-feed operation.

## BOOST PUMP SWITCHES.

Two guarded boost pump switches (figure 1-13), one for each fuel system, are located on the right subpanel of the front cockpit. All fuel pump circuit breakers ( figure 1-16) should be closed before operating boost pumps.

## BOOST PUMP INDICATOR LIGHTS.

Two boost pump indicator lights (figure 1-13), one for each boost pump, are located on the right subpanel of the rear cockpit. An indicator light illuminates when the corresponding boost pump switch is placed at OFF.

## FUEL PRESSURE CAUTION LIGHTS.

Two fuel low pressure caution lights, placarded LEFT FUEL PRESS, RIGHT FUEL PRESS (figure 1-13), are located on the right console of each cockpit. The caution light will illuminate when the warning system detects a low-pressure condition and will remain illuminated as long as the low-pressure condition exists. The caution lights may blink when afterburner power is selected. Various other conditions may cause the lights to blink; this blink is not an indication of boost pump failure.

# COCKPIT ARRANGEMENT — FRONT (TYPICAL)

A

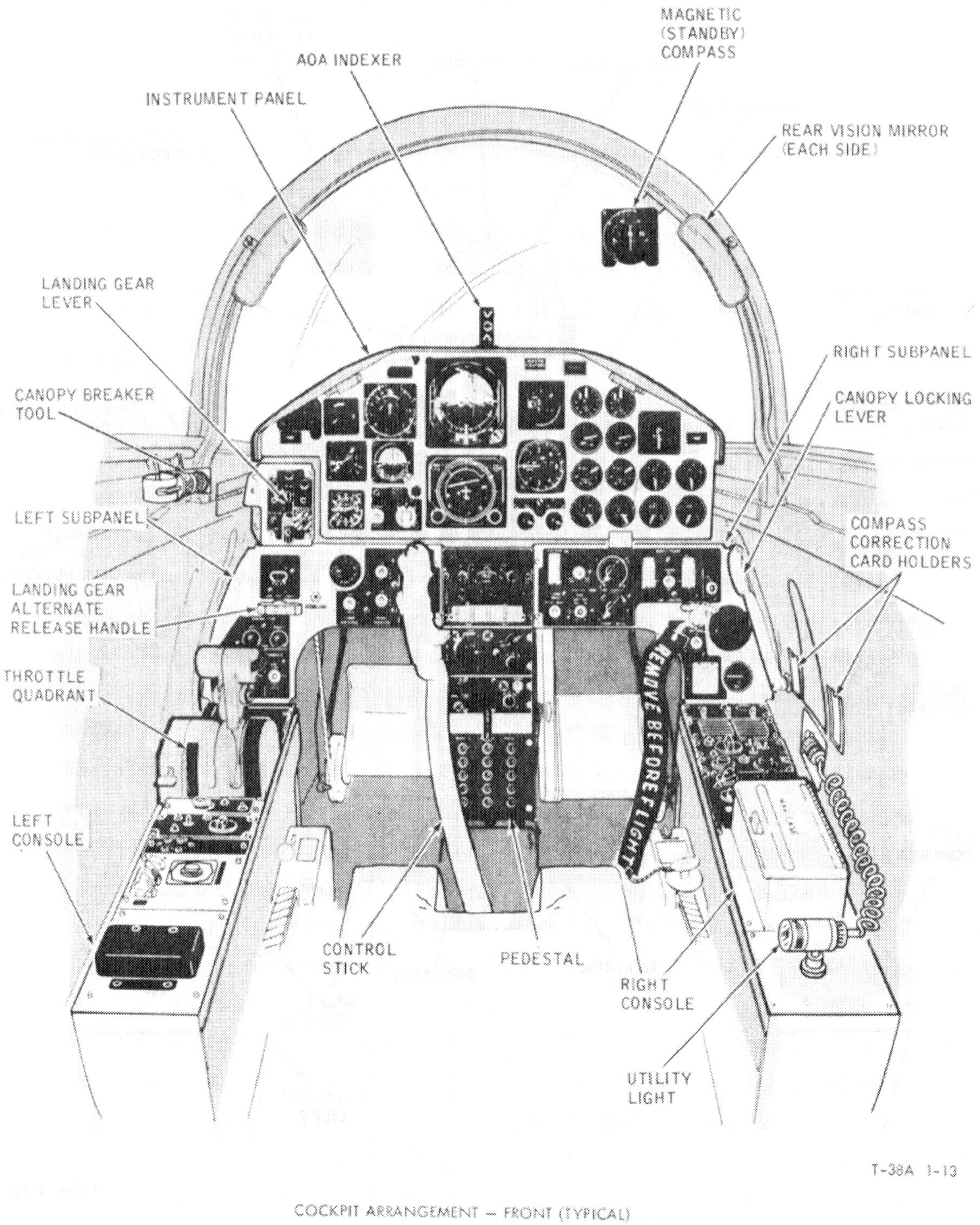


COCKPIT ARRANGEMENT — FRONT (TYPICAL)

**Figure 1-5. (Sheet 1 of 2)**

# COCKPIT ARRANGEMENT — FRONT (TYPICAL)

B

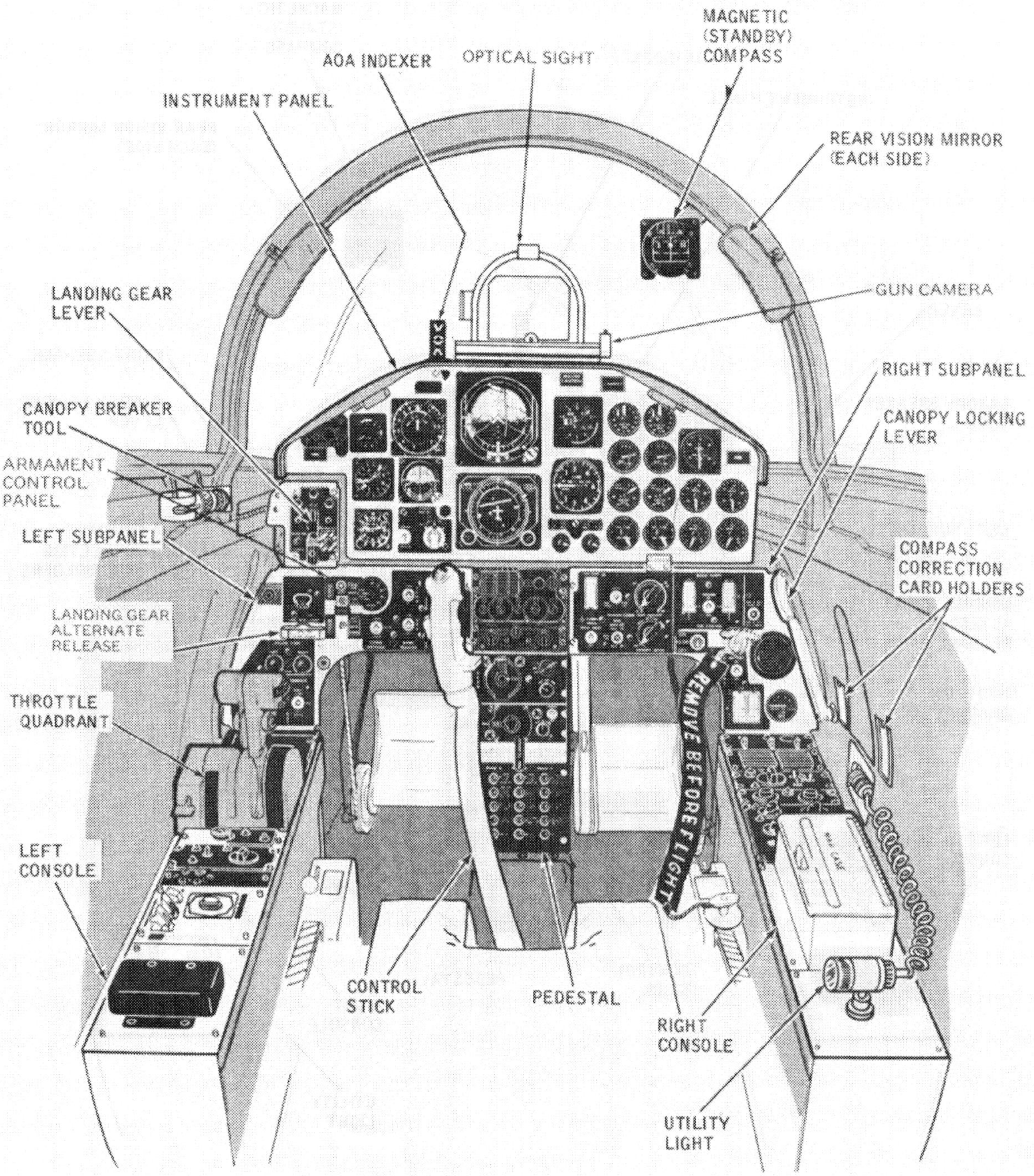


COCKPIT ARRANGEMENT — FRONT (TYPICAL)

**Figure 1-5. (Sheet 2 of 2)**

# COCKPIT ARRANGEMENT—REAR (TYPICAL)

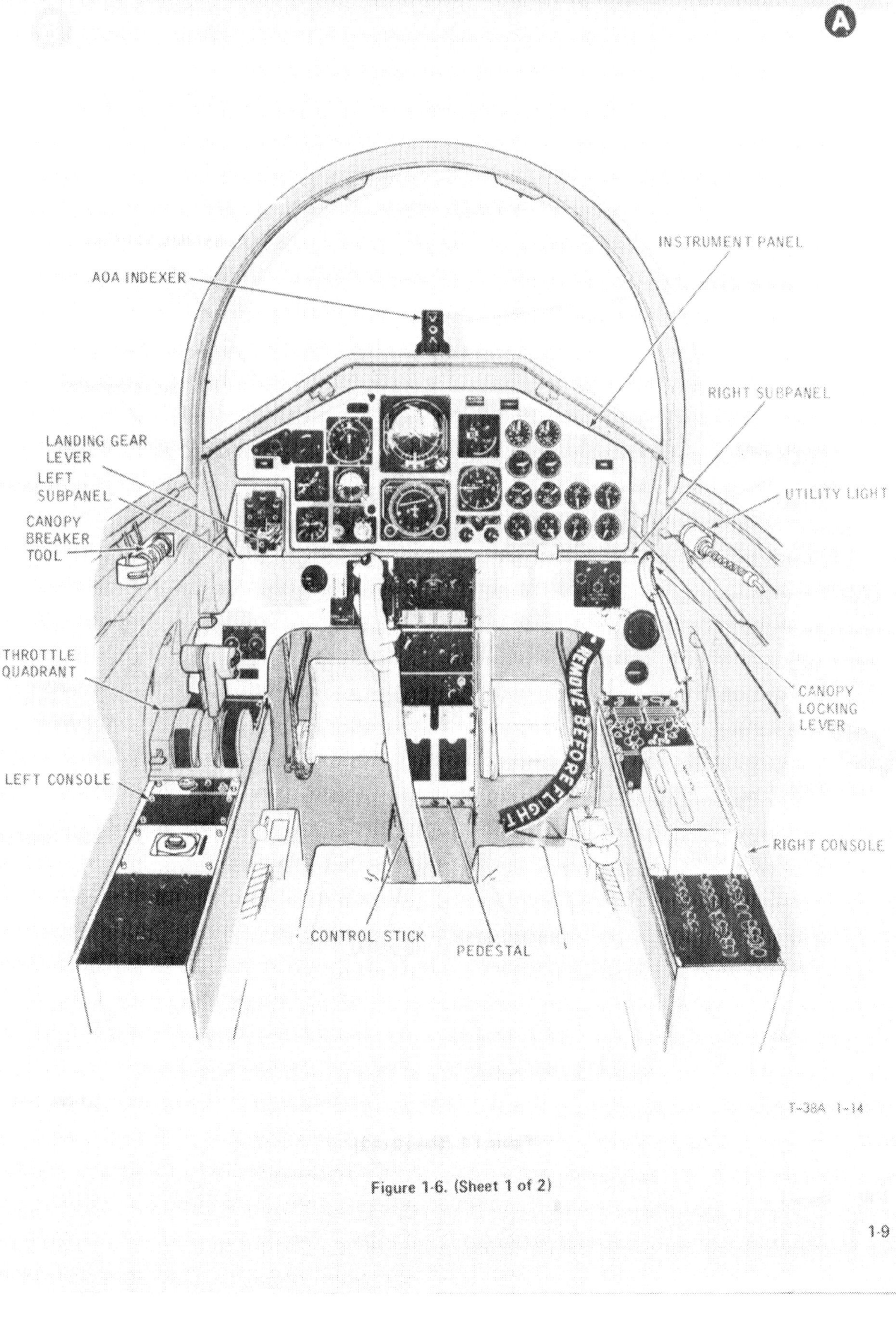


Figure 1-6. (Sheet 1 of 2)

# COCKPIT ARRANGEMENT — REAR (TYPICAL)

B

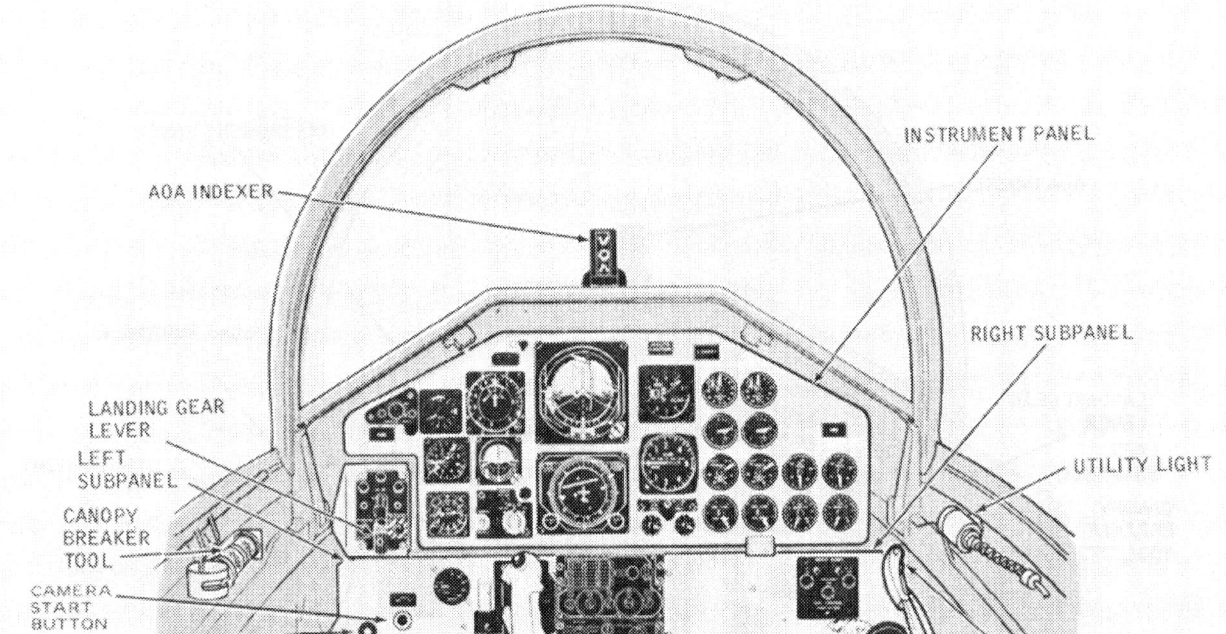


Figure 1-6. (Sheet 2 of 2)

# INSTRUMENT PANEL—BOTH COCKPITS (TYPICAL)

**AIRCRAFT WITH AOA SYSTEM**

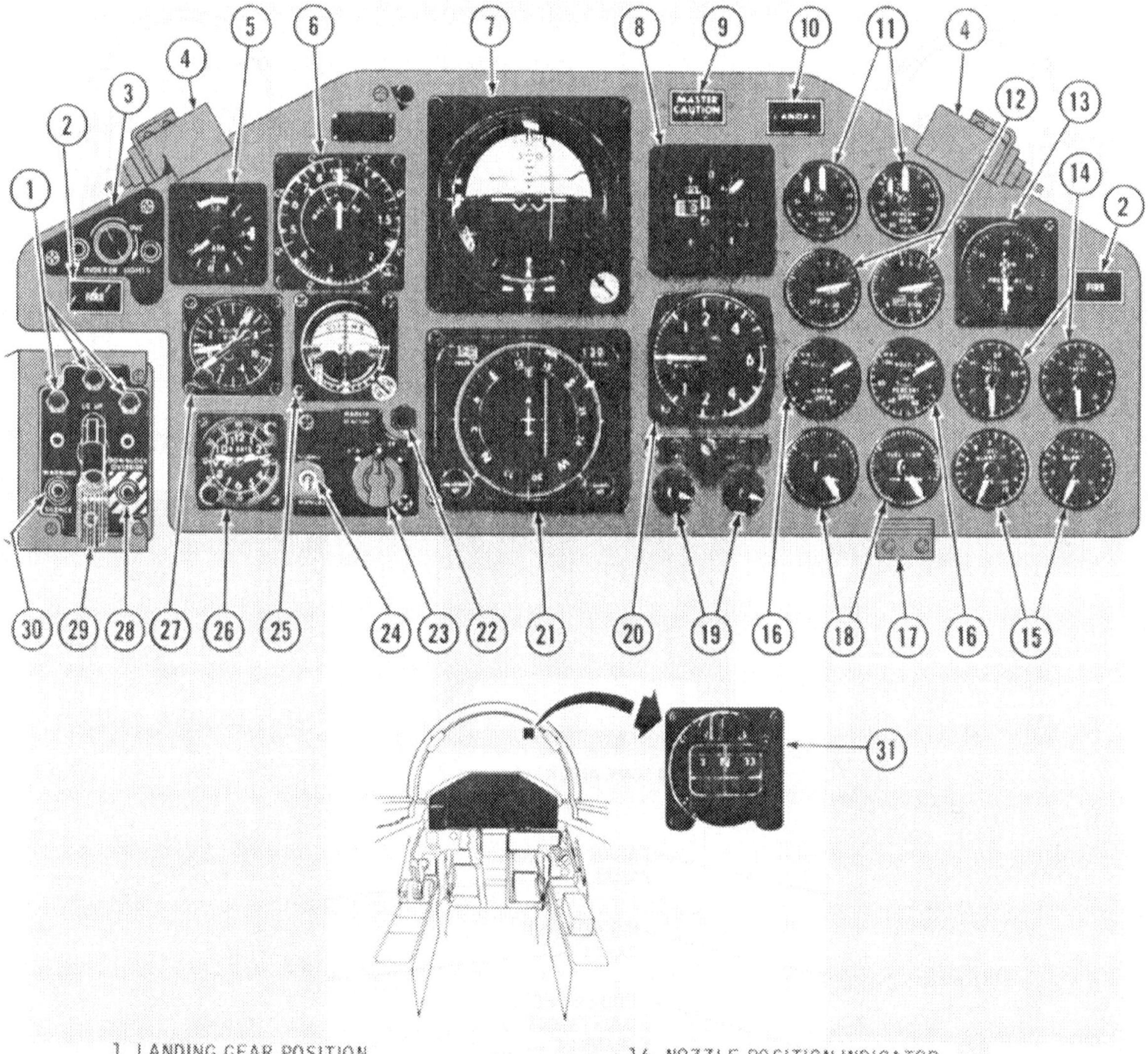

1 LANDING GEAR POSITION INDICATOR LIGHTS
2 ENGINE FIRE WARNING LIGHT
3 AOA INDEXER DIMMER
4 FLOODLIGHT
5 AOA INDICATOR
6 AIRSPEED/MACH INDICATOR
7 ATTITUDE DIRECTOR INDICATOR
8 ALTIMETER
9 MASTER CAUTION LIGHT
10 CANOPY WARNING LIGHT
11 ENGINE TACHOMETERS
12 EXHAUST GAS TEMPERATURE INDICATORS (Warning flags front cockpit only)
13 CABIN ALTIMETER (Front cockpit only)
14 OIL PRESSURE INDICATORS
15 FUEL QUANTITY INDICATORS
16 NOZZLE POSITION INDICATOR
17 CARD CLIP
18 FUEL FLOW INDICATORS
19 HYDRAULIC PRESSURE INDICATORS
20 VERTICAL VELOCITY INDICATOR
21 HORIZONTAL SITUATION INDICATOR
22 MARKER BEACON LIGHT
23 NAVIGATION MODE SWITCH
24 STEERING MODE SWITCH
25 STANDBY ATTITUDE INDICATOR
26 CLOCK
27 ACCELEROMETER
28 DOWNLOCK OVERRIDE BUTTON
29 LANDING GEAR LEVER
30 LANDING GEAR WARNING SILENCE BUTTON
31 MAGNETIC COMPASS (Front cockpit only)

T-38A 1-4

Figure 1-7.

# PEDESTALS (TYPICAL)

A B

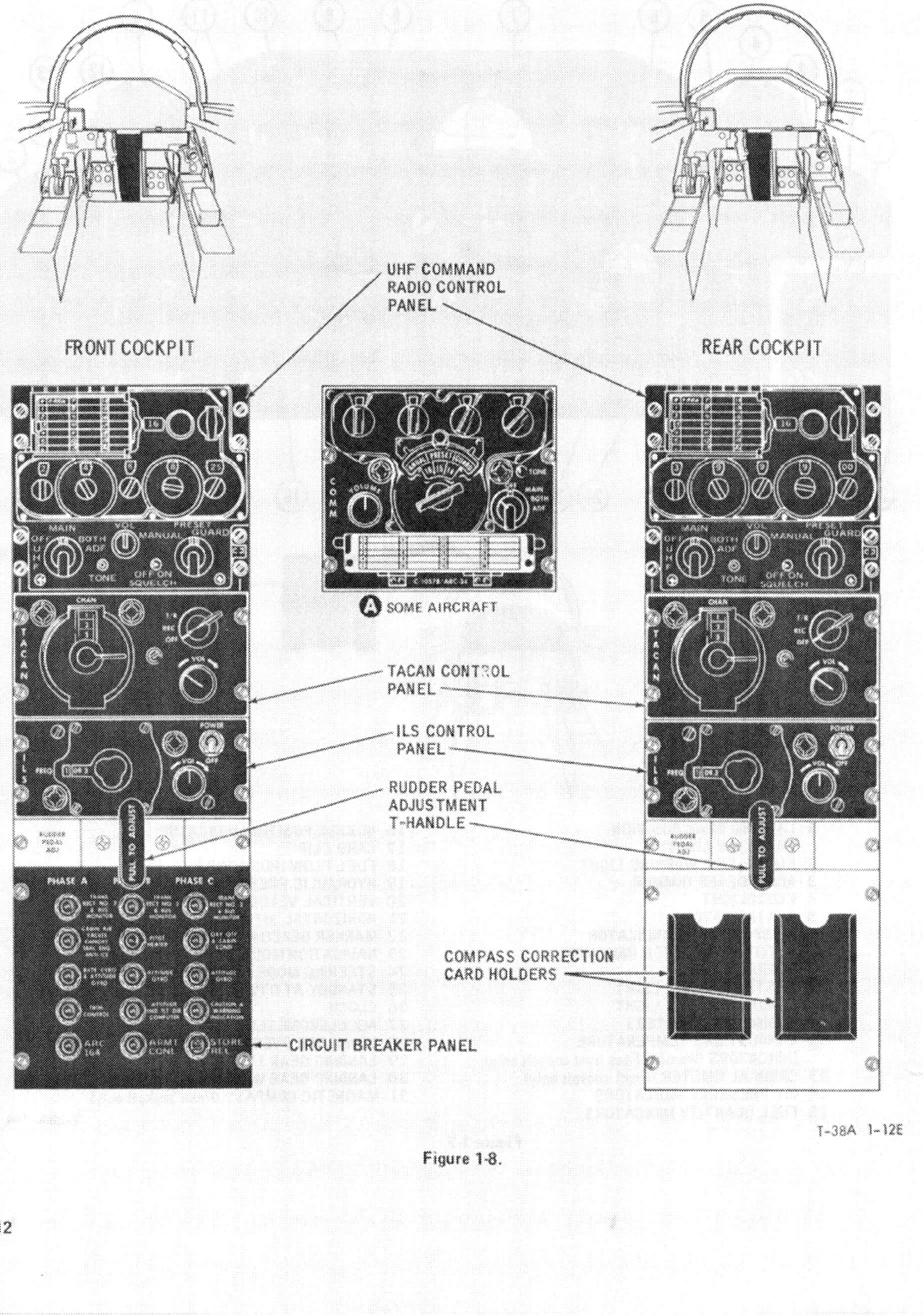


Figure 1-8.

# SUBPANELS—FRONT COCKPIT (TYPICAL)

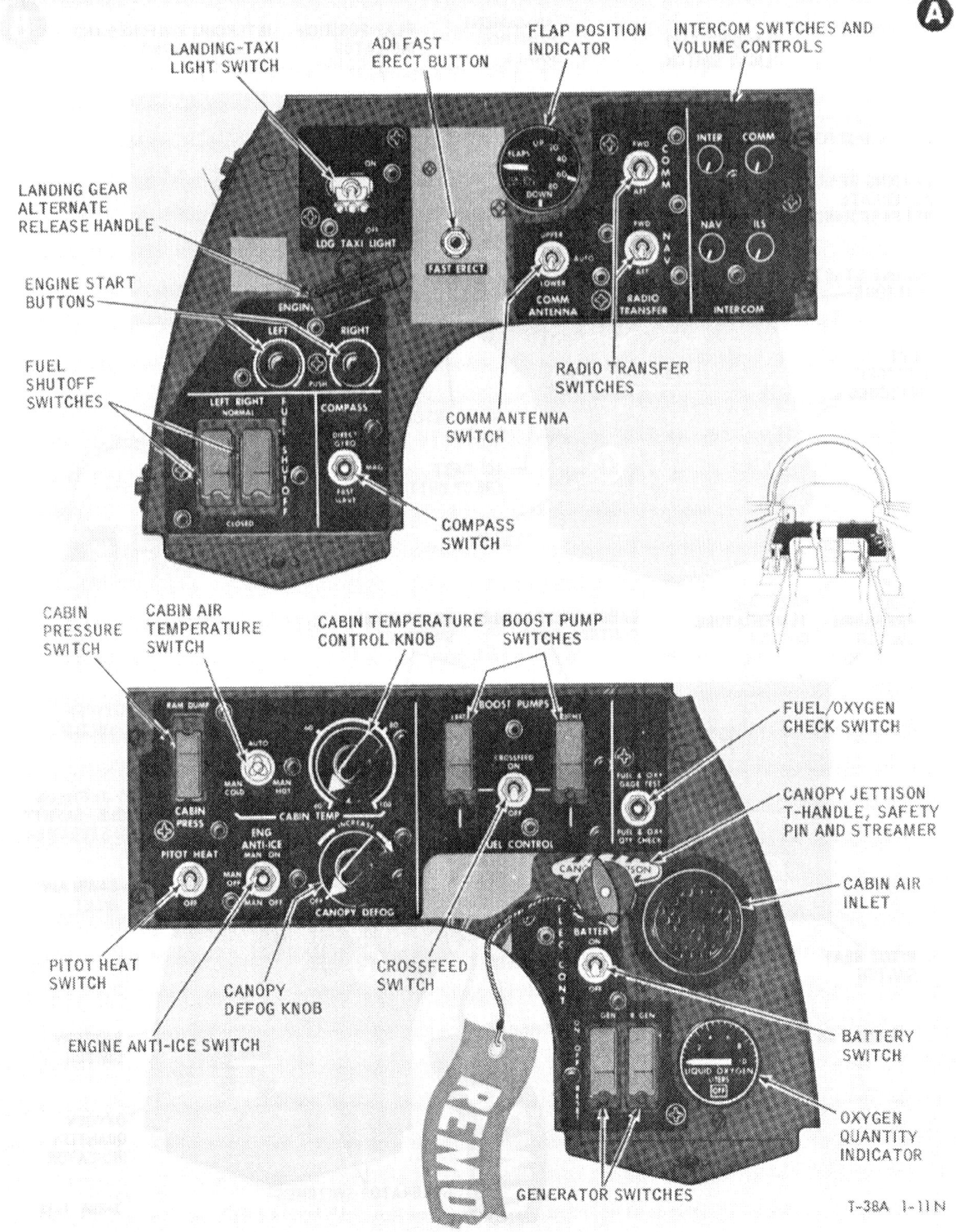


Figure 1-9. (Sheet 1 of 2)

# SUBPANELS—FRONT COCKPIT (TYPICAL)

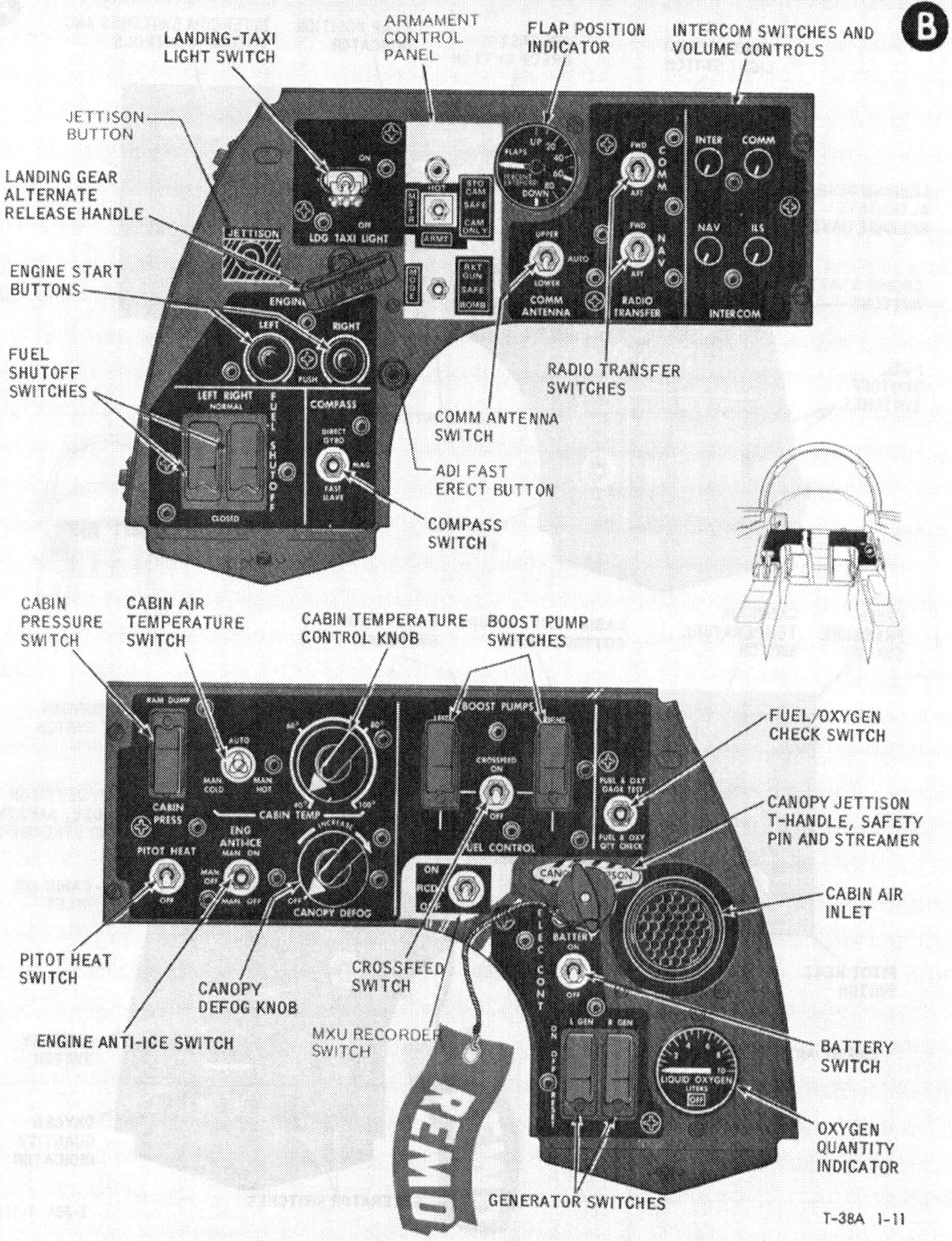


Figure 1-9. (Sheet 2 of 2)

# SUBPANELS—REAR COCKPIT (TYPICAL)

A

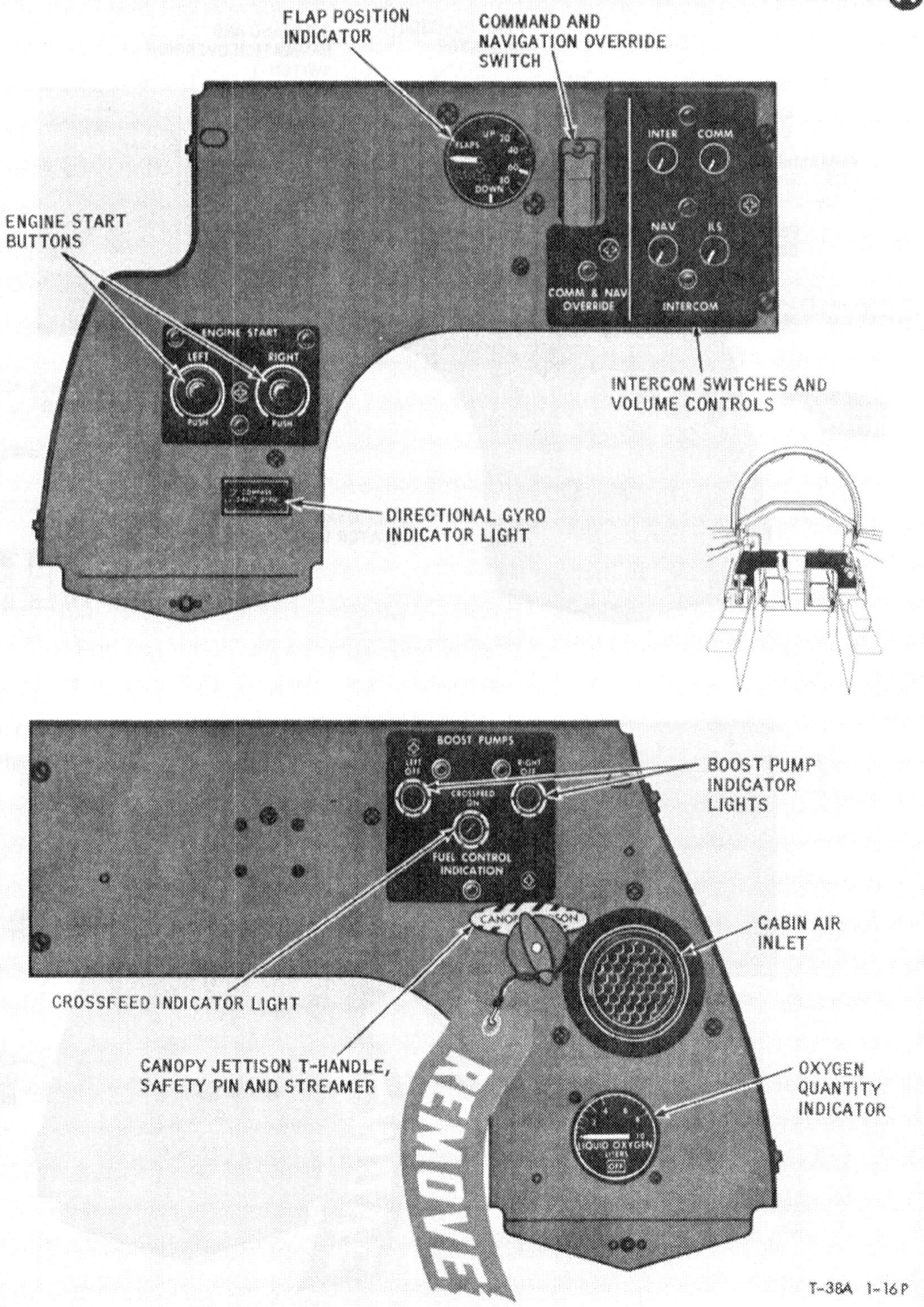


Figure 1-10. (Sheet 1 of 2)

# SUBPANELS—REAR COCKPIT (TYPICAL)

B

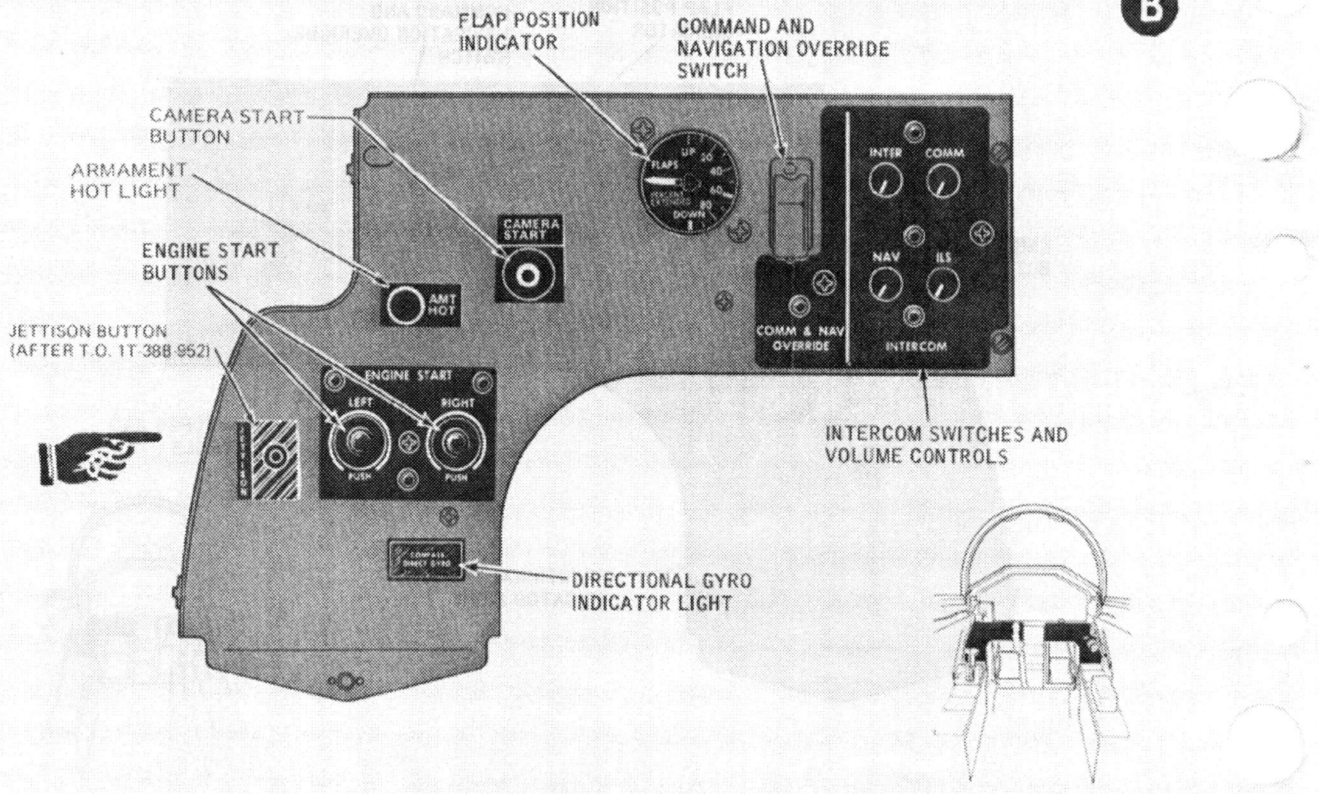


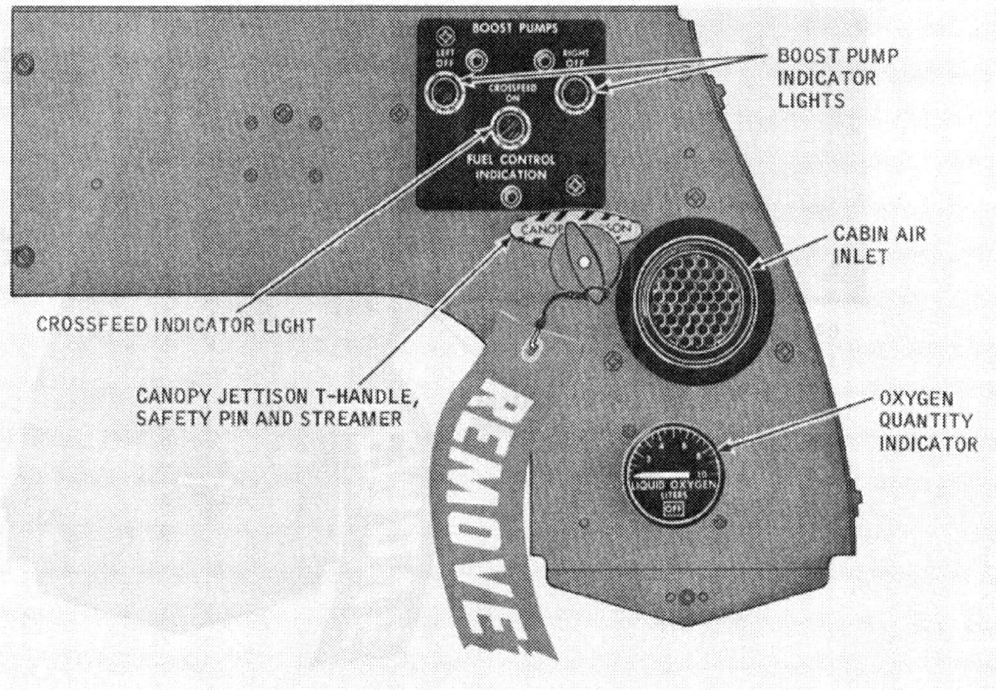


Figure 1-10. (Sheet 2 of 2)

# CONSOLE PANELS—FRONT COCKPIT (TYPICAL)

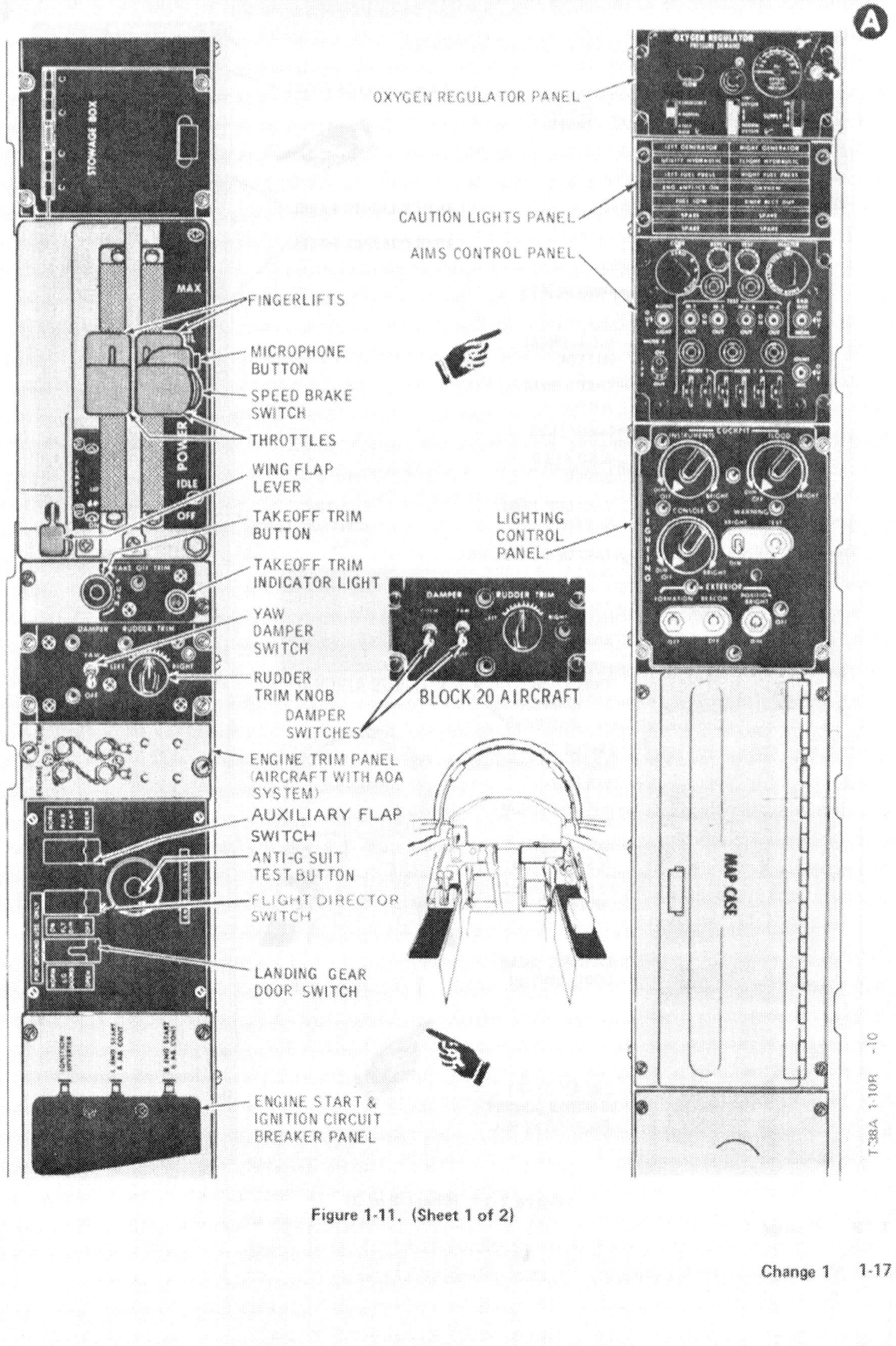


Figure 1-11. (Sheet 1 of 2)

# CONSOLE PANELS—FRONT COCKPIT (TYPICAL)

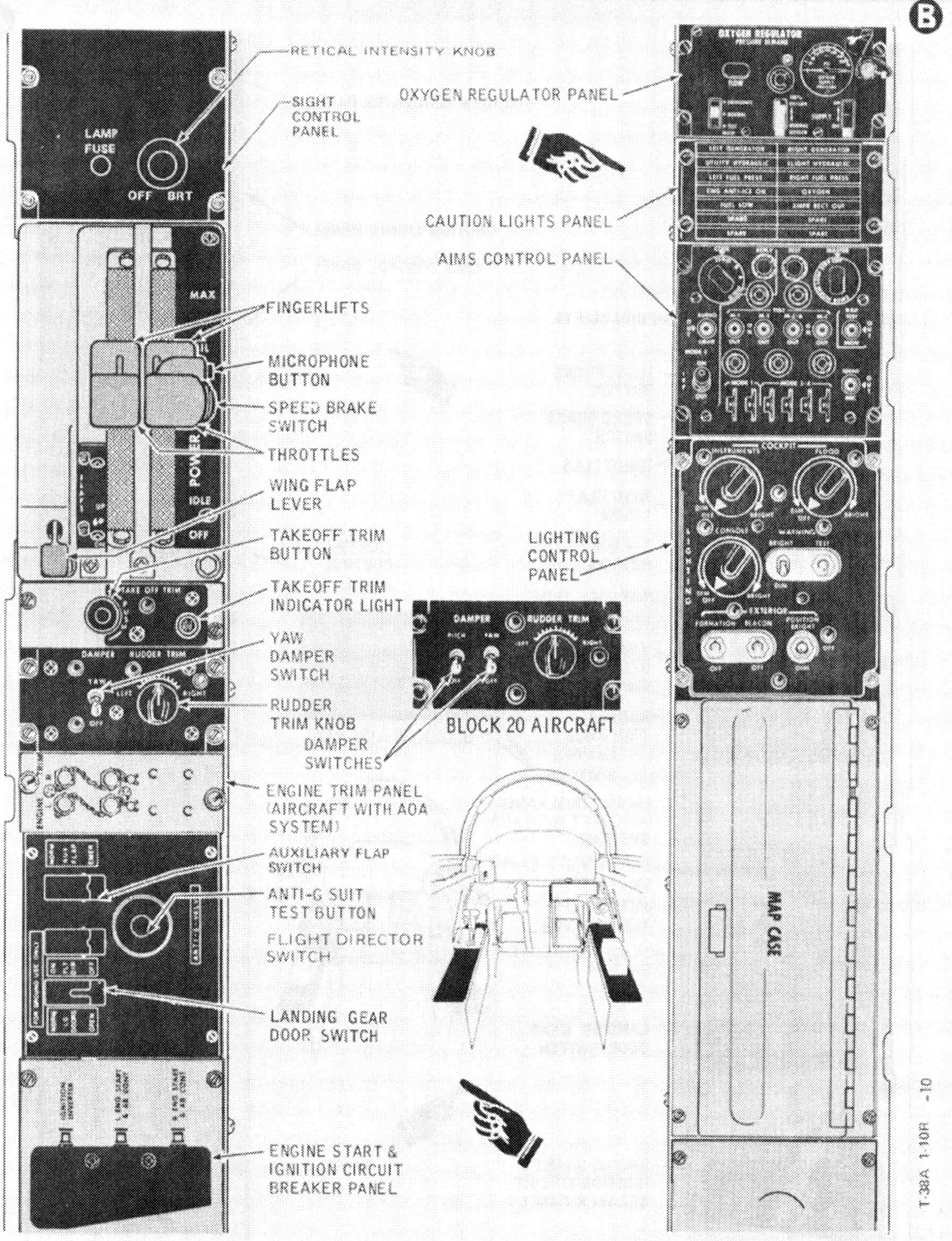


Figure 1-11. (Sheet 2 of 2)

# CONSOLE PANELS – REAR COCKPIT (TYPICAL)

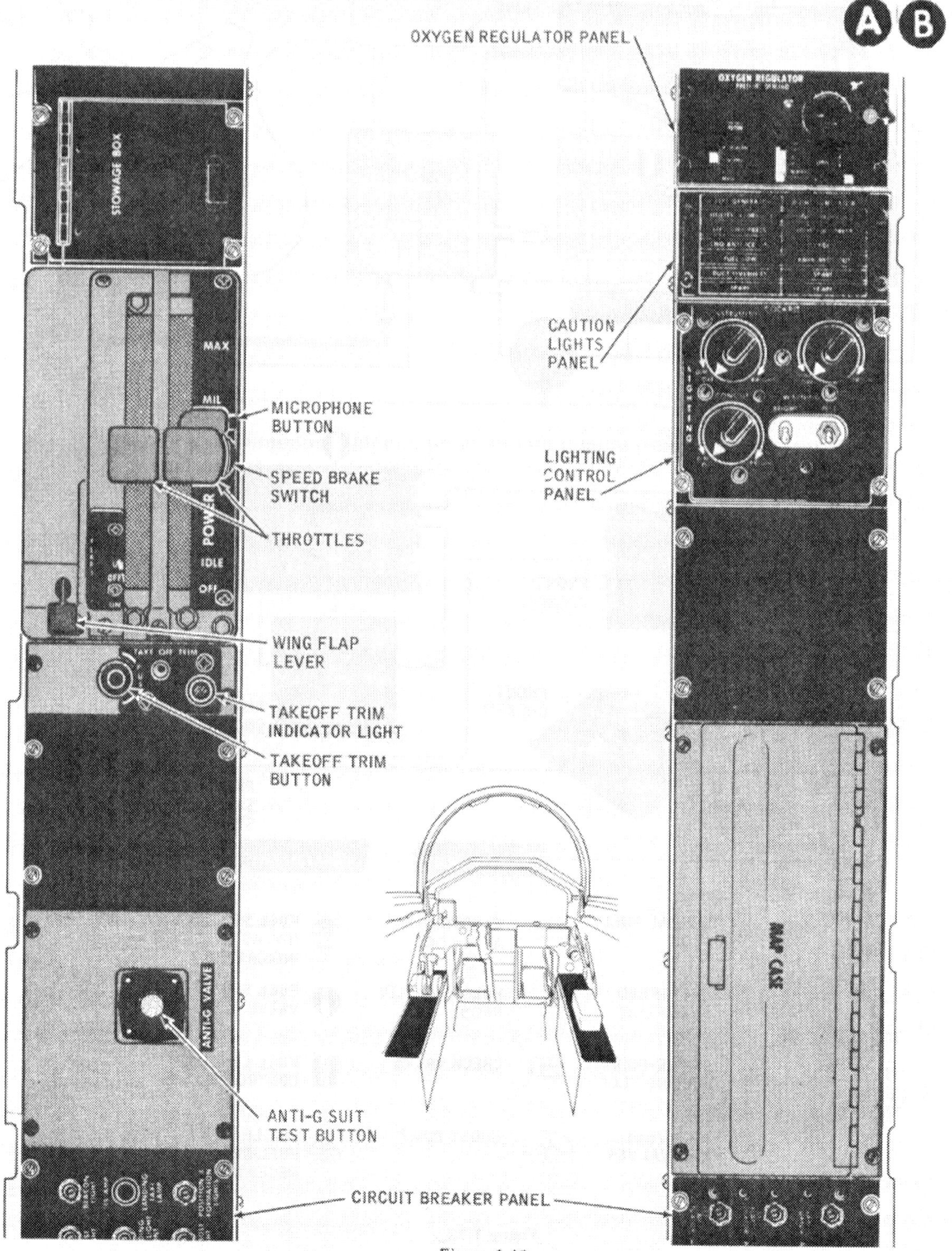


Figure 1-12.

# FUEL SYSTEM

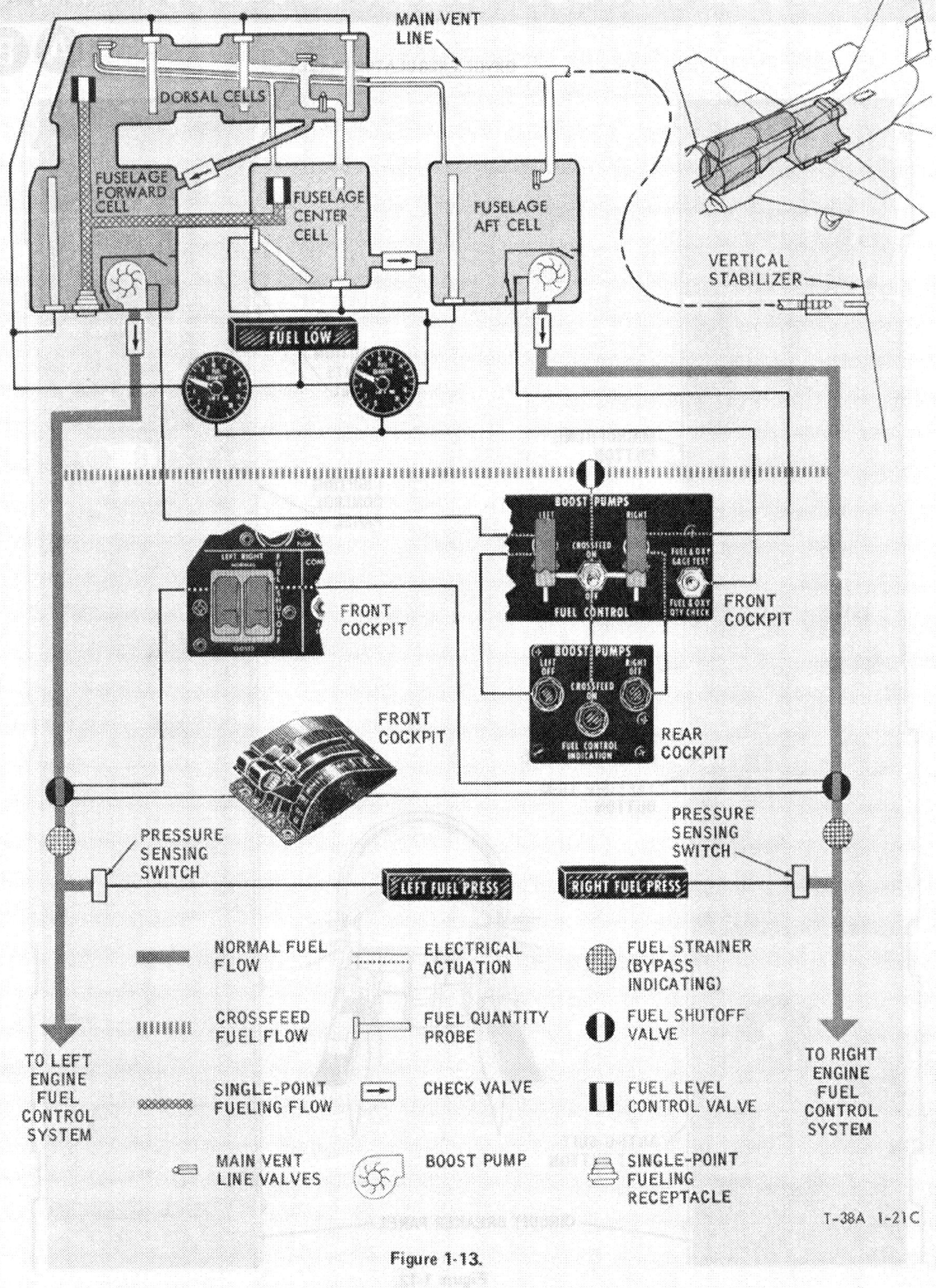


Figure 1-13.

CROSSFEED SWITCH.

A crossfeed switch (figure 1-13), operating on dc, is located on the right subpanel of the front cockpit. The switch is used to electrically open and close the crossfeed valve in the crossfeed fuel manifold that connects the two fuel systems. The switch is placed at ON to use the fuel from both systems to supply one engine or to operate both engines on fuel from one system under boost pump pressure.

WARNING

With the crossfeed switch ON and either both boost pumps ON - or both boost pumps OFF - a rapid fuel imbalance can occur. Fuel will feed predominantly from the left system.

CROSSFEED INDICATOR LIGHT.

An amber crossfeed indicator light (figure 1-13) is located on the right subpanel of the rear cockpit. When the crossfeed switch in the front cockpit is placed at ON, the crossfeed indicator light illuminates.

FUEL SHUTOFF SWITCHES.

Two guarded fuel shutoff switches (figure 1-13), one for each engine, are located on the left subpanel of the front cockpit. The fuel shutoff valves (dc operated) are normally controlled by the throttles, with the fuel shutoff switches in the NORMAL position. Placing either or both of these switches at the CLOSED position shuts off fuel flow to either or both engines in approximately 1 second without using the throttles.

CAUTION

The switches should be used only in an emergency, as damage to the engine-driven fuel pumps and main fuel control may occur.

FUEL QUANTITY INDICATORS.

Two fuel quantity indicators (figure 1-13), one for each fuel system, are located on each instrument panel. The indicators operate on ac and indicate in pounds the total usable fuel quantity in each fuel supply system.

FUEL QUANTITY CAUTION LIGHT.

A fuel quantity low-level caution light, placarded FUEL LOW, is located on the right console of each cockpit (figure 1-13). The caution light will illuminate after a 7.5-second delay when a fuel quantity indicator reads below 275 to 225 pounds. The left and right system fuel quantity indicators must be checked to determine which system is low.

FUEL/OXYGEN CHECK SWITCH.

Fuel and oxygen quantities and indicator operation can be checked by a switch on the right subpanel of the front cockpit (figure 1-9). The three-position switch is spring-loaded to the unmarked OFF position. With external or generator ac power, fuel and oxygen quantities are indicated when switch is at the OFF position. To check operation of fuel and oxygen quantity indicators, the switch is held at the FUEL & OXY GAGE TEST position. Indicator pointers should move counterclockwise. When the switch is released, each indicator pointer will return to indicate the fuel and oxygen quantities. With battery power only, the switch is held at the FUEL & OXY QTY CHECK position to read the fuel and oxygen quantities on board the aircraft. (The static inverter supplies ac power to the indicating circuits when the switch is actuated.)

## AIRFRAME-MOUNTED GEARBOX.

An airframe-mounted gearbox (figure 1-1) for each engine operates a hydraulic pump and an ac generator. A shift mechanism keeps ac generator output between 320 and 480 cycles per second. Gearbox shift occurs in the 65% to 70% RPM range.

## ELECTRICAL SYSTEMS.

Two alternating current systems and one direct current system (figure 1-15) supply electrical power to the aircraft. The 115/200-volt ac power supply systems consist of two identical engine-driven ac generating systems and an external power receptacle. The dc power supply system consists of a dc bus powered either by a 24-volt, 5-ampere-hour battery or two 28-volt dc transformer-rectifiers.

AC POWER SYSTEM.

AC power is normally obtained from two engine-driven ac generators. The power distribution is divided into a right system and a left system. The generators are cut in individually when engine speed accelerates to approximately 43% to 48% RPM. If one generator should fail or is turned off, the functioning generator will automatically supply electrical power to both systems through the bus contactor relay.

Generator Switches and Caution Lights.

Two guarded generator switches (figure 1-9), one for the left and one for the right generator, are located on the right subpanel of the front cockpit.

Generator caution lights, placarded LEFT GENERATOR and RIGHT GENERATOR (figure 1-14), are located on the right console of each cockpit. A caution light will illuminate when its generator switch is placed at OFF or when a generator malfunction occurs. A switch RESET position permits resetting the generators.

### DC POWER SYSTEM.

DC power is normally obtained thru two transformer-rectifiers which convert ac to dc power. If one transformer-rectifier fails, the other automatically supplies all dc requirements. If both transformer-rectifiers fail, the master caution light on the instrument panel and the XFMR RECT OUT (XMFR RECT OUT on 14-module panel) light on the right console will illuminate. Under this condition, the dc bus will revert to battery power.

**NOTE**

> The XFMR RECT OUT and master caution light may blink due to surge current developed by a high battery voltage overriding the dc bus voltage. This is a normal condition and does not indicate a failure.

#### Battery Switch.

A battery switch (figure 1-9) is located on the right subpanel of the front cockpit. Placing the switch at ON connects the battery to the dc bus. Under normal flight conditions, the battery switch should remain in the ON position to permit the battery to charge. A minimum battery voltage of 18 volts is required to close the battery relay.

### STATIC INVERTER.

A static inverter, powered by the dc bus, converts the dc bus voltage to 115 VAC. The inverter, when activated, provides an alternate source of ac power for the following:

a. Starting first engine on the ground or during flight.

b. Operation of right engine autosyn instruments during start of right engine.

c. Fuel and oxygen quantity indicators.

On the ground, with dc power only, the inverter is activated when either engine start button is pushed for engine start, or when the fuel/oxygen check switch is held at FUEL & OXY GAGE TEST or FUEL & OXY QTY CHECK position. During flight, with dc power only, the inverter is activated when either engine start button is pressed or either throttle is moved into MAX range for engine restarts, or when the fuel/oxygen check switch is held at FUEL & OXY GAGE TEST or FUEL & OXY QTY CHECK position. With normal ac/dc power or dc power only, an operational check of the static inverter can be accomplished by positioning the fuel/oxygen check switch to FUEL & OXY GAGE TEST and observing counterclockwise movement of fuel and oxygen quantity indicator pointers.

## CAUTION, WARNING, AND INDICATOR LIGHT SYSTEM.

### CAUTION LIGHT PANEL.

A 14- or 10-capsule word caution light panel (figure 1-14) on the right console of each cockpit is provided to alert the crewmember of individual system malfunction or status change. The 14-capsule

**CAUTION LIGHT PANEL**

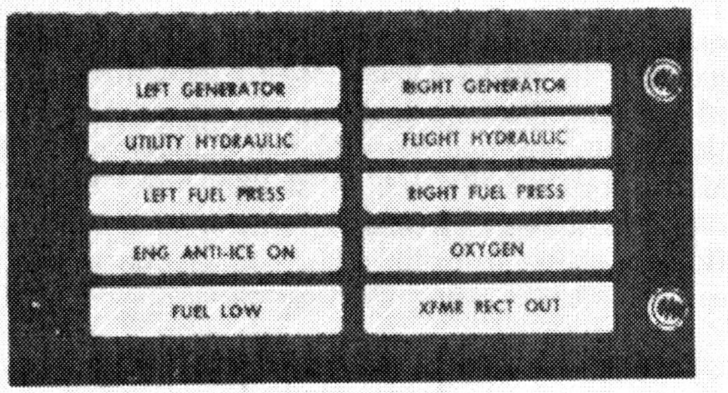


10-MODULE PANEL

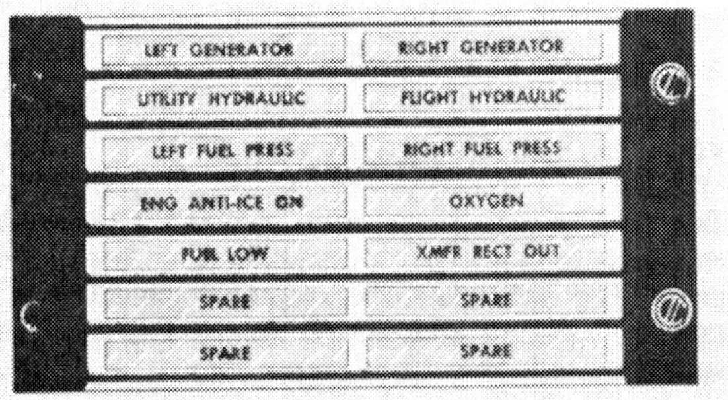


14-MODULE PANEL

T-38A 1-25A

Figure 1-14.

panel has 4 spare capsules. All capsule caution lights are yellow. Each caution light except the ENG ANTI-ICE ON light will remain illuminated as long as the malfunction exists or system status is unchanged. The caution lights will not go out if the master caution light is rearmed. The ENG ANTI-ICE ON light will illuminate when the engine anti-ice switch is turned on. Refer to the description of aircraft systems for operation of the applicable caution lights.

### MASTER CAUTION LIGHT.

A master caution light (figure 1-7), placarded MASTER CAUTION, is located on each instrument panel. When a light illuminates on the caution light panel, the master caution light will also illuminate. When the condition is corrected, the master caution light will automatically go out, but if the condition cannot be corrected, the master caution light may be pressed, causing it to go out and rearming it to provide warning of subsequent malfunctions.

### CAUTION, WARNING, AND INDICATOR LIGHT BRIGHT/DIM SWITCH.

A three-position switch (figures 1-11, 1-12) spring-loaded to neutral unmarked position is provided on the right console of each cockpit to dim all caution, warning, and indicator lights except the marker beacon light, (B) armament hot light, and the takeoff trim indicator light. With the instrument light control out of the OFF (detent) position, momentarily placing the switch in the DIM position will switch the power source from dc to ac, thus providing the DIM setting in that cockpit. Placing the switch momentarily to BRIGHT or placing the instrument light control to OFF will return the lights to bright.

### CAUTION, WARNING, AND INDICATOR LIGHT TEST SWITCH.

The landing gear audible warning signal, the fire detection system, the AOA indexer lights (aircraft with AOA system), and all caution, warning, and indicator lights except the takeoff trim indicator light, (B) armament hot light, and marker beacon light may be tested by placing the spring-loaded switch on the right console lighting control panel in each cockpit (figures 1-11, 1-12) at the TEST position.

NOTE

> If the warning test switches in both cockpits are actuated simultaneously, all fire warning lights will illuminate. The landing gear audible warning signal will not come on in either cockpit.

## FIRE WARNING AND DETECTION SYSTEM.

NOTE

> An illuminated fire warning light may be a valid fire indication even though the test circuit may be inoperative.

The fire warning and detection system is provided to give a warning of a fire or overheat condition in either engine bay. Heat detectors are located in the forward engine bay and boattail area for each engine. The system responds to an overall average temperature or to highly localized temperatures caused by impinging flame or hot gas. Operation of the system in each engine compartment is independent of the other except when testing the system using the caution, warning, and indicator test switch. Placing either cockpit test switch at TEST checks all system detectors and fire warning light bulbs/filaments (4) in each cockpit. For test purposes only, each bulb is connected to a detector. However, any fire or overheat condition in either engine compartment will illuminate both bulbs of the respective fire warning lights in both cockpits.

# ELECTRICAL SYSTEM

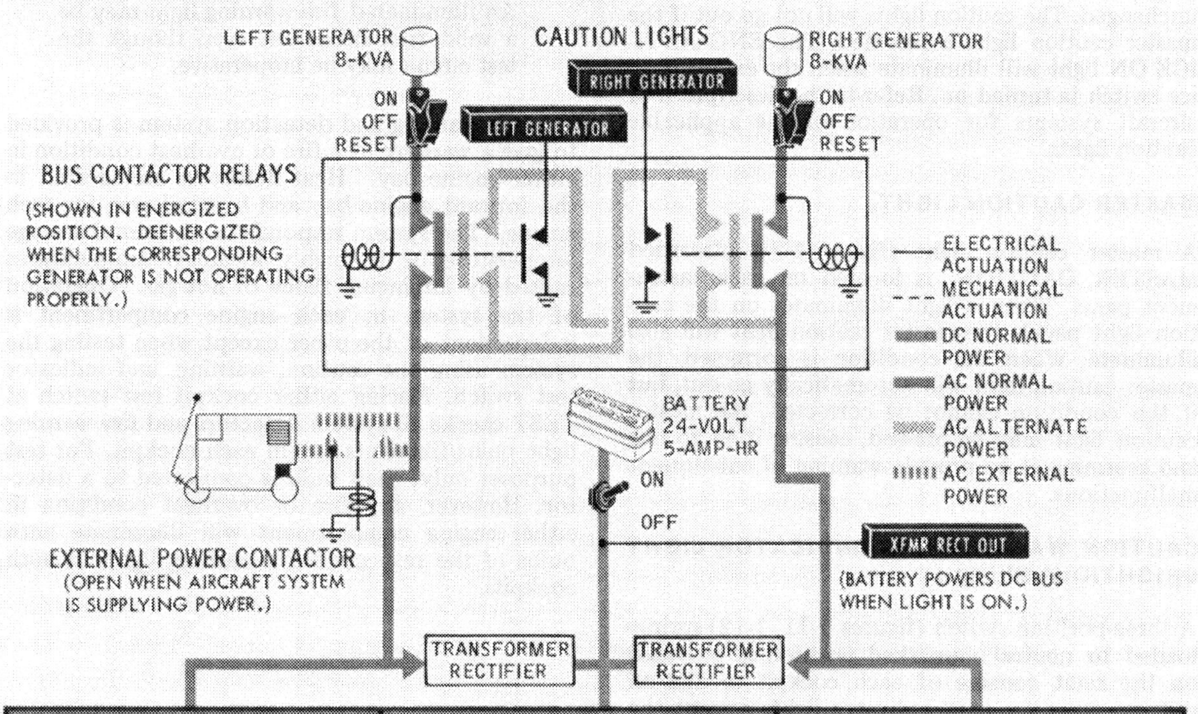


**LEFT 115/200-VOLT AC BUS**

- AIMS
- AOA COMPUTER
- AOA VANE HEATER
- CONSOLE LIGHTS
- FUEL LOW WARNING
- IFF-SIF
- INSTRUMENT LIGHTS
- LANDING-TAXI LIGHT (REQUIRES DC FOR EXTENSION)
- LEFT AUTOSYN INDICATORS
- LEFT FUEL BOOST PUMP
- MXU
- LEFT FUEL QUANTITY INDICATOR
- POSITION LIGHTS (6-VOLT AC)
- ROTATING BEACONS
- RUDDER TRIM CONTROL
- SIGHT
- STABILITY AUGMENTER
- STANDBY ATTITUDE INDICATOR

**28-VOLT DC BUS**

- AIR DIVERTER VALVE
- AOA INDEXER LIGHTS
- CAUTION, WARNING, AND INDICATOR LIGHTS–BRIGHT (DIM LIGHT IS AC)
- CROSSFEED VALVE
- ENGINE START AND IGNITION CONTROL
- FIRE DETECTORS
- FLAP CONTROL
- FLAP POSITION INDICATOR
- FLIGHT DIRECTOR MODE SELECTOR
- FLOODLIGHTS (EMERGENCY)
- FORMATION LIGHTS
- FUEL/OXYGEN CHECK SWITCH
- FUEL SHUTOFF VALVES
- ILS RECEIVER, INDICATORS, AND FLAGS
- INTERPHONE
- LANDING GEAR AUDIBLE WARNING SIGNAL
- LANDING GEAR CONTROL AND SAFETY CIRCUITS
- LANDING GEAR DOOR SWITCH
- MARKER BEACON LIGHT
- FLIGHT LOADS DATA RECORDER (MXU)
- MASTER CAUTION LIGHT CONTROL
- NOSEWHEEL STEERING CONTROL
- NOZZLE POSITION INDICATORS
- SPEED BRAKE CONTROL
- STATIC INVERTER
- TACAN
- TACAN ANTENNA SELECTOR
- UHF RADIO, ARC-164 (REAR COCKPIT)
- TURN NEEDLE
- VIBRATOR, AAU–19/A ALTIMETER

**RIGHT 115/200-VOLT AC BUS**

- ADI STEERING BARS
- ATTITUDE SPHERE
- BEARING POINTER
- CABIN AIR-CONDITIONING
- CANOPY SEAL
- CAUTION, WARNING, AND INDICATOR LIGHTS–DIM
- COMPASS CARD (SLAVING REQUIRES DC)
- ENGINE ANTI-ICE
- EQUIPMENT COMPARTMENT COOLING
- FLAP CONTROL
- FLAP MOTORS
- FLOODLIGHTS (NORMAL)
- FUEL LOW WARNING
- IGNITION EXCITERS
- ILS RECEIVER, INDICATORS, AND FLAGS
- LANDING GEAR AUDIBLE WARNING SIGNAL
- LANDING GEAR CONTROL AND SAFETY CIRCUITS
- NOSEWHEEL STEERING CONTROL
- OXYGEN QUANTITY AND LOW-LEVEL WARNING CIRCUIT
- PITOT HEATER
- RADIO TRANSFER CONTROL
- RIGHT AUTOSYN INDICATORS
- RIGHT FUEL BOOST PUMP
- RIGHT FUEL QUANTITY INDICATOR
- SEAT ADJUSTMENT
- SPEED BRAKE CONTROL
- TACAN
- TRIM CONTROL AND LIGHT
- UHF ANTENNA SELECTION
- UHF RADIO, ARC-34 (REAR COCKPIT)
- UTILITY LIGHTS

BLOCKS 20 THRU 30 AIRCRAFT
BLOCKS 20 THRU 35 AIRCRAFT
BLOCK 35 AND LATER AIRCRAFT
BLOCK 40 AND LATER AIRCRAFT
BLOCKS 20 THRU 70 AIRCRAFT
BLOCK 75 AND LATER AIRCRAFT
AIRCRAFT WITH AOA SYSTEM
AUTOSYN INDICATORS:
EXHAUST GAS TEMPERATURE INDICATORS
FUEL FLOW INDICATORS
HYDRAULIC PRESSURE INDICATORS
OIL PRESSURE INDICATORS

(ALSO REQUIRES AC)
(ALSO REQUIRES DC)

Figure 1-15.

# CIRCUIT BREAKER PANELS (TYPICAL)

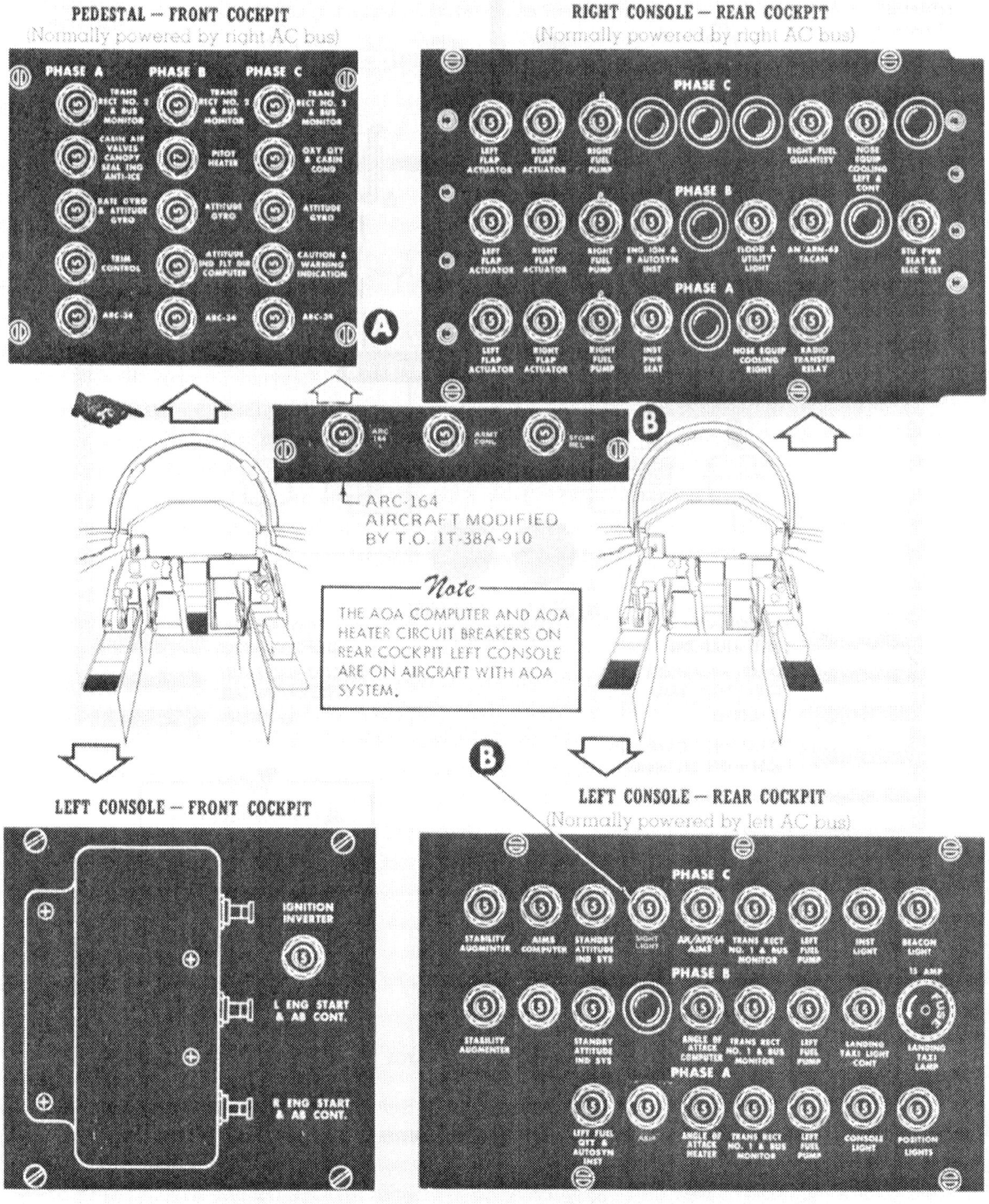


Figure 1-16.

# HYDRAULIC SYSTEMS

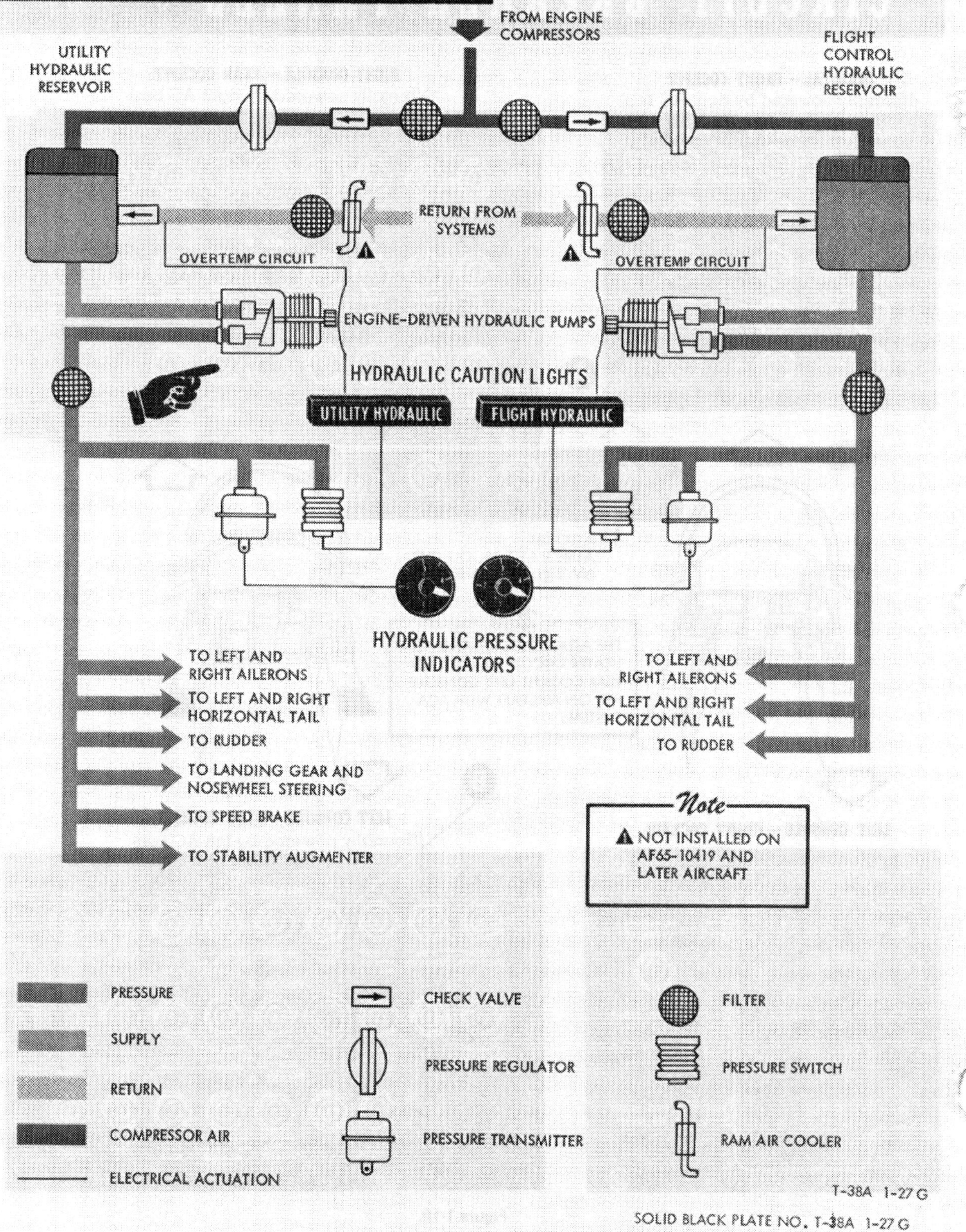


SOLID BLACK PLATE NO. T-38A 1-27 G
(+ THREE COLORS)

Figure 1-17.

### ENGINE FIRE WARNING LIGHTS.

Two red fire warning lights (figure 1-7), placarded FIRE, one for each engine, on the instrument panel in each cockpit, are provided to warn of an overheat or fire condition in either engine compartment. When the fire detection system senses an overheat condition or fire, the warning light for the respective engine will come on. This light will remain on until the condition is corrected and then will go out. Should the overheat condition or fire recur, the light will again come on. Each fire warning light contains two bulbs.

## HYDRAULIC SYSTEMS.

The aircraft hydraulic power supply systems (figure 1-17) include the 3000-psi utility system powered by the left engine and the 3000-psi flight control system powered by the right engine. No interflow can occur between the utility and flight control hydraulic systems. Separate pressure indicators and caution lights are provided for each system. On AF65-10419 and later aircraft, the ram air cooler has been deleted from the system. Refer to figure 1-25 for hydraulic fluid specification.

### HYDRAULIC PRESSURE INDICATORS.

Two ac powered hydraulic pressure indicators (figure 1-17), one for each hydraulic system, are located on the instrument panel in each cockpit.

### HYDRAULIC CAUTION LIGHTS.

A caution light for each hydraulic system placarded UTILITY HYDRAULIC and FLIGHT HYDRAULIC, is located on the right console of each cockpit (figure 1-17). The lights illuminate at approximately 1500 psi to indicate a low-pressure condition. The lights go out when a pressure of approximately 1800 psi is restored. The lights will also illuminate when the hydraulic fluid has excessively high temperatures. To determine which condition has caused the lights to illuminate, the hydraulic pressure indicators must be observed.

## FLIGHT CONTROL SYSTEM.

A hydraulically powered, irreversible flight control system is provided (airloads on the control surfaces can not cause control stick or surface movement). Conventional aerodynamic "feel" in the control stick is provided artificially by springs and bob weights. The springs progressively resist control stick displacement and the bob weight mechanism further resists aft stick travel during maneuvering flight. Lateral and longitudinal trim is provided by electric motors which change the neutral reference point of the feel springs and control stick position. Each control surface is moved by two hydraulic cylinders; one is powered by the UTILITY system, the other by the FLIGHT HYDRAULIC system.

### CONTROL STICK.

Each cockpit has a control stick with a standard stick grip (figure 1-18), which contains a flight trim switch and a nosewheel steering button. On Block 20 aircraft, a stability augmenter pitch damper cutoff switch is located below the stick grip.

### RUDDER PEDAL ADJUSTMENT T-HANDLE.

A mechanical rudder pedal adjustment T-handle (figure 1-8) is located on the pedestal of each cockpit. To adjust rudder pedals, pull T-handle out and hold until pedals are repositioned. Return the T-handle to the stowed position manually to lock the pedals in place.

**CONTROL STICK (TYPICAL)**

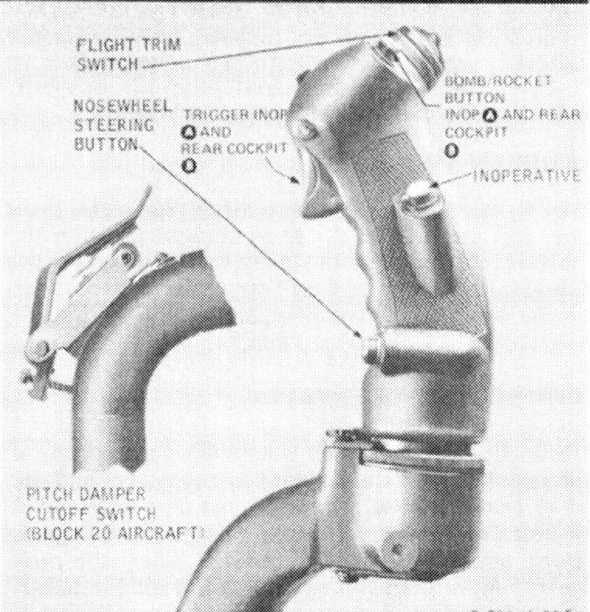


Figure 1-18.

**CAUTION**

Allowing the handle to snap back may trip or damage pedestal circuit breakers or ILS control and cause the cable to kink and wear excessively.

### TAKEOFF TRIM SYSTEM.

A takeoff trim system is installed to allow positioning of the horizontal tail for the optimum takeoff setting. The system uses the normal longitudinal trim system along with a push button and indicator light (figure 1-11, 1-12) installed on the left console in both cockpits. When the button is pushed and held, the trim motor moves the control stick to the required position at which point the motor stops and a green indicator light illuminates in the left console. The aircraft has external markings to visually confirm proper takeoff trim horizontal tail position.

### FLIGHT TRIM SYSTEM.

A conventional aileron/elevator trim switch is located on each stick grip. Operation of the switch (figure 1-18) causes operation of an AC motor, causing appropriate movement of the control stick. Limit and cutout switches are installed in the system which limit the range of stick travel obtainable through use of the trim system. Also, horizontal tail trim is interrupted when stick force is exerted against the direction of trim. These two cutout systems limit the effects of "runaway trim," since the aircraft can be flown with the control stick at either of the trim limit cutout positions; however, very heavy stick forces may be encountered.

### RUDDER TRIM KNOB.

An ac electrical rudder trim knob (figure 1-11) on the left console of the front cockpit provides the means of trimming the rudder. The yaw damper switch must be turned on before the rudder will assume the selected trim position.

### RUDDER LIMITER SYSTEM.

Deflection of the rudder is limited by a mechanical linkage between the rudder control system and the nose gear trunnion. When the nose gear is extended 3/4 or less, rudder deflection is limited to 6 degrees from neutral in either direction. When the nose gear is more than 3/4 extended, full rudder deflection of 30 degrees from neutral in either direction is available. The rudder limiter cannot be overcome by either crewmember.

### STABILITY AUGMENTER SYSTEM.

The stability augmenter system positions the rudder to reduce yaw oscillations. Manual rudder trim is accomplished thru the yaw damper. A yaw damper switch is located on the left console of the front cockpit (figure 1-11). The switch is spring-loaded to OFF position and is held in the YAW position by ac power. The yaw damper is disengaged by returning the switch to OFF. The augmenter will disengage automatically in the event of ac power failure or certain system malfunctions. On Block 20 aircraft, the stability augmenter system also positions the horizontal tail to reduce pitch oscillations when the pitch damper switch next to the yaw damper switch is positioned to the PITCH position. The switch is held in the PITCH position by ac power. The pitch damper is disengaged either by returning the switch to OFF or by actuating the pitch damper cutoff switch on the control stick (figure 1-18), and will disengage automatically in the event of ac power failure or certain system malfunctions.

### WING FLAP SYSTEM.

The wing flaps are electrically controlled by a flap lever. Two ac electric motors operate the flaps thru gear reduction units. The flaps are interconnected by a rotary flexible shaft. If one flap motor fails, both flaps are actuated thru the rotary shaft. Full flap extension or retraction takes from 10 to 17 seconds. Flaps are mechanically connected to the horizontal tail operating mechanism. This interconnect system operates progressively as wing flaps are extended (or retracted) and has three functions. First, as flaps are moved, the horizontal tail is automatically repositioned to essentially eliminate the pitch trim changes caused by flap movement. Second, as flaps are extended, the system progressively increases the amount of horizontal tail travel available in the nose down direction. Third, the interconnect system also changes the pitch authority of the control stick by increasing the amount of horizontal tail deflection per inch of stick travel. Besides reducing the pitch transients caused by flap extension, the interconnect system restores nose down pitch authority (amount of deflection) and sensitivity (slab/stick movement) lost due to flap extension. On some aircraft, the flap interconnect system incorporates a self-locking device which maintains the increased horizontal tail authority and greater pitch sensitivity if the interconnect system fails while the flaps are deflected. On other aircraft, the self-locking device is not installed; if the interconnect cable breaks, a spring will rapidly return the horizontal tail to the 0 percent flap position.

## WING FLAP LEVER AND POSITION INDICATOR.

A wing flap lever (figure 1-4) is located on the throttle quadrant of each cockpit. The two levers are mechanically interconnected by cables; however, the lever in the front cockpit actuates the electrical switch that operates the two flap motors. Sensing switches stop the flaps at 60% when the flap lever is placed in the 60% detent. When operating in the emergency mode, the flaps can be stopped at any position by placing the flap lever in the 60% detent. When UP or DOWN is selected, flap movement is stopped by limit switches at the fully retracted or extended position. The flap position indicator, which operates on dc, is located on the left subpanel of each cockpit (figures 1-9, 1-10). Flap extension is indicated as a percentage of full flap travel.

NOTE

If the wing flap lever is between the 0 - 60% position or the 60 - 100% position the flaps will not extend or retract.

## AUXILIARY FLAP CONTROL SWITCH.

The auxiliary flap control switch is located in the front cockpit (figure 1-11) on the left console. It is a two position switch. In the normal position, flap positions of full up, 60% down, and full down can be selected. In the emergency position, flaps can be set at any selection from full up to full down. In this mode of operation, flap extension or retraction is stopped by moving the lever to the 60% detent, which then functions as an OFF position when flaps have reached the desired position, or by limit switches when the flaps have fully extended or retracted.

## SPEED BRAKE SYSTEM.

A DC electrically controlled, hydraulically activated dual surface speed brake is located on the lower surface of the fuselage center section. Design of the activation system permits selection of intermediate speed brake positions other than fully extended.

### SPEED BRAKE SWITCH (figure 1-4).

A conventional, three position (UP-OFF-DOWN) speed brake switch (DC) is installed on the right throttle in each cockpit. The switch in the front cockpit has positive detents in each position. The switch in the rear cockpit has capability to override the position selected in the front cockpit and is spring-loaded to the center OFF position. Intermediate speed brake positions can be obtained by positioning the switch to the desired direction of movement and then returning to the OFF position. Speed brake creep will occur with the switch in the OFF position. Following override, control of the speed brake system is regained in the front cockpit by moving the switch to OFF. To prevent creep following actuation from the rear cockpit, the front cockpit switch should be placed in the position selected by the rear cockpit.

## LANDING GEAR SYSTEM.

Extension and retraction of the landing gear and gear doors are powered by the utility hydraulic system and electrically controlled by the landing gear levers. Landing gear extension or retraction normally takes approximately 6 seconds. The normal landing gear cycle may be reversed at any time. The normal extension sequence is doors open, gear extends, doors close. The retraction sequence is doors open, gear retracts, doors close.

## LANDING GEAR LEVER, WARNING SYSTEM, AND SYSTEM SILENCE BUTTON.

A landing gear lever (figure 1-7) is located on the instrument panel of each cockpit. The two levers are mechanically interconnected. A warning system consisting of an intermittent tone (beeper), audible thru the headset of each crewmember, and a red light within the wheel-shaped end of each landing gear lever will be activated if the landing gear is not down and locked and the following conditions exist:

a. The airspeed is 210 KIAS or less.

b. The altitude is 10,000 ± 750 feet or below.

c. Both throttles are below 96% RPM.

NOTE

Power required under single engine conditions may be in excess of that required to activate the landing gear warning system.

When airspeed is decreasing, the system is activated in the range of 210 to 180 KIAS. With the system activated and the aircraft accelerating, the light and tone may not go out until speed reaches approximately 240 KIAS. With the gear handle in the UP position, and the system not activated a red light in the landing gear lever indicates that the landing gear doors are not up and locked. The audible warning signal is not activated by an unlocked gear door condition. A landing gear warning silence button (figure 1-7) is located on the instrument panel of each cockpit. Pressing either button silences the audible warning signal.

CAUTION

Prior to incorporation of T.O. 1T-38-582, front cockpit pilot should not place the left foot outboard of the rudder pedal due to the possibility of striking the landing gear handle interconnect linkage causing uncommanded landing gear retraction.

**Landing Gear Lever Downlock Override Button.**

A landing gear lever downlock override button (figure 1-7) on the instrument panel of each cockpit enables either crewmember to raise the landing gear lever to the LG UP position if the locking solenoid fails to release the landing gear lever from the LG DOWN position. With button pressed the landing gear lever can be raised to the LG UP position during flight or on the ground. The rear cockpit downlock override button operates electrically; the front cockpit downlock override button operates mechanically.

## LANDING GEAR POSITION INDICATOR LIGHTS.

Three landing gear position indicator green lights (figures 1-7) on each instrument panel illuminate when the gear is down and locked.

NOTE

- There are separate contacts for each cockpit green light indicator. Good lights in either cockpit assure safe gear.
- With DC failure, the rear cockpit nose gear light will not illuminate due to gear relay wiring.

## LANDING GEAR ALTERNATE RELEASE HANDLE.

A landing gear alternate release handle (figure 1-9) on the left subpanel of the front cockpit permits gear extension without hydraulic pressure or electrical power. When the handle is pulled, the normal landing gear hydraulic and electrical systems are deenergized, and the gear uplocks and gear door locks are mechanically released, permitting the gear to extend by its own weight. No portion of the landing gear structure is under hydraulic pressure after extension by the alternate system. The handle must be held in the fully extended position (approximately 10 inches) until all three gears are unlocked. Extension of the main and nose landing gear will require approximately 15 seconds, but may take up to 35 seconds. If gear alternate extension was accomplished with the gear lever at LG UP, the lever must be placed at LG DOWN and then returned to LG UP to reactivate the normal system. After an alternate extension, the main gear doors will remain open and nosewheel steering will not be available until the system is reactivated. The nosewheel door assumes a spring-loaded closed position after alternate extension. A landing gear reset lever, located outboard of the left rudder pedal in the front cockpit, may be used to reset the landing gear switches.

NOTE

- During preflight, if the striker plate in the nose gear well is found in the extended position, check the reset lever in the reset (UP) position. This resets all gear switches, but will not raise the striker plate. The striker plate will remain extended until the nose gear retracts after takeoff.
- If the gear is lowered by the alternate release handle with the landing gear in the LG UP position, the red light in the landing gear lever will remain illuminated. In this situation, the illuminated red light indicates the gear door open condition normally associated with the gear retraction cycle. The landing gear green indicator lights will be illuminated and the warning signal silent, indicating a positive gear down and locked condition.

## LANDING GEAR DOOR SWITCH.

A guarded landing gear door switch is provided on the left console of the front cockpit (figure 1-11). With electrical and hydraulic power available, this switch permits opening and closing the landing gear doors when the landing gear lever is at LG DOWN. If the gear is extended in flight with the gear door switch at OPEN, the gear doors will remain open until the gear is retracted or the gear door switch is placed at NORMAL.

# NOSEWHEEL STEERING SYSTEM.

The nosewheel steering system provides directional control and shimmy damping. Hydraulic pressure for the system is supplied by the utility hydraulic system. Nosewheel steering is controlled by rudder pedal action and may be activated only when the weight of the aircraft is on the nosewheel. If the nosewheel position does not correspond to the position of the rudder pedals when steering is activated, the nosewheel will turn to correspond to the rudder pedal position.

NOSEWHEEL STEERING BUTTON.

Nosewheel steering is electrically controlled by the nosewheel steering button on the control stick (figure 1-18) in either cockpit. Steering is available only when the button is held in the pressed position. With button pressed, nosewheel steering is deactivated when one or both throttles are advanced

to MAX range and restored when both throttles are retarded below MAX range. Whenever the weight of the aircraft is not on the nose gear, the system automatically deactivates.

## NOSEWHEEL CENTERING MECHANISM.

A nosewheel centering cam mechanically streamlines the nosewheel whenever the nose gear strut is fully extended. Air pressure in the strut mechanism ensures that the nose gear strut remains fully extended during gear retraction.

## WHEEL BRAKE SYSTEM.

The main gear wheel brakes are the segmented rotor type and are powered by a separate, completely self-contained hydraulic system. The brake pedals are the conventional toe-operated type. Each brake pedal controls a hydraulic master cylinder. Control of the brakes transfers to the crewmember applying the greater pedal force.

## PITOT-STATIC SYSTEM.

The pitot-static system supplies both impact and static air pressure to the airspeed-mach indicator, the airspeed compensator of the stability augmenter system, and the airspeed and altitude pressure switch assembly that connects into the landing gear warning circuits. The altimeter and vertical velocity indicator receive only static pressure from the system.

## CANOPY.

Each cockpit contains a manually operated clamshell type canopy. The canopy is locked closed or unlocked by an individual locking lever in each cockpit, or by individual locking handles outside the left side of the front cockpit (figure 1-19). Each canopy is counter-balanced throughout its travel limits. The canopy opening mechanism is protected against excessive loads by a hydraulic canopy damper, which also restricts canopy opening and closing speeds. An inflatable pressurization seal installed on each canopy is inflated when both canopies are locked, the cockpit pressure switch is in the CABIN PRESS position, and an engine is operating.

**CAUTION**

- Canopy movement from the full open or closed and locked position must be initiated by the external or internal locking handle. Actual raising or lowering of the canopy must be done by hand pressure on the canopy frame. **Do not** apply pressure on the locking handle to raise or lower the canopy as damage to the mechanism may result.
- Use caution when opening the canopy under high or gusty wind conditions to avoid rapid canopy fly up.
- If an open canopy has been exposed to high winds or jet blast, it should be checked for normal operation, i.e., fully closed before taxi. If the canopy will not close, the aircraft should not be taxied or towed until cleared by qualified maintenance personnel. Aircraft movement may result in canopy separation.
- Damage and possible loss of canopy may occur if the hood is bunched between the drogue chute housing and canopy, and the seat is raised to the near full up position.
- If the canopy is closed with the shoulder harness on the drogue chute housing damage to the seat or canopy may occur.

## CANOPY WARNING LIGHT.

A canopy warning light, placarded CANOPY (figure 1-19), operating on dc (bright) or ac (dim), is located on the instrument panel of each cockpit. When either canopy is unlocked, both canopy warning lights illuminate.

## CANOPY JETTISON SYSTEM.

The canopy jettison system permits jettisoning each canopy individually from inside the cockpit or both canopies from outside the cockpit. From inside the cockpit, the canopy of the respective cockpit may be jettisoned independent of seat ejection by pulling the T-handle (figure 1-19) on the right subpanel of each cockpit. A safety pin is provided for each canopy jettison T-handle to prevent inadvertent jettison of the canopy. A spring clip on the bottom of the T-handle must be overcome to pull the T-handle out. To jettison the canopies from outside the cockpit, a canopy jettison D-handle (figure 1-19) is located externally on each side of the front cockpit, pointed out by the RESCUE decal. Opening either access door and pulling the D-handle jettisons both canopies. The front canopy jettisons first when the D-handle is pulled, followed 1 second later by jettison of the rear canopy. The canopy jettison system will function properly only with the canopy closed and locked. However, with the aircraft at rest, the canopy may not separate from the hinges if the canopy is in the fully

# CANOPY CONTROLS

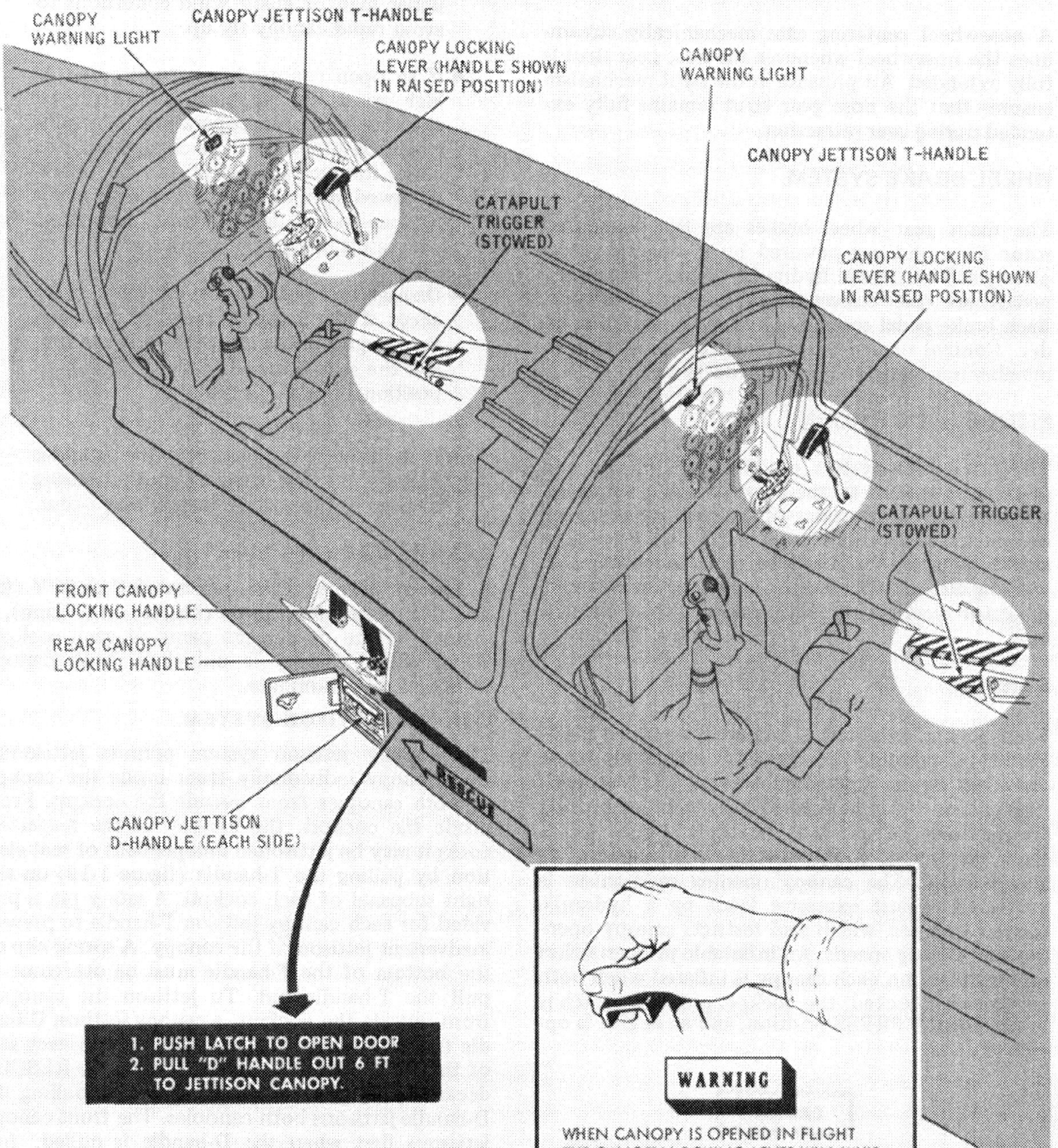


Figure 1-19.

open position. If the jettison system is activated with the canopy in a position other than fully open, the canopy will move to the full open position and probably separate from the aircraft.

**WARNING**

If the canopy jettison is activated with the canopy in other than the closed and locked position, the canopy could fall off its hinges and into the cockpit area.

### CANOPY BREAKER TOOL.

A canopy breaker tool (figures 1-5, 1-6) is stowed on the left canopy frame in each cockpit. The tool is used to break the canopy glass if other methods of opening the canopy fail.

## EJECTION SYSTEM.

The ejection system consists of an ejection seat with drogue chute and man-seat separator (figure 1-20), an automatic opening safety belt with 0.65 second delay initiator (figure 1-21 or 1-22), and an automatic opening parachute with 0.25 second delay initaitor or zero delay lanyard parachute with a one second delay initiator.

After ejection from the aircraft, the drogue chute deploys to stabilize the seat, the safety belt opens and actuates the man-seat separator forcing the crewmember from the seat. An aneroid delays parachute opening until 14,000 feet pressure altitude when free falling. Below 14,000 feet, parachute opening is initiated at 0.25 second (or one second) after seat separation. Low altitude capability (below 2000 feet AGL) is provided by the 0.25 second delay initiator or the zero delay lanyard connection. With the zero delay lanyard hooked to the parachute ripcord handle, the ripcord is pulled upon man-seat separation providing immediate parachute deployment. A stowage ring is provided on the parachute harness for the zero delay lanyard when not in use.

Refer to section II for proper connection of the zero delay lanyard and to section III for proper use of ejection equipment.

### EJECTION SEAT.

Each cockpit is equipped with a rocket catapult ejection seat (figure 1-20). A calfguard, hinge-mounted to the forward end of each seat, is pulled downward behind the crewmember's legs during ejection to prevent the crewmember's legs from being thrust backward beneath the seat by wind blast and to assist in man-seat separation. Controls for the ejection sequence are the catapult triggers. During the first part of seat ejection, initial seat movement simultaneously disconnects the oxygen, anti-G suit, and communication disconnects, pulls the calfguard down, fires the safety belt delay initiator, disconnects the seat adjuster power cable and initiates drogue gun operation. Each seat is equipped with a canopy piercer and will eject thru the canopy if canopy jettison malfunction is experienced. The front seat canopy piercer is attached to the seat and is raised and lowered with the seat. The rear seat canopy piercer is not attached to the seat and will remain in a fixed position when the seat is raised and lowered.

#### Legbraces.

Two legbraces (figure 1-20), terminating in handgrips, are attached to the ejection seat (one on each side) and are linked together mechanically so that they rise simultaneously. Each legbrace contains a handgrip and a catapult trigger. Raising either handgrip locks the shoulder harness and exposes the catapult triggers. Initial movement of either handgrip releases the downlock on both legbraces. When actuated, the legbraces are held in the raised position by an uplock and cannot be returned to the down stowed position by the crewmember.

# EJECTION SEAT

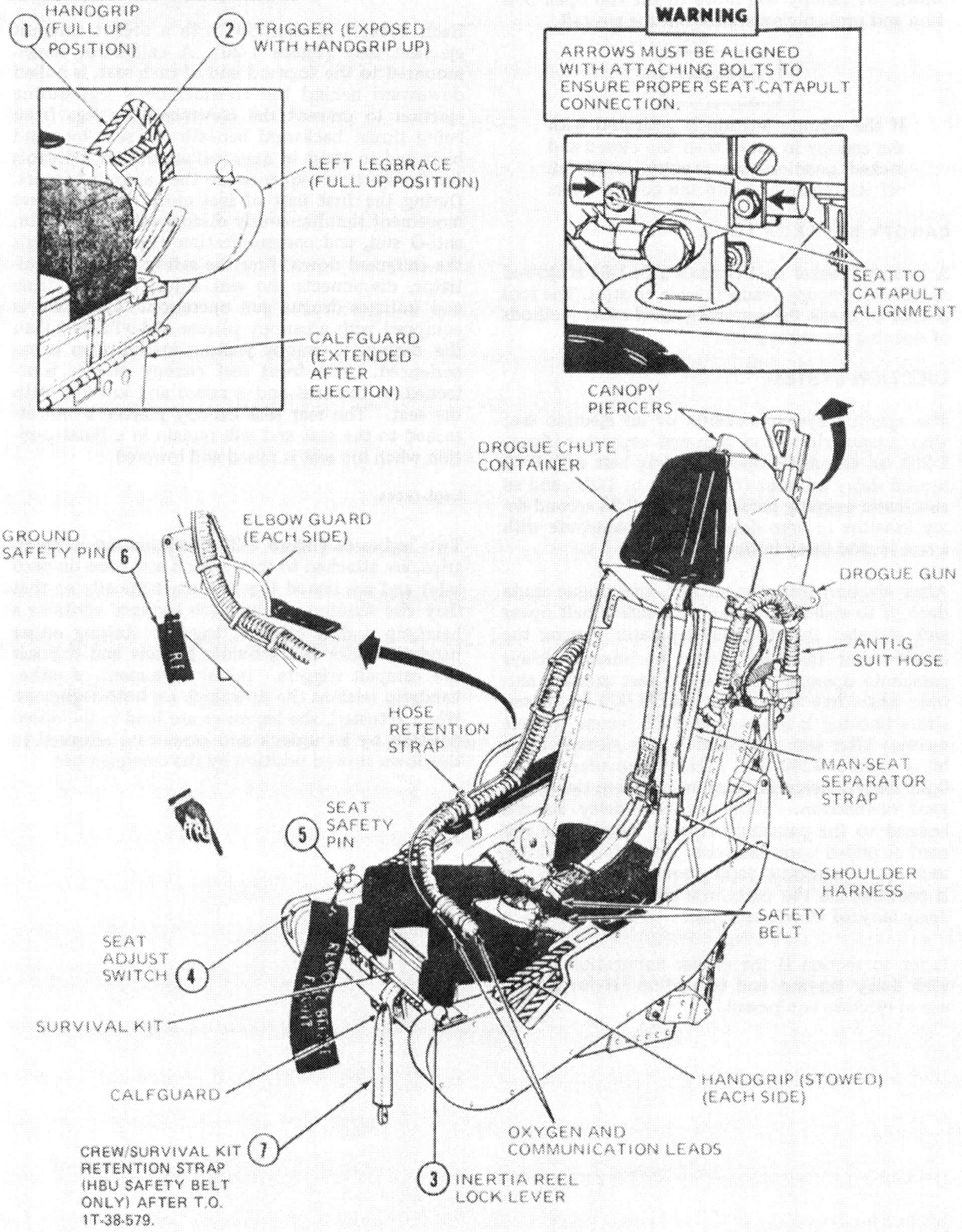


Figure 1-20

EJECTION SEAT CONTROLS (Figure 1-20)

| CONTROLS | | FUNCTION | |
|---|---|---|---|
| 1 | Handgrips (Yellow with black diagonal stripes) | Pulling either or both handgrips up to travel limits raises legbraces to fully up and locked position and exposes triggers. First 12 degrees of travel unlocks both legbraces. | |
| 2 | Firing Triggers (Yellow with black diagonal stripes) | Squeezing either or both triggers initiates canopy jettison and seat ejection. | |
| 3 | Inertial Reel Lock Lever | LOCK | —Locks shoulder harness. |
| | | AUTO | —Unlocks shoulder harness, freeing it to reel in and out. Harness will automatically lock during rapid 3-g acceleration and/or during seat ejection. |
| 4 | Seat Adjust Switch | Forward and Hold | —Lowers seat electrically. |
| | | Center | —Spring-loaded neutral position. |
| | | Aft and Hold | —Raises seat electrically. |
| 5 | Seat Safety Pin | Inserted | —Holds right legbrace handgrip down. The streamer is attached to the canopy jettison handle safety pin streamer. |
| 6 | Ground Safety Pin | Provides mechanical safing of the safety belt initiator during ground maintenance. | |
| 7 | Crew/Survival Kit Retention Strap (HBU Safety Belt only) | Retains crew and survival kit in position during zero and negative G maneuvers. | |

Ejection Seat Safety Pin.

The safety pin (figure 1-20), when inserted, holds the right legbrace handgrip down, preventing inadvertent seat ejection. The streamer for the ejection seat safety pin is attached to the streamer for the canopy jettison T-handle safety pin.

Catapult Triggers.

The catapult triggers (figure 1-20) are locked in the stowed position when the legbraces are down. When the handgrips are raised, the triggers move to the exposed position. Squeezing either or both catapult triggers jettisons the canopy of that cockpit, followed in 0.3 second by seat ejection.

Seat Adjustment Switch.

A seat adjustment switch (figure 1-20) on the right legbrace provides control of seat adjustment thru a vertical range of 5 inches. The adjustment switches operate on ac.

**CAUTION**

Hard items stored under the seat may puncture the cockpit floor when the seat is lowered resulting in loss of cabin pressurization.

Oxygen/Communication Block.

The oxygen/communication block (figure 1-20) is secured to the seat to prevent injury to the pilot during ejection. The oxygen hose retention strap (figure 1-20) effects positive hose disconnection after man-seat separation. A snap fastener on the retention strap allows individual adjustment of the oxygen hose to obtain freedom of movement without disconnecting the hose.

Anti-G Suit Hose.

The anti-G suit hose (figure 1-20) is located on the left side of the ejection seat next to the headrest. The hose is held in the stowed position by a flexible spring.

#### Inertia Reel Lock Lever.

A shoulder harness inertia reel lock lever (figure 1-20) is located on the left legbrace.

#### Automatic-Opening Safety Belt.

The automatic opening safety belt (HBU or MA-5) is equipped with a 0.65 second delay initiator. The use of either type automatic-opening safety belt greatly reduces the time for separation from the seat and deployment of the parachute and consequently reduces the altitude required for safe ejection. The MA-5 safety belt incorporates a manual release lever, which locks both halves of the safety belt together when positioned into the detent slot. The belt is manually opened by pulling the lever up. Refer to figure 1-21 for proper connection and operation of the MA-5 belt. The HBU safety belt has an improved design buckle assembly to prevent inadvertent opening of the belt. The buckle on the left half of the belt incorporates a rotary latch mechanism, consisting of a belt latch, lanyard latch, interlock device, and a serrated manual release handle spring-loaded to the locked position. The interlock device prevents fastening the safety belt without first attaching the automatic parachute arming lanyard into the lanyard latch. Actuation of the handle is not necessary when manually attaching the lanyard anchor and connecting the right half of the safety belt. The lanyard anchor, once inserted, remains locked in the latch mechanism during automatic opening, and can be released only manually. Full manual counterclockwise rotation of the release handle releases the lanyard anchor and the belt link. Refer to figure 1-22 for proper connection and operation of the HBU belt.

#### Man-Seat Separation System.

A man-seat separation system forcibly separates the crewmember from the ejection seat when the safety belt initiator fires after ejection. On ejection, man-seat separation is aided by full deployment of the drogue chute.

### PERSONNEL LOCATOR BEACON.

A personnel locator beacon installed in the parachute harness is used in locating crewmembers who have ejected. The beacon transmits a signal on 243.0 megahertz. The beacon will operate automatically upon parachute deployment when the actuator tab is snapped to the stud tab below the canopy release on the right-hand main lift web (figure 1-23). With the actuator tab unsnapped, the beacon will not operate automatically.

## SURVIVAL KIT.

The CNU-68/P or CNU-68A/P survival kit (figure 1-24) is designed to fit in the ejection seat and be used as a seat cushion with a back type parachute. The kit is divided into two sections, an aft section and a forward section. The aft section serves as a support for the back type parachute. The forward section contains a life raft attached to a 20-foot lanyard and a $CO_2$ bottle to inflate the life raft. The survival kit is attached to the crewmember's parachute harness by attaching straps on each side of the survival kit. An emergency release handle is located on the right side of the survival kit forward section. Pulling the emergency release handle during descent after ejection releases and inflates the life raft. Pulling the emergency release handle while seated in the aircraft releases both attaching straps from the survival kit, and the crewmember will be free of the unopened survival kit.

**WARNING**

Seats containing a survival kit or a spacer with a 4-inch thick seat cushion in place of a survival kit shall contain a cutout to prevent interference with control stick full aft movement.

### IMPROVED SURVIVAL KIT (AFTER T.O. 1T-38A-940)

The improved survival kit (figure 1-24A) fits in the ejection seat and is attached to the parachute harness by two quick-disconnect buckle/web strap assemblies. The forward section of the kit top is equipped with a seat cushion. The rear section of the kit top provides support for a back type parachute. Depending upon local command desires, kit contents will vary and may include a life raft.

### AUTOMATIC/MANUAL DEPLOYMENT

The kit will be automatically released during the ejection sequence or retained for normal release, depending upon the selected position of the survival kit AUTO/MANUAL selector before ejection. During parachute deployment, the parachute shroud lines pull the kit auto-release cable. If the AUTO/MANUAL selector is at AUTO, the kit auto-release cable pull will cause an initiator cartridge to fire, and after a 4-second delay, the survival kit is automatically released. If the selector is at manual, the cartridge is safetied and the kit must then be released manually by pulling the emergency release handle. When the kit is released, either automatically or manually, the quick-disconnect

# AUTOMATIC-OPENING SAFETY BELT MA-5

## LOCKED

1. INITIATOR HOSE TO AUTOMATIC RELEASE MECHANISM.
2. SHOULDER HARNESS LOOPS OVER SWIVEL LINK.
3. ANCHOR (FROM AUTOMATIC PARACHUTE ARMING LANYARD) SLIPPED OVER SWIVEL LINK.

**WARNING**

- ANCHOR MUST BE INSTALLED AS SHOWN OR PARACHUTE WILL NOT OPEN AUTOMATICALLY IF EJECTION IS NECESSARY.
- LANYARD MUST BE OUTSIDE PARACHUTE HARNESS AND NOT FOULED ON ANY EQUIPMENT, TO PERMIT CLEAN SEPARATION FROM SEAT.

4. MANUAL RELEASE LEVER CLOSED.

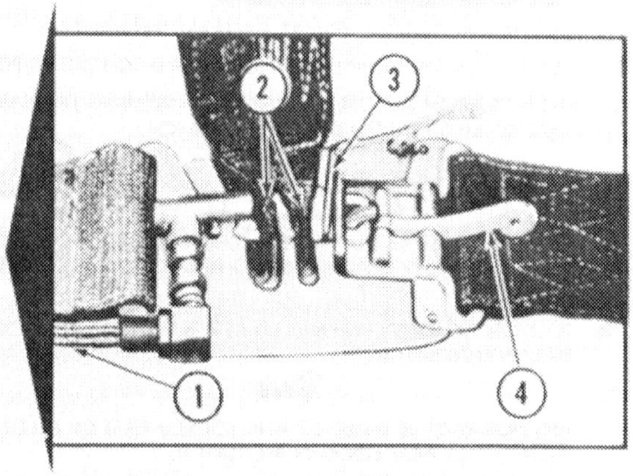


## AUTOMATICALLY OPENED

1. AUTOMATIC RELEASE MECHANISM ACTUATED BY GAS PRESSURE FROM INITIATOR; SWIVEL LINK DETACHED ON AUTOMATIC RELEASE SIDE.
2. ANCHOR (FROM AUTOMATIC PARACHUTE ARMING LANYARD) RETAINED BY SWIVEL LINK.
3. MANUAL RELEASE LEVER CLOSED.
4. SWIVEL LINK RETAINED BY MANUAL RELEASE LEVER.

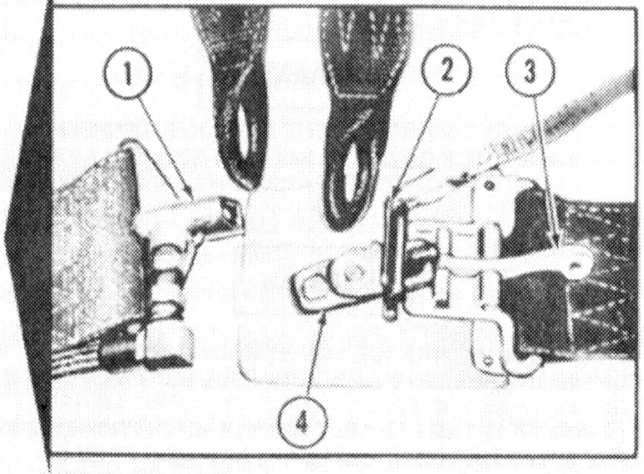


## MANUALLY OPENED

1. SWIVEL LINK RELEASED BY MANUAL RELEASE LEVER (AUTOMATIC RELEASE MECHANISM NOT ACTUATED).
2. ANCHOR (FROM AUTOMATIC PARACHUTE ARMING LANYARD) FREED FROM SWIVEL LINK.

**WARNING**

IF THE BELT IS MANUALLY OPENED DURING EJECTION, THE PARACHUTE WILL NOT OPEN AUTOMATICALLY UPON SEPARATION FROM THE SEAT.

3. MANUAL RELEASE LEVER OPENED.

*Note*

MANUAL RELEASE LEVER CAN BE USED TO UNLOCK BELT AT ANY TIME, EVEN IF AUTOMATIC-OPENING SEQUENCE HAS BEEN INITIATED.

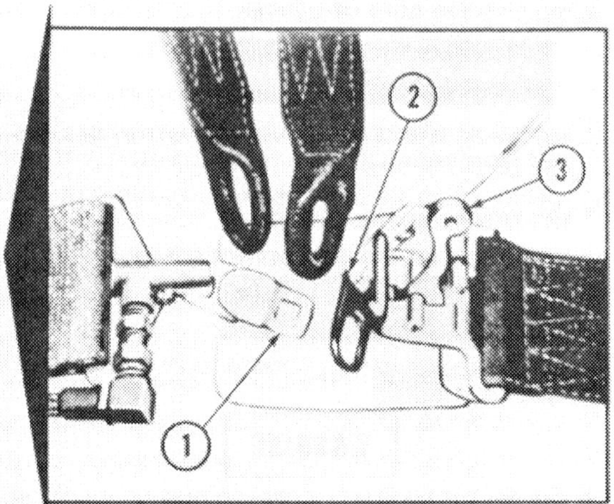


T-38A 1-33 E

Figure 1-21.

# AUTOMATIC-OPENING SAFETY BELT HBU

**LOCKED**

1. INITIATOR HOSE TO AUTOMATIC RELEASE MECHANISM.
2. MANUAL RELEASE HANDLE SPRING-LOADED TO LOCKED POSITION.
3. ANCHOR (FROM AUTOMATIC PARACHUTE ARMING LANYARD) INSERTED AND LOCKED IN LANYARD LATCH.

**WARNING**

- LANYARD MUST BE OUTSIDE PARACHUTE HARNESS AND NOT FOULED ON ANY EQUIPMENT, TO PERMIT CLEAN SEPARATION FROM SEAT.
- PULL ON ARMING LANYARD TO ASSURE ANCHOR IS LOCKED IN LANYARD LATCH.

*Note*

ANCHOR MUST BE INSERTED AND LOCKED FIRST OR BELT LINK (ITEM 5) WILL NOT LOCK IN BELT LATCH.

4. SHOULDER HARNESS LOOPS OVER BELT LINK.
5. CREW/SURVIVAL KIT RETENTION STRAP OVER BELT LINK AFTER SHOULDER HARNESS.

**WARNING**

FAILURE TO INSTALL THE CREW/KIT RETENTION STRAP LOOP ON BELT LINK LAST MAY DELAY OR NEGATE SEAT/MAN SEPARATION DURING EJECTION.

6. BELT LINK INSERTED AND LOCKED IN BELT LATCH.

**WARNING**

THE HBU-2B/A LAP BELT CAN BE INADVERTENTLY OPENED BY FULL AFT AND THEN FULL LEFT MOVEMENT OF THE CONTROL STICK. THIS CONDITION CAN BE ALLEVIATED IF THE LAP BELT BUCKLE IS POSITIONED SO THE SHOULDER HARNESS LOOPS ARE CENTERED ON THE INDIVIDUAL WHICH WILL OFFSET THE LAP BELT BUCKLE TO THE LEFT OF CENTER.

**AUTOMATICALLY OPENED**

1. AUTOMATIC RELEASE MECHANISM ACTUATED BY GAS PRESSURE FROM INITIATOR; BELT LINK AUTOMATICALLY RELEASED.
2. ANCHOR (FROM AUTOMATIC PARACHUTE ARMING LANYARD) RETAINED IN LANYARD LATCH.
3. MANUAL RELEASE HANDLE DOES NOT ROTATE.

**MANUALLY OPENED**

1. MANUAL RELEASE HANDLE ROTATED FULLY COUNTERCLOCKWISE.

**WARNING**

IF THE BELT IS MANUALLY OPENED DURING EJECTION, THE PARACHUTE WILL NOT OPEN AUTOMATICALLY UPON SEPARATION FROM THE SEAT.

2. ANCHOR (FROM AUTOMATIC PARACHUTE ARMING LANYARD) RELEASED FROM LANYARD LATCH.
3. BELT LINK RELEASED FROM BELT LATCH.

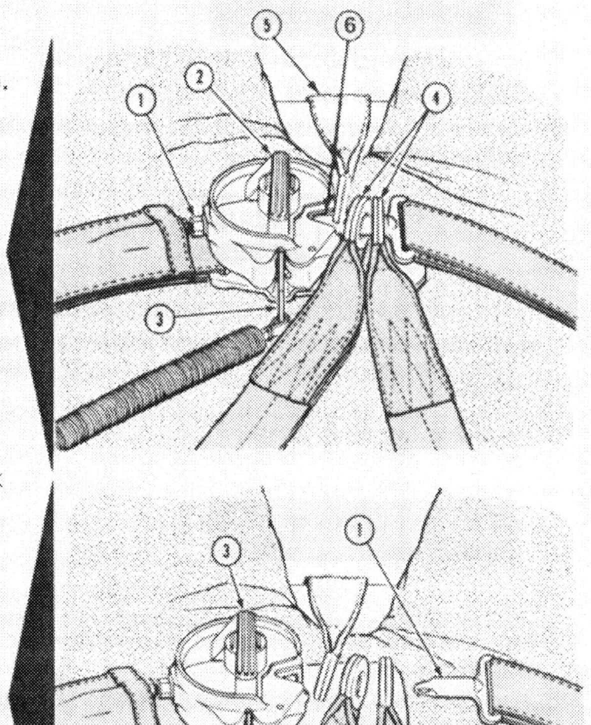


Figure 1-22.

# STRAP-IN CONNECTIONS

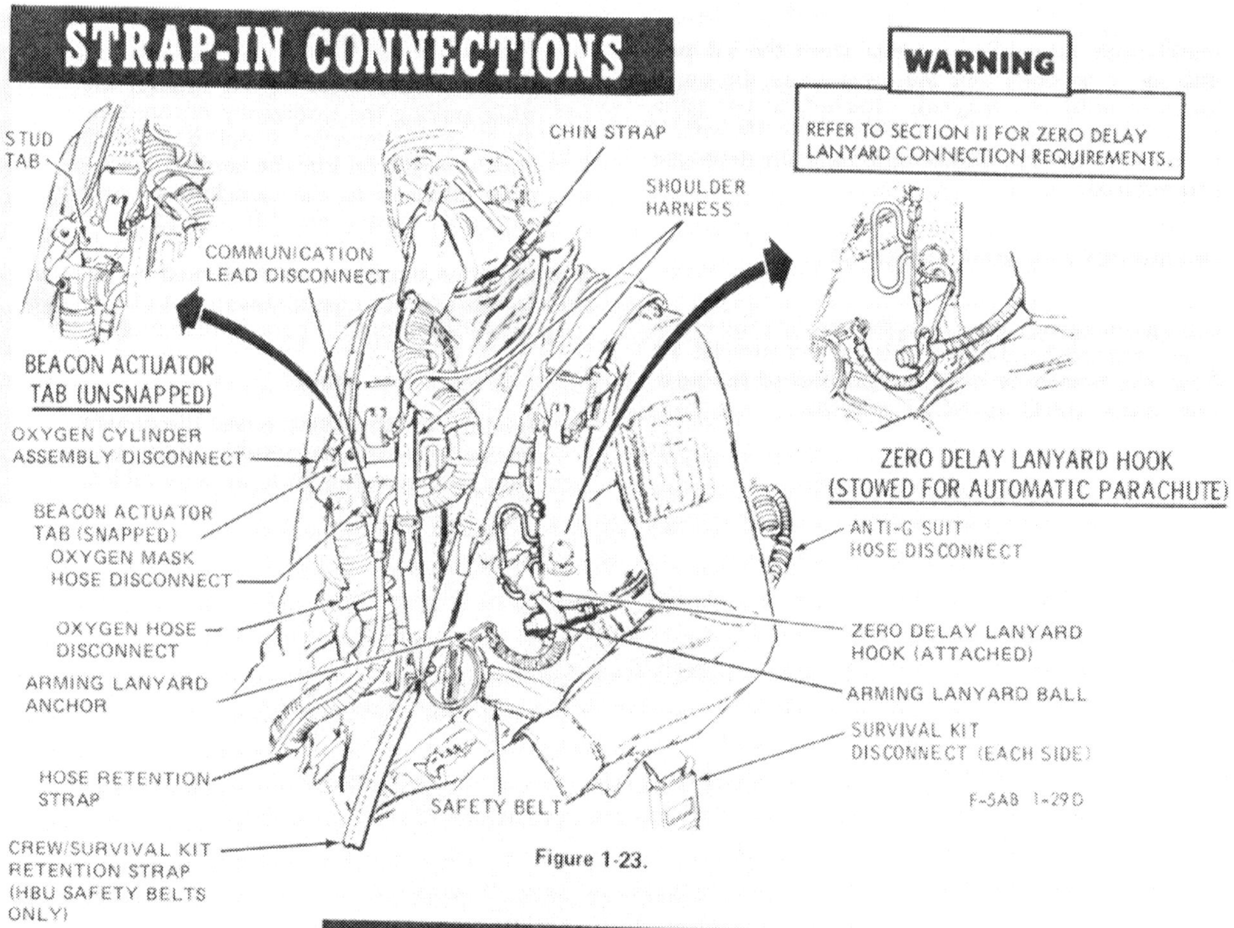


Figure 1-23.

# SURVIVAL KIT

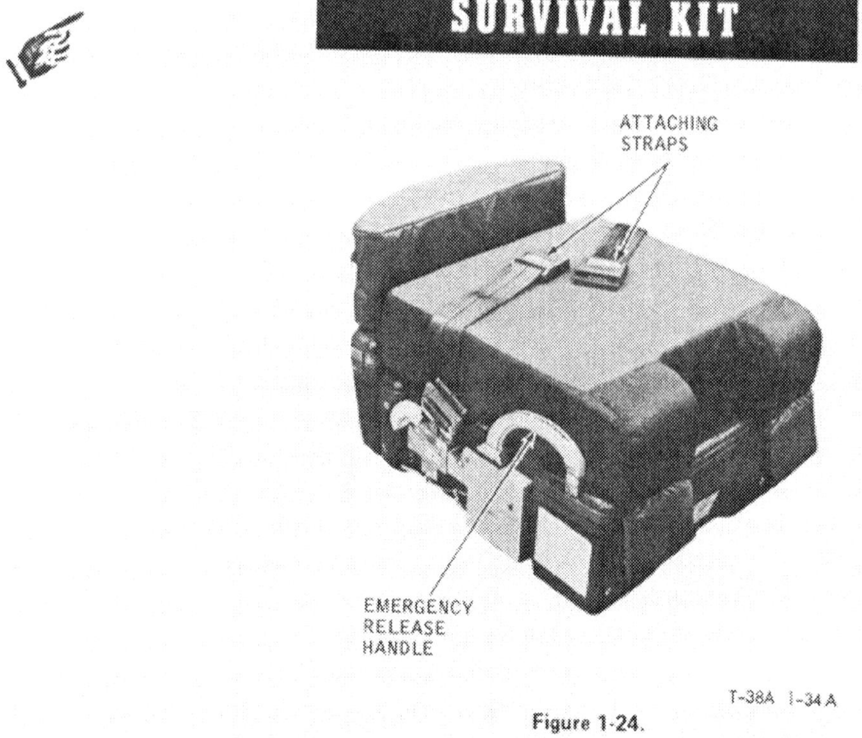


Figure 1-24.

buckle/web assemblies separate from the kit permitting it to open and fall away from the crewmember until the lanyard, attached to the parachute harness, is fully extended. The life raft, if included in the kit, will be automatically deployed and inflated.

## EMERGENCY AND NORMAL EGRESS

During emergency ground egress, pulling the emergency release handle will release the survival kit from the parachute harness regardless of the position of the AUTO/MANUAL selector.

**NOTE**

Pilot's weight must be on survival kit while pulling the emergency release handle. If the handle is pulled without pilot's weight on kit, the lanyard will remain attached to the parachute harness and could cause egress difficulties.

Normal egress should be accomplished by manually disconnecting both quick-disconnect buckle/web strap assemblies from the parachute harness.

**NOTE**

Binding of the right quick disconnect buckle/web strap is possible while manually disconnecting from the survival kit.

# SURVIVAL KIT IMPROVED

AFTER T.O. 1T-38A-940

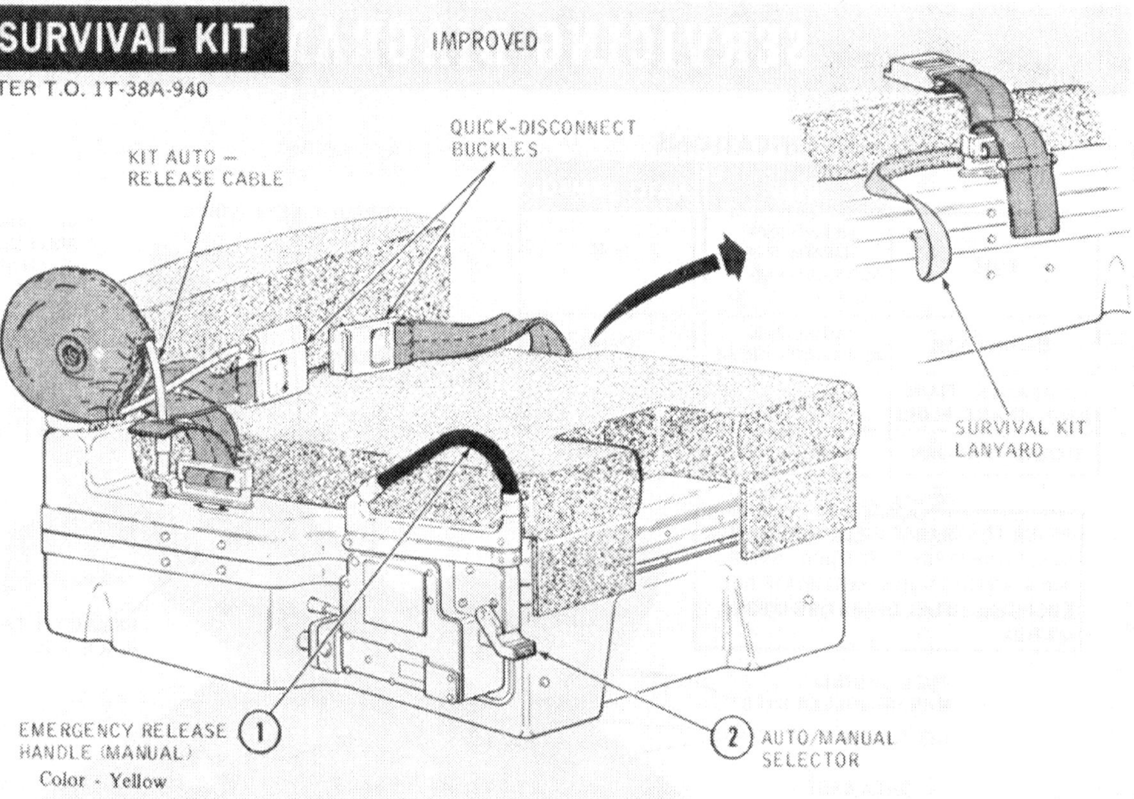


Figure 1-24A.

**SURVIVAL KIT — IMPROVED**

| CONTROLS | | FUNCTION |
|---|---|---|
| 1 Emergency Release Handle | Pull | — a. After ejection, with AUTO/MANUAL selector at MANUAL; releases kit.<br>b. While seated on survival kit, regardless of the position of the AUTO/MANUAL selector; releases both quick-disconnects from kit. |
| 2 AUTO/MANUAL Selector | AUTO (Up) | — Permits automatic deployment of survival kit 4 seconds after parachute shroud lines are fully stretched. |
| | MANUAL (Down) | — Permits manual deployment of survival kit when emergency release handle is pulled. |

# SERVICING DIAGRAM

## FLUID SPECIFICATIONS

| | USAF | NATO SYMBOL |
|---|---|---|
| FUEL | MIL-T-5624 GRADE JP-4 ALTERNATE: * EMERGENCY: * | F-40 |
| ENGINE OIL | MIL-L-7808 ALTERNATE: NONE | O-148 |
| HYDRAULIC FLUID AND BRAKE FLUID | MIL-H-5606 | H-515 |
| LIQUID OXYGEN | MIL-O-27210 | NONE |

*Note

REFER TO STRANGE FIELD PROCEDURES IN SECTION II FOR ELECTRICAL UNITS, AIR-STARTING UNITS, ALTERNATE AND EMERGENCY FUEL OPERATING INFORMATION.

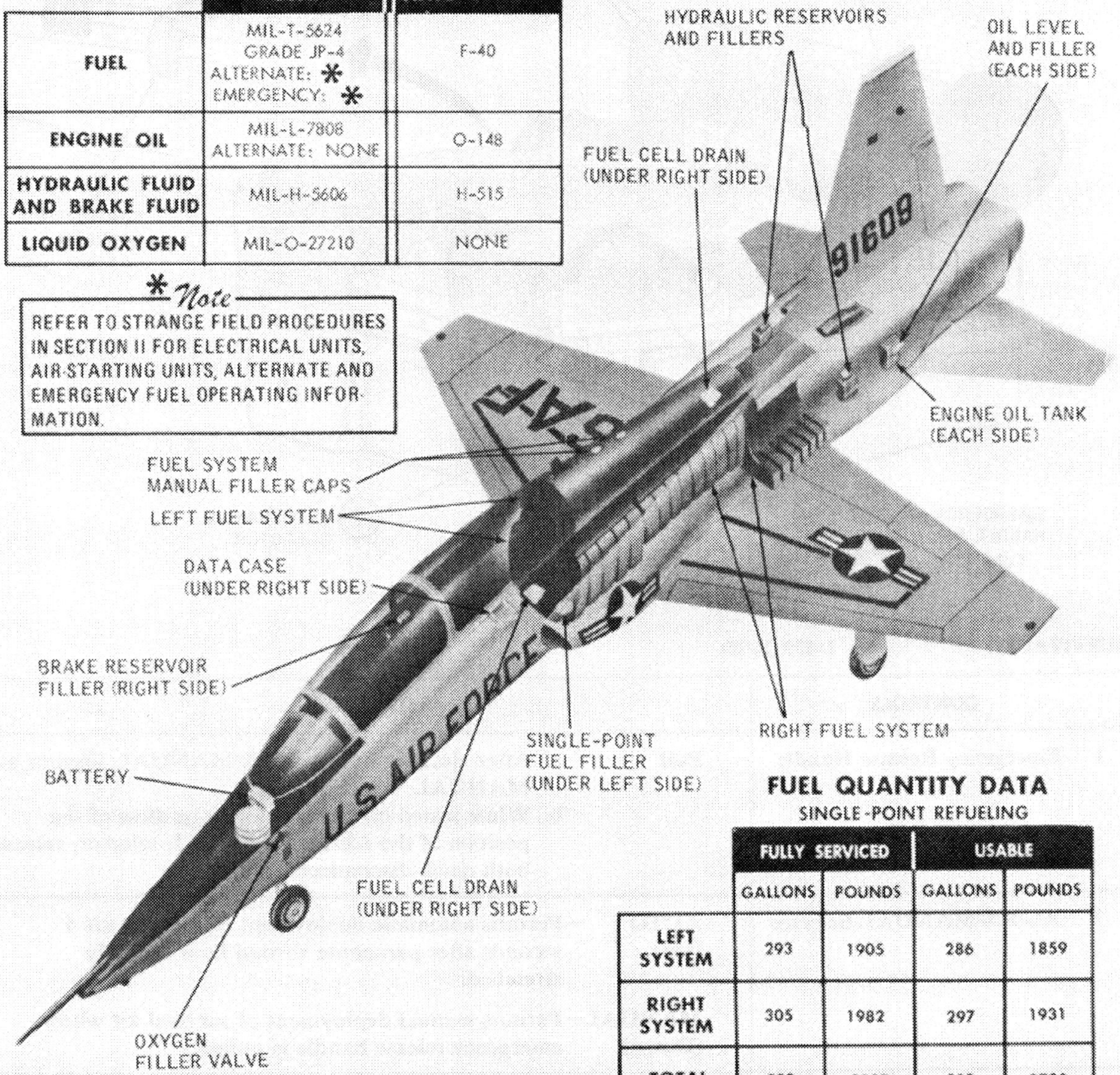


## FUEL QUANTITY DATA

SINGLE-POINT REFUELING

| | FULLY SERVICED | | USABLE | |
|---|---|---|---|---|
| | GALLONS | POUNDS | GALLONS | POUNDS |
| LEFT SYSTEM | 293 | 1905 | 286 | 1859 |
| RIGHT SYSTEM | 305 | 1982 | 297 | 1931 |
| TOTAL | 598 | 3887 | 583 | 3790 |

Note

- FUEL — JP-4 6.5 LB PER GALLON
- SUBTRACT 6 GALLONS (39 POUNDS) FROM TOTALS IF MANUALLY REFUELED.

DATA BASIS: **ACTUAL**
DATE: 1 JULY 1978

T-38A 1-35R

Figure 1-25.

# NORMAL PROCEDURES

## SECTION II

T-38A 1-101

TABLE OF CONTENTS



### PREFLIGHT CHECK.

#### BEFORE EXTERIOR INSPECTION.

(B) 1. Pylon Safety Pin — Installed (If Applicable).

2. Form 781 — Check for both aircraft status and proper servicing.

3. Seat and Canopy Safety Pins — Installed. If safety pins other than seat and canopy pins are installed, do not remove until the status of the ejection system has been checked with maintenance personnel.

4. Seat Pack Tiedown Straps — Security and Condition (if applicable).

5. Oxygen Hose Retention Strap — Check Security and Adjust.

6. Seat Attach Bolts — Check.

**WARNING**

The two attach bolts must be aligned with reference line (or shoulder) of catapult head. (See EJECTION SEAT illustration, Sec. I.)

7. Publications — Check to ensure that all required navigational publications are on board.

8. Fuel and Oxygen Quantity — Check.

9. Wing Flap Lever — UP.

NOTE

If the flaps are other than full up the flap switch must be set to correspond with the actual flap position. Otherwise, inadvertent flap extension/retraction will occur when AC power is applied.

10. Landing Gear Lever — LG DOWN.

(B) 11. Armament Switches — Safe.

(B) 12. Reticle Intensity Knob — OFF.

13. Radio Transfer Switch — FWD.

## EXTERIOR INSPECTION.

Conduct the exterior inspection as shown in figure 2-1.

## INTERIOR INSPECTION.

### Rear Cockpit (Solo Flights).

1. Seat and Canopy Safety Pins — Check Installed, Streamers Fastened Together.

2. Survival Kit/Seat Pack — Remove, or Secure.

WARNING

Seat safety belt and shoulder harness do not provide adequate restraint for survival kit/seat pack during zero or negative-G maneuvers.

3. Safety Belt, Shoulder Harness, Crew/Kit Retention Strap, Oxygen Hose and Man-Seat Separator Straps — Secure and Lock.

WARNING

If these items are not secured, they may become entangled with the control stick. In securing the man-seat separator straps, do not twist the upper portion which is reeled into the back of the ejection seat.

4. Stowage Box Cover — Closed and Secured.

5. Communication and Navigation Equipment — Check.

   a. Command radio: Function Switch — BOTH; Manual/Preset/Guard Switch — GUARD.

   b. TACAN: Function Switch — T/R; Channel Selector Knobs — Desired Channel.

   c. ILS: Steering Mode Switch — NORMAL; Navigation Mode Switch — LOCALIZER; Power Switch — POWER; Channel Selector — Desired Channel.

6. Command and Navigation Override Switch — OFF.

7. Loose Equipment — Check Securely Stowed.

8. Circuit Breakers — Check.

9. Lights — OFF.

10. Oxygen — NORMAL — 100% — ON.

11. Instrument Hood — Remove or Secure. Check all bungee cords connected.

12. Canopy — Closed and Locked.

CAUTION

- The pilot will personally close and lock the rear canopy. To check for a locked condition, the pilot will push up on the canopy.
- While stowing the outside handle, do not apply clockwise pressure after the canopy is locked.

### Cockpit (All Flights).

On dual flights, all items marked with an asterisk should also be checked in the rear cockpit.

1. Battery/External Electrical Power — As Required.

CAUTION

If external power is connected the battery should be turned off to prevent battery damage.

# EXTERIOR INSPECTION

DURING THE EXTERIOR INSPECTION, THE AIRCRAFT SHOULD BE CHECKED FOR GENERAL CONDITION, WHEELS CHOCKED, ACCESS DOORS, PANELS, AND FILLER CAPS SECURED, GROUND WIRES REMOVED, FOR HYDRAULIC, OIL, AND FUEL LEAKS, AS WELL AS FOR THE FOLLOWING SPECIFIC ITEMS:

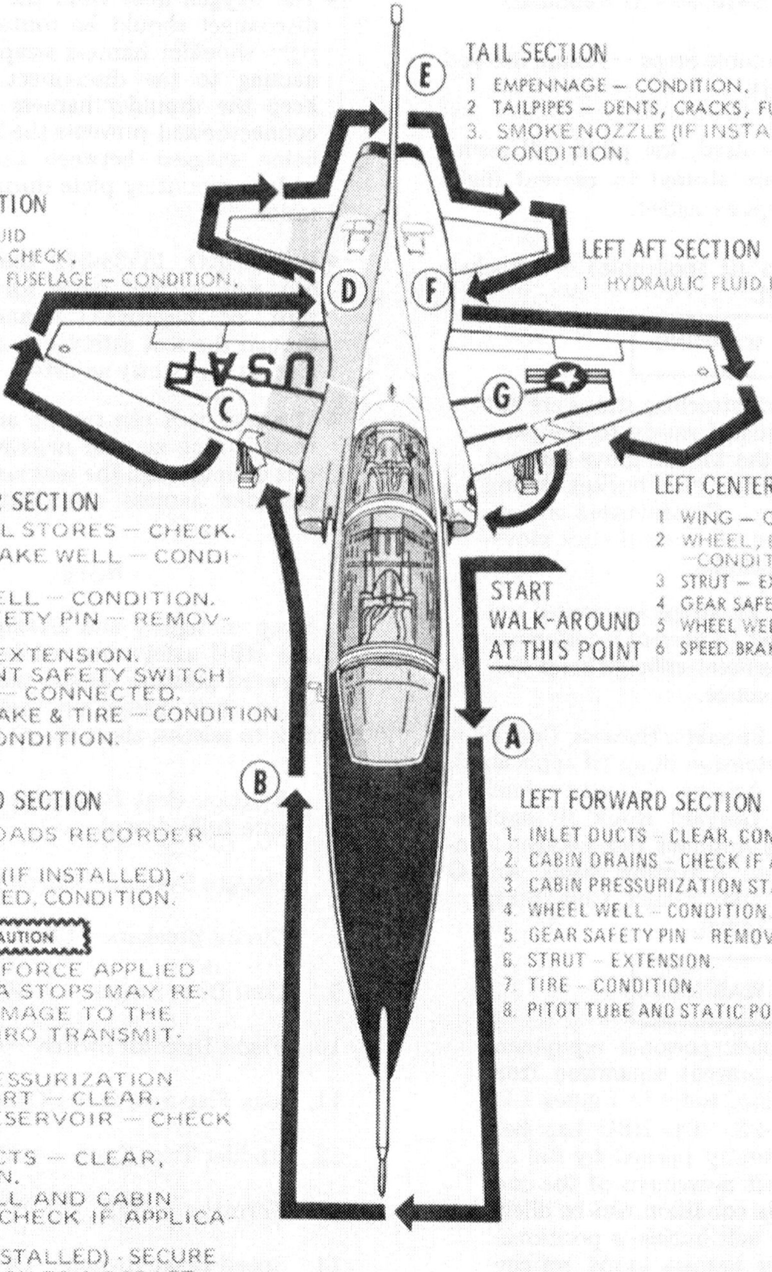


**TAIL SECTION**

1. EMPENNAGE — CONDITION.
2. TAILPIPES — DENTS, CRACKS, FUEL.
3. SMOKE NOZZLE (IF INSTALLED)-CONDITION.

**RIGHT AFT SECTION**

1. HYDRAULIC FLUID INDICATORS — CHECK.
2. UNDERSIDE OF FUSELAGE — CONDITION.

**LEFT AFT SECTION**

1. HYDRAULIC FLUID INDICATORS — CHECK.

**RIGHT CENTER SECTION**

(B) 1. EXTERNAL STORES — CHECK.
2. SPEED BRAKE WELL — CONDITION.
3. WHEEL WELL — CONDITION.
4. GEAR SAFETY PIN — REMOVED.
5. STRUT — EXTENSION.

(B) 6. ARMAMENT SAFETY SWITCH LINKAGE — CONNECTED.
7. WHEEL, BRAKE & TIRE — CONDITION.
8. WING — CONDITION.

**LEFT CENTER SECTION**

1. WING — CONDITION.
2. WHEEL, BRAKE & TIRE — CONDITION.
3. STRUT — EXTENSION.
4. GEAR SAFETY PIN — REMOVED.
5. WHEEL WELL — CONDITION.
6. SPEED BRAKE WELL — CONDITION.

**RIGHT FORWARD SECTION**

(B) 1. FLIGHT LOADS RECORDER SET.
2. AOA VANE (IF INSTALLED) - PIN REMOVED, CONDITION.

CAUTION

EXCESSIVE FORCE APPLIED TO THE AOA STOPS MAY RESULT IN DAMAGE TO THE AOA SYNCHRO TRANSMITTER.

3. CABIN PRESSURIZATION STATIC PORT — CLEAR.
4. BRAKE RESERVOIR — CHECK LEVEL.
5. INLET DUCTS — CLEAR, CONDITION.
6. STICK WELL AND CABIN DRAINS — CHECK IF APPLICABLE.
7. WSSP (IF INSTALLED) - SECURE CHECK PYLON TO AIRCRAFT ATTACH BOLTS AND POD NOSE SECTION FOR SECURITY.

**LEFT FORWARD SECTION**

1. INLET DUCTS - CLEAR, CONDITION
2. CABIN DRAINS - CHECK IF APPLICABLE.
3. CABIN PRESSURIZATION STATIC PORT - CLEAR.
4. WHEEL WELL - CONDITION.
5. GEAR SAFETY PIN - REMOVED.
6. STRUT - EXTENSION.
7. TIRE - CONDITION.
8. PITOT TUBE AND STATIC PORTS - CLEAR.

T-38A 1-36N

Figure 2-1.

NOTE

If the aircraft will not accept external AC power and the APU checks good, cycling the battery switch ON then OFF may actuate necessary relays to allow the aircraft to accept external AC power.

*2. Intercomm Switches - As Required.

3. Crew Retractable Steps — Assure Stowed (If Required).

If steps are used, the pilot will assure that they are stowed to prevent flight with the steps extended.

*4. Survival Kit (if applicable) — Attached and adjusted.

WARNING

- If the survival kit attaching straps are not attached or adjusted snugly to the parachute harness, the kit can move forward and up out of the seat bucket during zero or negative - G maneuvers and restrict or prohibit aft control stick movement in both cockpits.
- Ensure survival kit straps are routed under the safety belt to prevent interference and possible man-seat entanglement during ejection sequence.

*5. Safety Belt, Shoulder Harness, Crew/Survival Kit Retention Strap (if applicable), Parachute Arming Lanyard Anchor, Zero-Delay Lanyard Hook (if applicable), Beacon Actuator Tab, Oxygen Connectors, Hose Retention Strap, Anti-G Suit Hose, and Helmet Chin Strap — Fasten and Adjust.

WARNING

- Failure to attach personal equipment correctly may prevent separation from seat after ejection. Refer to figures 1-21 or 1-22 and 1-23. The HBU Lap Belt can be inadvertently opened by full aft and then full left movement of the control stick. This condition can be alleviated if the lap belt buckle is positioned so the shoulder harness loops are centered on the individual which will offset the lap belt buckle to the left of center.
- Assure that hose retention strap is adjusted to preclude hose separation from oxygen-disconnect on parachute harness.
- Do not route the anti-G suit hose under the safety belt or in any manner which would interfere with disconnecting the hose if required.
- The oxygen hose from the mask to the disconnect should be routed under the right shoulder harness strap before connecting to the disconnect. This helps keep the shoulder harness clear of the connector and prevents the harness from being snagged between the connector and its mounting plate during seat separation.
- Before T.O. 1T-38-579 the survival kit can rise up and move forward during zero or negative-G maneuvers even though the seat safety belt and shoulder harness are tightly adjusted.
- The seat pack can rise up and move forward during zero or negative-G maneuvers even through the seat safety belt and shoulder harness are tightly adjusted.

NOTE

Strap in tightly and attempt to release the HBU safety belt while maintaining forward pressure against the belt. If the safety belt hangs up momentarily or fails to release, abort the aircraft.

*6. Ejection Seat Handgrips — Push (to ensure fully down).

*7. Oxygen System — Check (PRICE).

*8. Circuit Breakers — Check.

9. Gear Door Switch — NORMAL.

10. Flight Director Switch — ON.

11. Aux Flap Switch — NORMAL.

12. Rudder Trim Knob — CENTERED.

*13. Throttles — OFF.

14. Speed Brake Switch — OPEN.

15. Compass Switch — MAG.

16. Fuel Shutoff Switches — NORMAL (guarded position).

17. Landing Gear Alternate Release Handle — IN.

18. Landing-Taxi Light Switch — OFF.

*19. Landing Gear Lever — LG DOWN.

*20. Airspeed-Mach Indicator — Check.

*21. Accelerometer — Check.

*22. Clock — Set.

*23. Steering Mode Switch — As Required.

*24. Navigation Mode Switch — As Required.

*25. Marker Beacon Light — Test.

*26. Radio Transfer Switches — As Required.

27. Comm Antenna Switch — AUTO.

*28. UHF, TACAN, ILS — ON.

29. Magnetic Compass — Check.

*30. Altimeter — Set.

*31. Vertical Velocity Indicator — Check.

32. Cabin Altimeter — Check.

33. Cabin Pressure Switch — CABIN PRESS.

34. Cabin Air Temperature Switch — AUTO.

35. Pitot Heat Switch — OFF.

36. Engine Anti-Ice Switch — As Required.

37. Fuel Boost Pump Switches — ON.

38. Crossfeed Switch — OFF.

39. Generator Switches — ON.

40. IFF/SIF — STBY.

(B) 41. MXU — ON.

*42. Warning Test Switch — TEST.

(Without ac power on Block 30 and earlier aircraft, no landing gear audible warning signal.)

NOTE

- When the test switches in both cockpits are actuated simultaneously, all fire warning lights will illuminate. The landing gear audible warning signal will not come on in either cockpit.
- All four fire warning light bulbs in both cockpits must illuminate during TEST. Failure of any bulb to illuminate may indicate an inoperative fire detector.

*43. Interior and Position Lights — As Required.

44. Rotating Beacon — ON.

*45. Forms/Publications — Stowed.

(B) 46. Sight Dust Cover — Stowed.

## STARTING ENGINES.

### RIGHT ENGINE.

Start the right engine first, using the following procedure:

1. Signal for air supply.
2. Engine Start Button — Push Momentarily.
3. Throttle — Advance to IDLE at 14% minimum RPM.

CAUTION

- Prior to moving either throttle to IDLE, assure that the respective OFF flag is out of view or the ON flag is in view as applicable (front cockpit only) otherwise an engine start cannot be properly monitored.
- If ignition does not occur before fuel flow reaches 360 LB/HR, retard throttle to OFF. Maintain airflow to permit fuel and vapors to be purged from engine. Wait at least 2 minutes to permit fuel to drain before attempting another start.
- If EGT does not begin to rise within 12 seconds after the first indication of fuel flow, abort the start. If engine light is normal but RPM do not reach generator cut-in speed before termination of the start cycle, push the engine start button to assure aircraft electrical power is available to monitor the start.

NOTE

Engine speed must reach a minimum of 14% RPM within 15 seconds. If engine speed does not reach 14% RPM within 15 seconds, abort the start and change start carts before attempting another start.

4. Engine Instruments — Check.
5. Hydraulic Pressure — Check.
6. Caution Light Panel — Check.

### LEFT ENGINE.

1. Left Engine — Start Same As Right Engine.

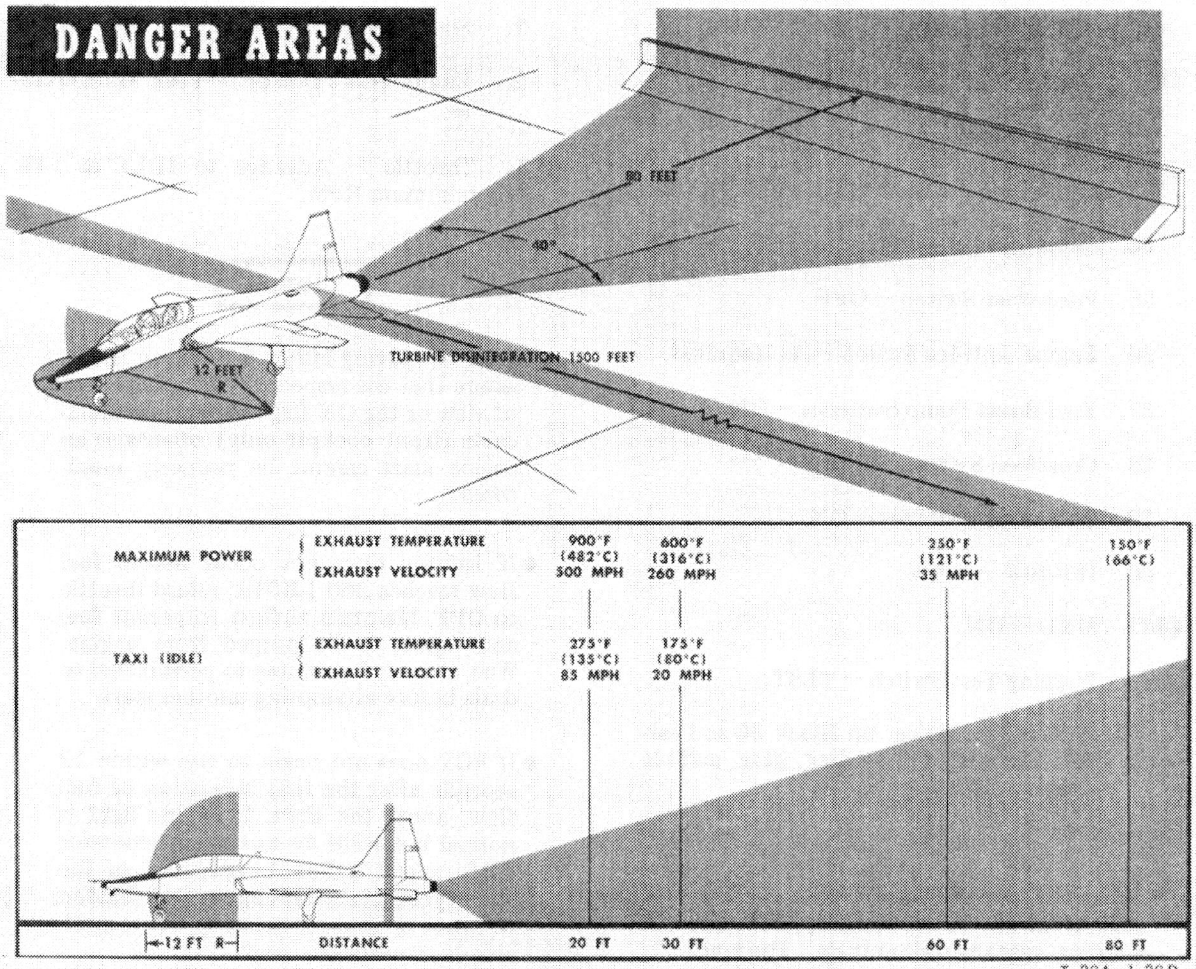


Figure 2-2.

**CAUTION**

Do not push left engine start button until a minimum of 30 seconds has elapsed after right engine start button has been pushed. The left engine start cycle will be shortened and may result in a hot start due to loss of external air to the engine.

2. Signal ground crew to disconnect external power and/or air supply.

3. Battery Switch — Check ON.

## BEFORE TAXIING

**CAUTION**

Allow 3 1/2 minutes to elapse after power has been applied to the attitude gyro control assembly before taxiing.

1. Canopy Defog, Cabin Temp and Pitot Heat — Check (check pitot heat if required).

**WARNING**

For night or anticipated weather operation with conditions of high humidity and narrow temperature-dewpoint spread, the canopies should be closed and the cockpit temperature increased to the 100° AUTO position to preheat all flight instruments and canopy surfaces. Return temperature control to a comfortable inflight setting after completion of the line-up check.

*2. Circuit Breakers — Check.

3. Yaw Damper Switch — YAW.

4. (Block 20 aircraft) Pitch Cutoff Switch — Check.

a. Pitch Damper Switch — PITCH.

b. Pitch Cutoff Switch — Actuate.

c. Pitch Damper Switch — Moves to OFF.

d. Pitch Damper Switch — PITCH; check that horizontal tail does not move.

e. If horizontal tail moved, Pitch Damper Switch — OFF.

5. Flight Trim System — Check.

Verify proper operation of fore and aft trim. Press takeoff trim button until indicator light illuminates.

6. Aileron Trim — Neutral (Check Visually).

7. Flight Controls — Check.
With normal movement, hydraulic pressure should not drop below 1500 PSI. Check visually for proper displacement and freedom of movement. Rudder deflection may be checked with mirrors when canopy is open.

8. Speed Brake — Closed.

9. Wing Flaps — Down to 60%, full down, then retract to 60%.

Check visually that trailing edge of horizontal tail moves down as flaps are lowered, up as flaps are retracted. The 60% flap setting should be checked for a steady gauge reading of 60 ± 5% as flaps are lowered and retracted. To ascertain proper horizontal tail position for takeoff trim, verify with the crew chief that the leading edge of the horizontal tail lies within the upper index marked Takeoff Trim 60% Flap.

**WARNING**

If horizontal tail does not reposition to the proper index, do not attempt flight.

*10. Communication and Navigation Equipment — Check.

Refer to Section IV for description of proper system operation.

B 11. Optical Sight/Camera — Check/Set.

*12. Altimeter — Check (as required).

Refer to Section IV for altimeter operation.

**CAUTION**

Do not rotate the barometric set knob at a rapid rate or exert force to overcome momentary binding. If binding should occur, the required setting may be established by rotating the barometric set knob a full turn in the opposite direction and then approaching the desired setting carefully.

13. Fuel/Oxygen Check Switch — FUEL & OXY GAGE TEST.

B 14. MXU — OFF.

*15. Survival Kit (AUTO/MANUAL) — As Required.

*16. Seat and Canopy Safety Pins — Remove, display to ground crew, and stow.

**CAUTION**

Care should be taken to prevent inadvertent pulling of the canopy Jettison T-handle when removing the safety pin.

*17. Brakes — Check Pedal Pressure.

18. Chocks — Removed.

## TAXIING.

**WARNING**

If carbon monoxide contamination is suspected during ground operation, use 100% oxygen.

**CAUTION**

- To prevent possible damage to the canopy downlock mechanism, taxi with either both canopies open or both closed and pressurized whenever practical.
- If brake drag is encountered or suspected, the aircraft should be aborted.
- Simultaneous use of wheel brakes and nose wheel steering to effect turns results in excessive nosewheel tire wear. Nosewheel tires are severely damaged when maximum deflection turns are attempted at speeds in excess of 10 knots.

**CAUTION**

A low nose gear strut indicates insufficient strut pressure and may result in a cocked nosewheel and/or damage to the nosewheel well during retraction. Do not fly the aircraft if the nose gear strut is deflated or if the strut "bottoms" during taxiing.

*1. Flight Instruments — Check (as required).

## BEFORE TAKEOFF.

*1. Takeoff Data — Review.

2. Battery Switch — Check ON.

3. Crossover Relay Check — Right Generator Switch — OFF then ON (for Flight in IMC).

4. Canopy Defog, Cabin Temp — As Required.

5. Engine anti-Ice — As Required.

*6. Parachute Arming Lanyard, Zero-Delay Lanyard (if applicable) and survival kit (if applicable) — Check Attached.

*7. Cockpit Loose Items — Check Secured.

*8. Helmet Visors — As Required.

9. Flight Controls — Check for free and proper movement.

10. Takeoff Trim Button — Press. Check that indicator light illuminates.

*11. Canopy — Closed, Locked; Warning light — Out.

**WARNING**

Adjust seat height to ensure ability to assume ejection position.

**CAUTION**

- Before lowering the canopy, ensure the hood is not bunched between the ejection seat drogue chute housing and the canopy or damage may occur to the seat or canopy.
- Should the canopy jam in the fully open position, the aircraft should not be taxied or towed until cleared by qualified maintenance personnel. Efforts to close the canopy or vibrations set up by aircraft movement could result in canopy separation.

## LINEUP CHECK.

1. Pitot Heat — As Required.

2. IFF/SIF — As Required.

3. Nosewheel Steering — Check Disengaged.

4. Throttles — MIL.

5. Master Caution Light — Out.

6. Engine Instruments — Check.

7. Hydraulic Pressure — Check.

## TAKEOFF.

The following takeoff procedure, and that given in figure 2-3 forms the basis of the performance predictions in Appendix I. Conditions such as weight, wind, single engine performance considerations, etc., may make it prudent to delay rotation and lift-off above the speeds shown in the figure. However, tire limit speeds should be observed.

**WARNING**

Avoid wake turbulence. Allow a minimum of two minutes before takeoff behind any heavier type aircraft or helicopter and a minimum of four minutes behind jumbo jet aircraft. With effective crosswind of over 5 knots, the interval may be reduced. Attempt to remain above and upwind of the preceding aircraft's flight path. See Section VI.

1. Brakes — Release.

2. Throttles — MAX.

3. Instruments — Check.

**CAUTION**

The takeoff should be aborted if either afterburner fails to light within 5 seconds or if the light off is abnormal.

**NOTE**

The acceleration check speed is the only means by which actual aircraft (engine) performance can be referenced to the computed values. Less than predicted acceleration will invalidate all computed speeds and associated distances.

# NORMAL TAKEOFF (TYPICAL)

BASED ON GROSS WEIGHT OF 12,500 LBS

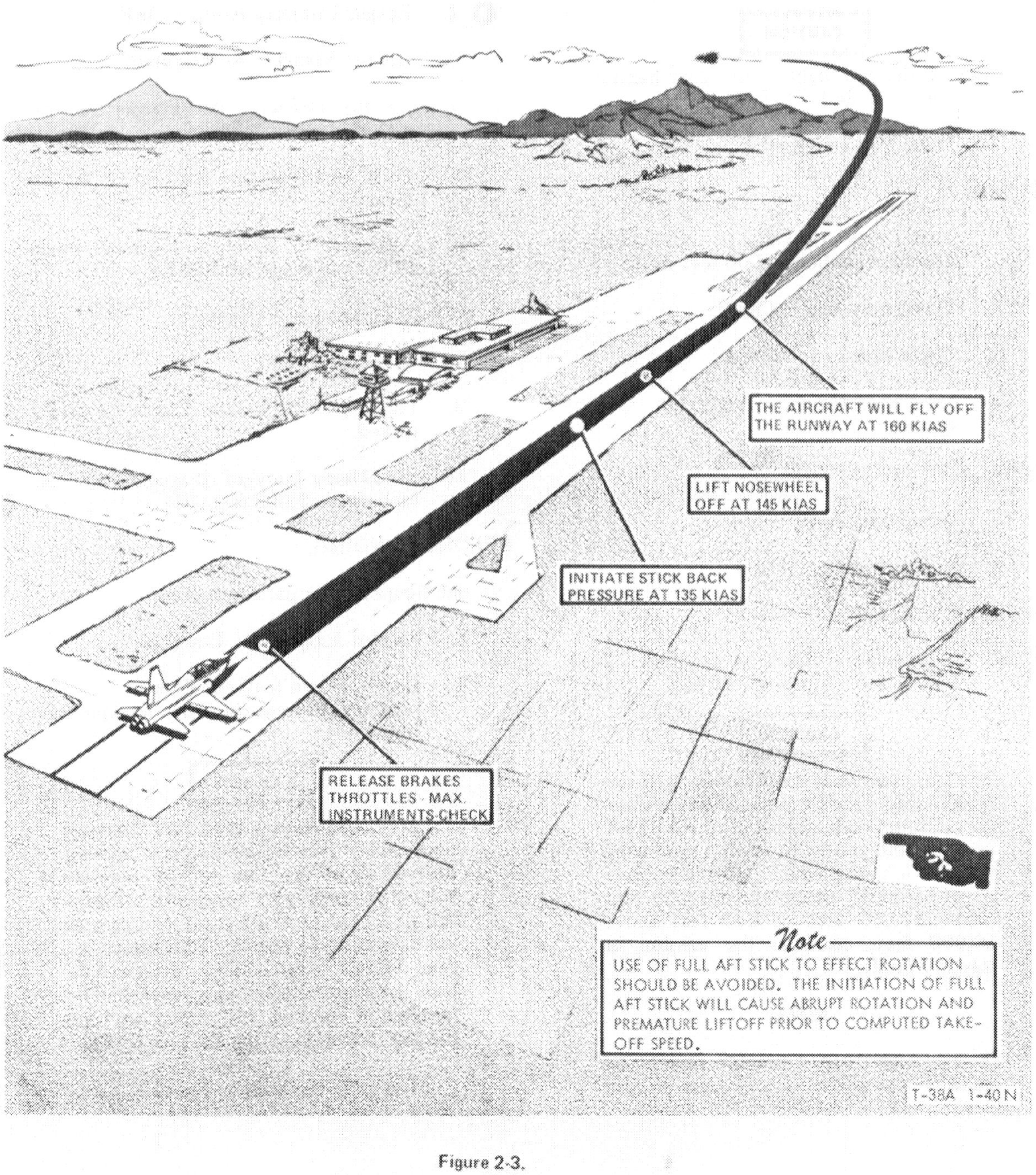


Figure 2-3.

## CROSSWIND TAKEOFF.

Aileron into the wind will aid in directional control and help in preventing compression of the downwind strut. The aircraft should be allowed to crab into the wind as rotation occurs.

## AFTER TAKEOFF.

1. Landing Gear Lever — LG UP, when definitely airborne.

**CAUTION**

Check the red light in the gear handle out prior to 240 KIAS.

2. Wing Flap Lever — UP.

## CLIMB.

*1. Zero-Delay Lanyard (if applicable) — Disconnect above 2,000 feet AGL.

*2. Oxygen System — Check.

3. Cabin Pressure — Check.

4. Canopy Defog and Cabin Temp — As Required.

## LEVEL-OFF AND CRUISE.

*1. Oxygen System — Check.

2. Cabin Pressure — Check.

3. Fuel Quantity — Check.

*4. Altimeter — Reset as required, check STBY and return to RESET.

**CAUTION**

The altimeter may malfunction without reverting to standby mode. If any altimeter malfunction is suspected, check STBY position and return to RESET. If during any check the difference between primary and standby mode exceeds 150 feet below 10,000 feet or 250 feet above 10,000 feet, continue the mission in standby mode.

**NOTE**

If the AAU-19/A altimeter reverts to standby operation at any time during flight, attempt to return to the servoed mode of operation by placing the altimeter function switch momentarily to the RESET position. If the altimeter will not reset or reverts to standby after a few seconds, continue mission in the standby mode.

## DESCENT.

(B) *1. Armament Safety Check — Completed.

(B) 2. Reticle Intensity Knob — OFF.

*3. Helmet Visors — As Required.

4. Canopy Defog, Cabin Temp — As Required.

5. Pitot Heat, Engine Anti-Ice — As Required.

*6. Altimeter — Reset as required, check STBY and return to RESET.

*7. Fuel Balance — Check.

8. Crossfeed — OFF.

9. Landing and Position Lights — As Required.

*10. Zero-Delay Lanyard (if applicable) — Attach above 2,000 feet AGL.

## BEFORE LANDING.

See figure 2-4 for pattern speeds:

*1. Pattern Airspeeds — Compute.

*2. Gear — Down & Check Down. (Physically press the front cockpit lever full down.)

**CAUTION**

Failure of the landing gear lever interconnect cable while the landing gear is being lowered from the rear cockpit may result in normal gear extension without full down travel of the front landing gear lever, leading to possible uncommanded gear retraction on landing. To preclude this, the front landing gear lever shall be physically checked full down anytime the gear is lowered from the rear cockpit.

*3 Hydraulic Pressures — Check.

# LANDING AND GO-AROUND PATTERN
(TYPICAL)

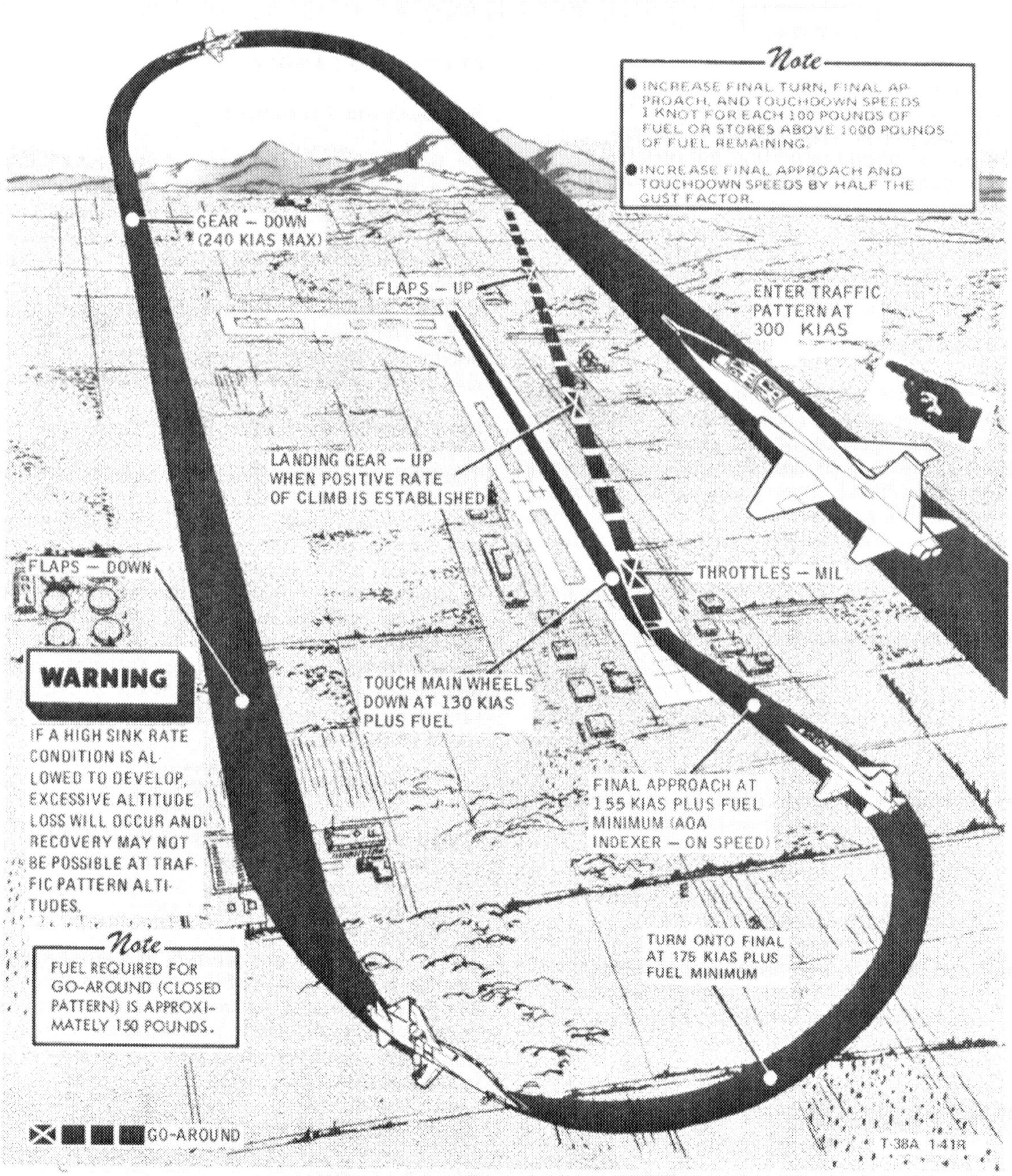


Figure 2-4.

*4. Flaps — As Required.

## LANDING.

WARNING

Avoid wake turbulence. Allow a minimum of two minutes before landing behind any heavier type aircraft or helicopter and a minimum of four minutes behind jumbo jet aircraft. With effective crosswinds of over 5 knots, the interval may be reduced. Attempt to remain above and upwind of the preceding aircraft's flight path. See Section VI.

## NORMAL LANDING.

Refer to figure 2-4 for recommended landing and go-around pattern. After touchdown, continue to increase back-pressure on the stick to obtain the highest possible nose-high attitude without flying the aircraft off the runway. Just prior to loss of elevator authority, lower the nosewheel to the runway. After the nosewheel is lowered to the runway, a single, smooth brake application should be used to stop. This technique could increase landing distance as much as 50 percent from that computed from the landing distance chart in Part 7 of Appendix.

CAUTION

- Extreme CAUTION must be exercised when applying wheel brakes above 120 KIAS as locked wheels or tire skids are difficult to recognize. If tire skids are detected, immediately release both brakes and cautiously reapply.
- Ⓑ Attempting to aerobrake using full back stick until the nose can no longer be held up will produce a hard nosewheel impact at approximately 100 KIAS.

## MINIMUM ROLL LANDING.

Decrease airspeed 10 knots below normal landing final approach airspeed to assure touchdowns at speeds noted in the landing distance chart in Part 7 Appendix. The landing distance chart shows data for landing at computed touchdown speed at approximately 12 degrees nose high attitude. Just prior to loss of elevator authority, lower the nosewheel to the runway and apply optimum wheel braking. For wet runways, a firm touchdown will tend to reduce the effects of hydroplaning.

## CROSSWIND LANDING.

### Approach and Touchdown.

On final approach, counteract drift by crabbing into the wind, maintaining flight path alignment with the runway. The crab should be held thru touchdown. When the crosswind component exceeds 15 knots, touchdown should be planned for the center of the upwind side of the runway. Maintain precise airspeed control throughout the final approach; in gusty conditions, increase the indicated airspeed by one-half of the gust increment above the wind velocity. Refer to Section V for landing rate of descent.

### After Touchdown.

The landing attitude should be maintained by continuing to increase back pressure on the stick. Aileron into the wind will aid in directional control, will help in preventing compression of the downwind strut, and will prevent the upwind wing from becoming airborne. Maintain directional control of the aircraft with the rudder. A too rapid increase in the back stick pressure may cause the aircraft to become airborne and drift across the runway. Drift will create a high probability of tire damage. Just prior to loss of elevator authority, lower the nosewheel to the runway.

CAUTION

Ⓑ Attempting to aerobrake using full back stick until the nose can no longer be held up will produce a hard nosewheel impact at approximately 100 KIAS.

Lowering the nose prematurely in a crosswind will produce a compression of the downwind strut. This hampers directional control and may be minimized by use of aileron. Early downwind strut compression combined with weathervaning usually results in damage to the downwind tire.

## USE OF WHEEL BRAKES.

### Wheel Brake Operation.

To minimize brake wear, the brakes should be used as little and as lightly as possible. If the first application of brakes does not provide adequate pressure or if the brakes feel spongy, normal pressure might be regained by pumping the brake pedals. The pedals should be allowed to return to the full up position between strokes. Failure of certain brake components within a cockpit may result in complete failure of one or both brakes. Should this occur, braking might be gained by operating the brakes in the other cockpit. Full advantage of the length of runway should be taken during landing or aborted takeoff. Minimize use of brakes during turns and avoid dragging the brakes during taxiing. When there is considerable lift on the wings, such as immediately after touchdown, heavy brake pressure will lock the wheel more easily than when the same pressure is applied after the full weight of the aircraft is on the tires. Once a wheel is locked, it may be necessary to completely release brake pressure to allow wheel rotation.

### Optimum Braking Action.

The physical limitations of the tire and brake system make it extremely difficult to consistently achieve optimum braking action, particularly at high speeds (above 120 KIAS), where the weight component is reduced due to lift. A single, smooth application, increasing as airspeed decreases, offers the best braking opportunity. Great caution should be used when braking at speeds above 100 KIAS. Locked brakes are difficult to diagnose until well after the fact. Braking should be discontinued at the first indication of directional problems and then cautiously reapplied. At speeds below 100 KIAS, the chances of approaching optimum braking action are greatly increased.

**WARNING**

- Braking required for high speed, heavy gross weight abort may result in extremely hot brakes or brake failure and the possibility of tire fire should be anticipated.
- If hot brakes are suspected, the aircraft should not be taxied into a congested area. Ensure all personnel remain clear of the main wheels until they have cooled.

**NOTE**

All stopping distances computed from the appendix are based on optimum braking. Optimum braking is difficult to achieve. Variables such as brake and tire condition, pilot technique, etc. may increase computed distances.

## GO-AROUND

Make the decision to go-around as early as possible. Military power is normally sufficient for go-around, but do not hesitate to use maximum power if necessary.

**WARNING**

If conditions do not permit an aerial go-around, do not try to hold the aircraft off the runway; continue to fly the aircraft to touchdown and follow the go-around procedure.

1. Throttles — MIL (MAX if necessary).
2. Landing Gear Lever — LG UP, when definitely airborne.
3. Wing Flap Lever — UP.

**NOTE**

If touchdown is made, lower the nose slightly to accelerate. Establish takeoff attitude to allow the aircraft to fly off the runway at takeoff speed.

## TOUCH-AND-GO LANDINGS.

To make a touch-and-go landing, perform the desired approach and landing. After touchdown, follow the normal go-around procedure.

**WARNING**

Touch-and-go landings encompass all aspects of the landing and takeoff procedures in a relatively short time span. Be constantly alert for possible aircraft malfunctions and/or unsafe operator technique during these two critical phases of flight.

## AFTER LANDING.

1. Pitot Heat — OFF.
2. Cabin Altimeter — Check.
   If reading is below field elevation, place cabin pressure switch at RAM DUMP before opening either canopy.

*3. Cockpit Loose Items — Check Secured (before opening canopy).

*4. Seat and Canopy Safety Pins — Install.

5. Gear Door Switch — OPEN.
*6. Takeoff Trim Button — Press.
*7. Wing Flaps — Up.
8. Speed Brake — Open.
9. Landing-Taxi Light — As Required.
*10. TACAN, ILS, IFF/SIF — OFF.

## ENGINE SHUTDOWN.

*1. Canopy — Unlocked.

NOTE

The canopy seals will remain inflated if engines are shut down with both canopies locked.

2. Cabin Pressure Switch — CABIN PRESS.
3. Operate engines at 70% RPM or below for a minimum of 1 minute.
4. Position Lights — OFF.

NOTE

Allow 10 seconds for landing-taxi light retraction and/or closure of ram dump door prior to engine shutdown.

5. Rotating Beacon — OFF.
6. Throttles — OFF.
*7. Oxygen — 100%.
*8. All Unguarded Switches — OFF.
9. Wheels — Chocked.
10. Battery — OFF.
11. Flight Director Switch — OFF.
(B) 12. Sight Dust Cover — Replace.

## STRANGE FIELD PROCEDURES.

The following information provides guidance for operation at fields that do not normally support the T-38A:

1. Oil: Use MIL-L-7808 (NATO 0-148). Alternate: None.

   Check oil level immediately after flight.

2. Fueling: Use MIL-T-5624, Grade JP-4 (NATO F-40).

### Alternate Fuels.

Alternate fuels can be used continuously with a possible loss of efficiency. The use of these fuels might result in increased maintenance.

a. Use of an alternate fuel requires adjusting of the density setting on the engine main and afterburner fuel controls as well as retrimming the engine. If this has not been accomplished, any fuel listed as an alternate is downgraded to emergency status.

b. Any fuel used as an alternate fuel must contain an anti-icing inhibitor. If it does not, it must be downgraded to emergency status.

c. Flight is restricted to altitudes where temperatures below the fuel freeze points will not be encountered.

d. With or without fuel control changes and engine trim, use of JP-5, Jet A-1, Jet A, and Jet A-50 may cause slow but satisfactory starts below 22,000 feet and/or temperatures down to -29°C. Below -29°C and/or above 22,000 feet starting capability is significantly reduced and failure to start can be expected. Do not initiate afterburner operation above 25,000 feet with JP-5, Jet A-1, Jet A, or Jet A-50 since afterburner lights will be very hard possibly causing compressor stall and engine flameout.

### Emergency Fuels.

Emergency fuels may cause significant damage to the engine or other systems. Examples of conditions that might warrant use of emergency fuels are an accomplishment of an important mission and emergency evacuation flights.

a. Use of emergency fuels is restricted to a one time subsonic flight with minimum maneuvers and power changes. Engine RPM and EGT must be closely monitored to prevent exceeding operating limits during throttle movement. Rapid throttle movements and afterburner lights in flight are allowed only under emergency conditions.

b. Without the fuel control density setting changed, idle speed (minimum thrust) may be increased, acceleration may be faster causing the engine to stall, max RPM and EGT may be exceeded, and finally afterburner fuel flow may be high and cause the engine to stall.

c. When an emergency fuel is used the conditions in paragraphs c and d for alternate fuels must also be followed.

Single-Point — Use a 45 - 55 PSI system no flow pressure. After fuel flow starts, expect a drop in pressure. Do not increase fuel flow pressure during refueling. Start fuel flowing and then move the precheck valve handle, located adjacent to the single-point fueling adapter, to the PRIM (primary) position. Fuel flow should stop within 10 seconds. Stoppage is indicated by fuel flow not greater than 10 gallons per minute at fuel truck meter. Return precheck valve handle to OFF. Allow fuel flow to continue for a short duration and then place precheck valve handle in the SEC (secondary) position. Fuel flow should stop within 10 seconds. Return precheck valve handle to OFF position and continue refueling. If fuel flow fails to stop in both check positions, do not use single-point refueling.

Manual — Service left system first or aircraft may settle on tail.

3. Oxygen: Use MIL-O-27210.

4. Hydraulic Fluid and Brake Fluid: Use MIL-H-5606 (NATO H-515).

5. Tire Pressure:

   Main — 250 PSI. Nose — 75 PSI.

6. Loose Fasteners: Use Torq-Set bit.

7. Air Starting Units:

   Air Force — MA-1, MA-1A, MA-1MP, MA-2, MA-2MP, M32A-60, MA-3MP, and 502-70.

   Navy — GTC-85, MA-1E, WELLS AIR START SYSTEM, and RCPP/RCPT/NCPP-105.

8. Electrical Units: (115/200 volts, 3-phase, 400-cycle required).

| FUEL SPECIFICATIONS | | | |
|---|---|---|---|
| GRADE | MIL SPEC | COMMERCIAL DESIGNATION | NATO SYMBOL |
| SPECIFIED FUEL | | | |
| JP4 | MIL-T-5624 | . . . . | F40 |
| ALTERNATE FUELS | | | |
| . . . . | . . . . . . . . | JET B | F40 |
| JP5 | MIL-T-5624 | . . . . | F44 |
| . . . . | . . . . . . . . | JET A-1 | F34 |
| . . . . | . . . . . . . . | JET A | F30 |
| . . . . | . . . . . . . . | JET A-50 | . . . |
| EMERGENCY FUEL | | | |
| AVGAS | . . . . . . . . | AVGAS | . . . |

NOTE

- When aviation gasoline is used a 3 percent lubricating oil, specification MIL-L-22851, type II, must be added to improve its lubricity characteristics.
- When an alternate or emergency fuel is used enter the fact in the aircraft AFTO Form 781.

# EMERGENCY PROCEDURES

## SECTION III

T-38A 1-102

TABLE OF CONTENTS

GROUND-OPERATION EMERGENCIES

## IN-FLIGHT EMERGENCIES



## LANDING EMERGENCIES



NOTE

- A Critical Procedure is an emergency procedure that must be performed immediately without reference to printed checklist and that must be committed to memory. These critical procedures appear in **BOLDFACE** capital letters. Non-critical Procedures are all other steps wherein there is time available to consult the checklist.
- In the event of multiple emergencies, the pilot is required to exercise sound judgment as to the appropriate action. A thorough knowledge of the correct procedures and aircraft systems is essential to analyze the situation correctly and determine the best course of action.
- To assist the pilot when an emergency occurs, three basic rules are established, which apply to most emergencies occurring while airborne. They should be remembered by each aircrew member.

1. Maintain Aircraft Control.
2. Analyze the Situation and Take Proper Action.
3. Land as soon as conditions permit.

# GROUND-OPERATION EMERGENCIES

## ENGINE FIRE DURING START.

If a fire warning light illuminates, or if there is other indication of a fire, proceed as follows:

1. Throttles — OFF.
2. Battery/APU—OFF.

## DEPARTING PREPARED SURFACE.

Any time the aircraft departs a hard surface (taxiway or runway), immediately shut down both engines. REFER TO EMERGENCY EXIT ON THE GROUND TO ABANDON THE AIRCRAFT.

## EMERGENCY EXIT ON THE GROUND.

When a situation develops which requires a crewmember to abandon the aircraft, place the throttles at OFF, battery switch at OFF, insert the ejection seat safety pin, release the survival kit, if one is carried, by pulling the survival kit emergency release handle (figure 1-24), disconnect personal leads and release safety belt. Crewmembers should consider removing the parachute when disconnecting equipment to facilitate exit from cockpit. Open the canopy. If either canopy cannot be opened by the normal procedure, pull the canopy jettison T-handle. If either canopy fails to open or jettison, break thru the canopy, using the canopy breaker tool (figure 3-1).

**WARNING**

To avoid kit deployment and possible pilot-survival kit entanglement during emergency exit, the survival kit must be seated firmly in position before the survival kit emergency release handle is pulled.

**CAUTION**

The canopy seals will remain inflated if engines are shut down with both canopies locked making the canopies more difficult to open.

## USE OF CANOPY BREAKER TOOL.

To break the canopy, grasp the canopy breaker tool with both hands and use your body weight behind an arm swinging thrust. Aim the point of the tool to strike perpendicular to the canopy surface. The blade alignment will determine the direction of the cracks. No set pattern of blows is necessary on the front canopy. Several minutes of chopping may be required to open an adequate hole in the rear canopy.

**WARNING**

To preclude personal injury, the curved edge of the blade must be towards you. This will allow glancing blows against the canopy to deflect away from you.

## SMOKE, FUMES, OR ODORS IN COCKPIT.

Do not take off if smoke, fumes, or unidentified odors are detected. Refer to Smoke, Fumes, or Odors in Cockpit procedure under Inflight Emergencies, this section.

**CANOPY BREAKER TOOL**

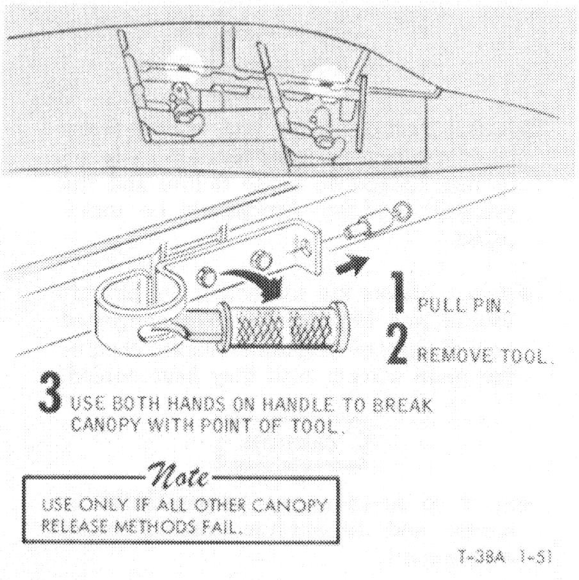


Figure 3-1.

# TAKEOFF EMERGENCIES

## ABORT/BARRIER ENGAGEMENT.

If the decision is made to ABORT during a takeoff or a touch-and-go-landing, maximum deceleration can be obtained by optimum braking while in a three point attitude. However, optimum wheel braking is difficult to achieve as gross weight and /or airspeed increases. It is therefore recommended that aerodynamic braking be used to the maximum extent possible. When the nosewheel is lowered to the runway, immediately commence moderate braking. Optimum braking should not be attempted in excess of 100 KIAS. This will minimize the possibility of skidding, blown tires, etc. Aerodynamic braking performed with less than full flaps or a 12 degree pitch attitude becomes progressively less efficient, although in some circumstances it may not be possible to attain the optimum configuration. Aerodynamic braking at intermediate pitch attitudes versus cautious wheel braking are probably equally effective in most cases. If runway length is insufficient to completely stop the aircraft, decelerate as much as possible and prepare to depart the hard surface or engage the barrier (if available). After barrier engagement, actuation of the controls or changing aircraft configuration may cause damage to the aircraft.

1. THROTTLES—IDLE.
2. WHEEL BRAKES—AS REQUIRED.

**WARNING**

- Braking required for high speed, heavy gross weight abort may result in extremely hot brakes or brake failure and the possibility of tire fire should be anticipated.
- If hot brakes are suspected, the aircraft should not be taxied into a congested area. Ensure all personnel remain clear of the main wheels until they have cooled.

**CAUTION**

- Steer so as to engage perpendicular to barrier and discontinue braking before engagement.
- Heavy braking above 100 KIAS may cause skidding, tire failure, and loss of directional control.
- Extreme caution must be exercised when applying wheel brakes above 120 KIAS as locked wheels or tire skids are difficult to recognize. If tire skid is detected, immediately release both brakes and cautiously reapply.
- MA1A barrier engagement is unlikely with the WSSP/External Stores installed or speed brake open.

Refer to Takeoff/Abort charts in Part 2 of Appendix.

## ENGINE FAILURE/FIRE WARNING DURING TAKEOFF.

If an engine fails on takeoff prior to reaching decision speed, use the procedure in this section titled ABORT/BARRIER ENGAGEMENT. If an engine fails on takeoff above the computed decision speed, it is possible to continue the takeoff. Limited excess thrust is available for takeoff, acceleration, and climb-out when operating on a single engine. The available runway should be used to accelerate the aircraft above single-engine takeoff speed (SETOS). The computed single-engine takeoff speed is the minimum speed at which the aircraft will takeoff and be able to fly out of ground effect. Thrust predictions of the takeoff factor can be verified only by an accurate acceleration speed check. A significant relationship exists between airspeed and initial climb performance: between takeoff speed and SETOS + 20 KIAS, single-engine climb performance increases at the rate of 100 feet per minute for each knot of airspeed above SETOS. Best acceleration occurs with the aircraft in a three point attitude, with the stick at, or slightly aft of, the takeoff trim setting. The nosewheel should not be allowed to "dig-in", nor should it be permitted to lift off. This attitude must be maintained until the airspeed reaches a minimum of SETOS; however, SETOS + 10 KIAS is the optimum rotation speed. Initial pitch attitude is shallower than normal. Climb should be restricted to only that required to avoid obstacles until the airspeed reaches 190 KIAS and flaps are retracted. The gear should be retracted as soon as the aircraft is airborne above SETOS + 10 KIAS. Gear door drag is not a factor during retraction above SETOS + 10 KIAS. The flaps should be raised after gear retraction and above 190 KIAS.

Due to the critical nature of airspeed and altitude and the ejection envelope, the decision made by the pilot may vary.

If Decision Is Made To Stop:

1. ABORT.

NOTE

If the Abort was made as a result of an engine fire, place the throttle of the affected engine to OFF once the aircraft is under control. If the fire is confirmed, accomplish the Emergency Exit On The Ground procedures once the aircraft is stopped.

If Takeoff Is Continued:

1. THROTTLES–MAX.

NOTE

Depending on airspeed and altitude, it may be necessary for the pilot to leave the throttle of the affected engine at a high power setting until reaching a safe airspeed and/or altitude for ejection.

2. FLAPS–60%.

(B) 3. EXTERNAL STORE – JETTISON IF NECESSARY.

4. ATTAIN AIRSPEED ABOVE SETOS, 10 KNOTS DESIRED.

5. Gear — UP (when airborne above SETOS + 10 KIAS).

6. Flaps — UP (at 190 KIAS minimum).

WARNING

- Continuing a takeoff on a single engine should be attempted only at maximum thrust.

- With TOF above 4.5 (gross weight 11,800 lbs) or above 4.2 (gross weight 12,500 lbs) single-engine takeoff is not considered possible.

- With other than 60% flaps, single-engine capability is impaired to such an extent that high takeoff factors coupled with heavy gross weights may make takeoff impossible.

- If engine failure occurs after rotation, it will probably be necessary to lower the nose to attain speed above single-engine takeoff speed. If engine failure occurs after takeoff, it may be necessary to allow the aircraft to settle back to the runway.

NOTE

- If the left engine is inoperative but windmilling, generally gear retraction may be accomplished but will require an extended time period; however, gear doors may not completely close. Gear retraction, when initiated between single-engine takeoff speed plus 10 knots and 190 KIAS, may require up to one minute.

- If unable to retract the landing gear, best level flight/climb capability is obtained at 190 KIAS with 60 percent flaps or at 220 KIAS with the flaps up. At high gross weight, flap retraction should not be initiated prior to 220 KIAS.

- After flaps are set at 60% the flap indicator should be checked to insure flaps are within 60% range.

## TIRE FAILURE DURING TAKEOFF.

Tire failure on takeoff may present a greater problem than tire failure on landing. Directional control is more difficult, and braking effectiveness is greatly reduced at higher gross weights.

If Decision Is Made To Stop:

1. ABORT.

If Takeoff Is Continued:

1. DO NOT RETRACT GEAR.

## LANDING GEAR RETRACTION FAILURE.

If the warning light in the landing gear lever remains illuminated after the lever has been moved to the LG UP position, proceed as follows:

1. Airspeed — Maintain below 240 KIAS.

2. Landing Gear Lever — LG DOWN.

# INFLIGHT EMERGENCIES

## SINGLE-ENGINE FLIGHT CHARACTERISTICS.

Single-engine directional control can normally be maintained at all speeds above the stall. Very little rudder is required because of the close proximity of the thrust lines to the centerline of the aircraft. In high-drag, high-thrust, low-airspeed conditions, rudder must be used to coordinate flight to obtain optimum aircraft performance. There are conditions under which the aircraft will not maintain altitude in takeoff configuration or landing configuration with one engine operating at either MIL or MAX thrust. For fully fueled takeoffs, single engine takeoff speed should be attained to insure excess thrust is available. At other fuel weights, final approach speed will insure excess thrust is available for go-around. Single-engine performance in a landing configuration with 60% flaps is shown in the Thrust Required and Available chart and the Effect of Bank Angle on Vertical Velocity charts in Part 7 of the appendix. Minimum single-engine flying speed for any condition occurs where the thrust available and thrust required lines cross. If the airspeed is less than the minimum speed, altitude must be sacrificed to attain this minimum and/or the configuration must be changed to reduce the drag. Every effort should be made to immediately attain a speed that will give excess thrust. It is imperative that the speed brake be closed during all single-engine flight to obtain the performance stated in the single-engine charts. The single-engine service ceiling can be attained by following the climb schedule shown in the Single-Engine Service Ceiling chart in Part 3 of Appendix. Ⓑ In some single-engine flight conditions consideration should be given to the jettison of the external store. However, external store/airplane separation tests have not been conducted.

## ENGINE FAILURE DURING FLIGHT.

If an engine operates abnormally or fails during flight, reduce drag to a minimum and maintain airspeed and directional control while investigating to determine the cause. Failure of the left engine may deactivate speed brake, normal landing gear extension and retraction, nosewheel steering, and the stability augmenter system. However, left engine windmilling rpm under this condition may supply sufficient hydraulic pressure to operate these systems. Use the following procedure for shutting down an engine in flight:

1. Safe single-engine airspeed — Maintain.
2. Ⓑ External Stores—Jettison if necessary.
3. Throttle (inoperative engine) — OFF for 10 seconds before attempting a start if conditions permit.
4. Crossfeed — As Necessary.

**WARNING**

With the crossfeed ON and either both boost pumps ON — or both boost pumps OFF — a rapid fuel imbalance can occur. Fuel will feed predominantly from the left system.

Refer to Single-Engine Diversion Range Summary Table in Part 4 of Appendix.

## Ⓑ EXTERNAL STORE EMERGENCY JETTISON.

The external store may be released as follows:

1. Emergency Jettison Button — PUSH.

## DUAL ENGINE FAILURE AT LOW ALTITUDE.

If both engines fail during flight at low altitude and with sufficient airspeed, the aircraft should be zoomed (approximately 20 degrees nose up attitude) to exchange airspeed for altitude and to allow additional time to accomplish subsequent emergency procedures. ALTERNATE AIRSTART should be attempted immediately upon detection of dual engine flameout. If the decision is made to eject, ejection should be accomplished during the zoom while the aircraft is in a nose high positive rate of climb. It is imperative that the ejection sequence be initiated prior to reaching a stall or rate of sink.

**WARNING**

Do not delay ejection by attempting airstarts at low altitude if below the optimum airstart airspeed and below 2,000 feet AGL.

## RESTART DURING FLIGHT.

Airstarts can be expected at or below 26,000 feet between 250 KIAS and 310 KIAS. Optimum restart capability at higher altitudes occurs at 270 KIAS. At lower altitudes, an engine RPM of 16% or greater affords sufficient airflow for restart. If engine flameout is experienced, use the following procedure:

1. Altitude — 26,000 feet or below.

2. Airspeed — 250 KIAS to 310 KIAS.

3. Battery Switch — Check ON.

4. Boost Pump Switches — Check ON. (Circuit Breakers — IN, if applicable)

NOTE

When the rear cockpit is occupied, the fuel pump circuit breakers should be checked in.

5. Engine Start Button — Push Momentarily.

6. Throttle (windmilling engine) — Advance to slightly above IDLE, then retard to IDLE.

NOTE

- Leave throttle at IDLE for 30 seconds before aborting a start.
- If dual engine flameout occurs, right engine should be attempted first as right engine instruments will operate normally as soon as engine start button is pushed.

7. If Restart Attempt Fails — Place throttle in OFF position for approximately 10 seconds, turn crossfeed switch ON, check engine start and ignition circuit breakers, and attempt another start.

NOTE

- The RPM may hang up during restart after combustion occurs at low airspeeds. RPM hangup during an airstart may be eliminated by increasing airspeed.
- If it appears that a boost pump has failed, remain below 25,000 feet. Turn crossfeed OFF to avoid having to use an abnormal fuel balancing procedure.
- If it appears that a boost pump has failed and flight below 25,000 feet is impractical, engine operation above 25,000 feet with gravity fuel flow is possible at reduced power settings. If a reduced power setting is also impractical, use crossfeed operation to insure boost pump pressure and minimize the possibility of fuel flow interruption. Monitor the fuel balance and descend as soon as practical. Flight at lowest practical altitude and reduced power setting will minimize probability of fuel flow interruption.

## ALTERNATE AIRSTART.

The alternate airstart is primarily designed for use at low altitude when thrust requirements are critical. An airstart may be accomplished by advancing the throttle to MAX range. This energizes normal and afterburner ignition for approximately 30 seconds (if throttle remains in MAX range). If the engine does not start after 30 seconds, additional starts may be attempted by retarding the throttle out of MAX range to reset the circuit and again advancing the throttle into MAX range to reactivate the ignition cycle. After engine start, the throttle may be left in MAX range if afterburner operation is desired.

If alternate airstart is required, proceed as follows:

1. THROTTLE(S) – MAX.

WARNING

- If throttle is already in MAX, recycle throttle MIL to MAX.
- With dual engine failure, battery switch must be at ON to provide ignition.

NOTE

If the throttle is in the MAX range, pushing the start button will also provide ignition; however, only for that period of time which the button is held.

## COMPRESSOR STALL.

If an engine compressor stalls, proceed as follows:

1. Throttle — IDLE.

2. Increase airspeed and advance throttle slowly.

NOTE

If engine FOD is suspected, slow throttle advance is necessary to regain sustained engine power.

3. Throttle — OFF (if engine will not recover).

NOTE

- After experiencing a compressor stall, the engine may not recover to the full range of operation. If normal instrument indications can be achieved for a given power setting, the engine should not be shutdown unless other circumstances dictate.
- If the engine is shut down, an airstart may be attempted as applicable.
- Rapidly retarding the throttle to IDLE and immediately pushing the engine start button may permit the engine to recover and prevent complete flameout.

## FIRE WARNING DURING FLIGHT. (Affected Engine).

If a fire warning light illuminates, use the following procedure:

1. THROTTLE — IDLE.

When a fire warning light is preceded or accompanied by a pop, bang or thump it usually indicates a serious engine malfunction and/or fire. Consideration should be given to shutting down the engine.

2. THROTTLE — OFF IF FIRE WARNING LIGHT REMAINS ON.

WARNING

If engine cannot be shut down with the throttle, the fuel shutoff switch (affected engine) should be closed.

CAUTION

Do not attempt to restart the affected engine if the fire is extinguished. Make a single-engine landing.

NOTE

If the fire warning light goes out, check light by positioning warning test switch to TEST.

3. IF FIRE IS CONFIRMED — EJECT.

Any time the fire warning lights illuminate, verify the condition by other indications before abandoning the aircraft. It is possible that the warning system may have malfunctioned and given an erroneous indication. Fire is usually accompanied by one or more of the following indications: excessive EGT, erratic or vibrating engine operation, fluctuating fuel flow, smoke trailing the aircraft, and/or smoke in the cockpit.

## ELECTRICAL FIRE.

If an electrical fire occurs, proceed as follows:

1. Battery and Generator Switches — OFF.

NOTE

With boost pumps inoperative, engine flameout may occur if above 25,000 feet.

2. All Electrical Equipment — OFF.
3. Battery, Generator(s), electrical equipment — ON, as necessary for flight and landing.

## SMOKE, FUMES, OR ODORS IN COCKPIT.

All odors not identifiable shall be considered toxic. If smoke, fumes, or odors are encountered in the cockpit, proceed as follows:

1. Oxygen — 100%.
2. Check for Fire.
3. Cabin Pressure Switch — RAM DUMP, Below 25,000 Feet, If Possible.
4. If Smoke Becomes Severe — Jettison Canopy, Below 300 KIAS, if possible.

## LOSS OF CANOPY.

If either or both canopies are lost, immediately slow the aircraft to 300 KIAS or less to minimize turbulence and noise. Minimum drag occurs at approximately 225 KIAS. Reestablish intercockpit communications. Land as soon as conditions permit.

Figure 3-2.

## LOSS OF CABIN PRESSURE.

1. Descend immediately — Maintain aircraft at or below FL250.
2. Oxygen system — 100% and Emergency. Below 25,000 feet oxygen system operation may be returned to normal.
3. Land as soon as conditions permit.

## OXYGEN SYSTEM EMERGENCY OPERATION.

Should either pilot detect symptoms of hypoxia or hyperventilation, proceed as follows:

1. Supply Lever — Check ON.
2. Diluter Lever — 100% OXYGEN.
3. Emergency Lever — EMERGENCY.
4. Connections — Check security.

**WARNING**

If positive pressure is not felt after completing Step 4 or oxygen system contamination is suspected, use of the emergency bailout bottle should be considered. If oxygen system contamination is suspected, further consideration should be given to disconnecting the aircraft oxygen hose after activating the emergency bailout bottle.

5. Breathe at a rate and depth slightly less than normal until symptoms disappear.

6. Descend below 10,000 feet MSL (cabin pressure) and land as soon as conditions permit.

## OIL SYSTEM MALFUNCTION.

Abnormal engine oil pressure indications frequently are an early indication of some engine trouble. The engine oil pressure indicators are marked for normal operating conditions on the ground or in the air. If engine oil pressure exceeds the operating limits, proceed as follows:

1. Throttle—IDLE.

2. Throttle—OFF, if 5 psi minimum oil pressure cannot be maintained at idle RPM or if engine seizure appears imminent.

**CAUTION**

Consideration should be given to shutting down the engine for zero oil pressure or failure of both the engine RPM and oil pressure gauge which may be an indication of a sheared oil pump shaft.

**NOTE**

If the operating engine requires shutdown, the engine previously shutdown for oil system malfunction may be restarted.

## ENGINE OVERTEMPERATURE.

If excessive exhaust gas temperature occurs, immediately retard throttle to the setting at which the exhaust gas temperature of the affected engine decreases and remains within limits.

## NOZZLE FAILURE.

If nozzle failure occurs in closed range, excessive EGT is possible. If this condition occurs, follow the engine overtemperature procedure. If a nozzle fails in the open position, low EGT will result. The affected engine will operate from idle to mil, but with a much lower thrust output. Afterburner may not be available. Depending on the severity of either condition, consideration should be given to recovering the aircraft in accordance with single engine landing procedures. If the nozzle is closed, EGT may increase above acceptable limits during landing rollout or taxi. If this occurs, the engine should be shut down.

## EJECTION VS FORCED LANDING.

Ejection is preferable to landing on an unprepared surface. Do not land the aircraft with both engines flame out.

## DITCHING.

Ejection is to be accomplished in preference to ditching the aircraft.

## EJECTION PROCEDURE.

Escape from the aircraft should be made with the ejection seat, using the sequence shown in figure 3-3. After ejection, the safety belt automatically opens and a man-seat separation system forcibly separates the crewmember from the ejection seat (figure 3-7). A rapid deployment escape system is provided to improve low-altitude escape capability. The zero-delay lanyard if present on the parachute may be disconnected for completely controlled ejections if time and altitude permit. If the zero-delay lanyard is present it should be connected in accordance with present directives. However, if the crewmember knows he is going to eject at more than 2000 feet AGL (Above Ground Level) in a controlled condition, he should disconnect the zero-delay lanyard to reduce chances of seat-chute-man involvement. There is no evidence to indicate that one should attempt to connect the zero-delay lanyard after deciding to eject. The time lost in connection is greater than any advantages which may be gained.

**WARNING**

- Do not delay ejection below 2000 feet in futile attempts to start the engines, or for other reasons that may commit you to an unsafe ejection or a dangerous flame-out landing. Accident statistics emphatically show a progressive decrease in successful ejections as altitude decreases below 2000 feet above the terrain.

- Under uncontrollable conditions, eject at least 15,000 feet above the terrain whenever possible.

# EJECTION

**WARNING**

**ASSUME PROPER POSITION**

- SIT ERECT, HEAD FIRMLY AGAINST HEADREST, FEET HELD BACK AGAINST SEAT.
- POSITION ELBOWS CLOSE TO BODY WITHIN ELBOW GUARDS, TO PROTECT ELBOWS WHEN LEGBRACES ARE RAISED, AND DURING EJECTION.
- THE CREWMEMBER IN REAR COCKPIT SHOULD EJECT FIRST IF ALTITUDE PERMITS. THIS WILL PREVENT POSSIBLE INJURY FROM FRONT SEAT ROCKET BLAST.
- MAINTAINING AIRCRAFT CONTROL MAY REQUIRE USE OF ONE HAND TO INITIATE THE EJECTION SEQUENCE.

## 1 HANDGRIPS – RAISE

RAISE HANDGRIPS UNTIL LOCKED TO EXPOSE TRIGGERS, USING ONE OR BOTH HANDS.

## 2 TRIGGERS – SQUEEZE

SQUEEZING ONE OR BOTH TRIGGERS JETTISONS CANOPY AND EJECTS SEAT. SEAT WILL EJECT THRU CANOPY IF CANOPY FAILS TO JETTISON.

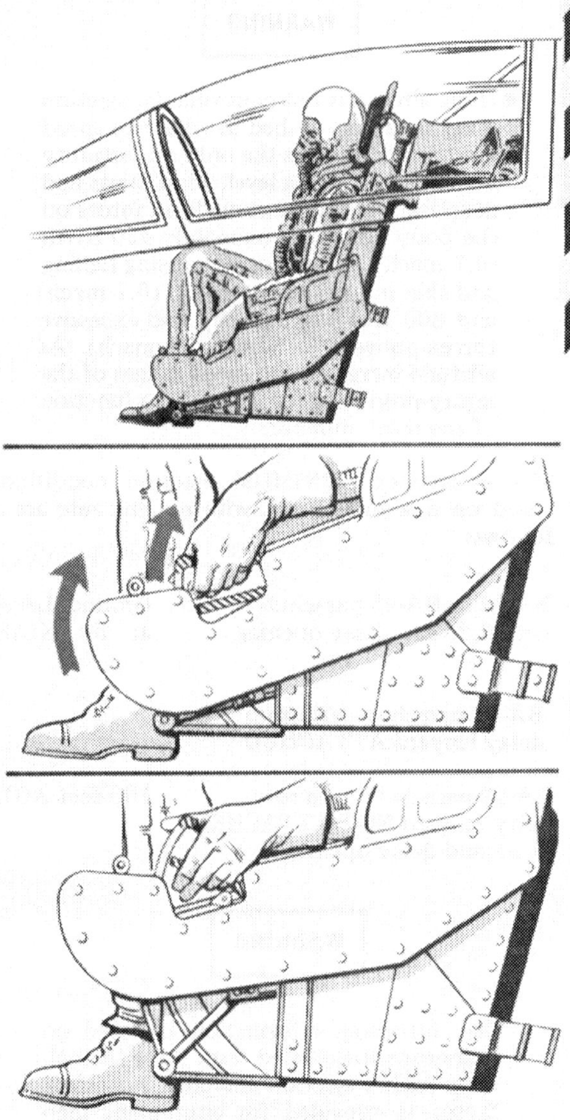

T-38A 1-50K

Figure 3-3.

During any low-altitude ejection, the chances for successful ejection can be greatly increased by pulling up to exchange airspeed for altitude if airspeed permits. Ejection should be accomplished while in a positive rate of climb with the aircraft approximately 20 degrees nose-up, and before the start of any sink rate. Ejection while the nose of the aircraft is above the horizon and in a positive rate of climb will result in a more nearly vertical trajectory for the seat, thus providing more altitude and time for seat separation and parachute deployment. See figure 3-5 for safe minimum ejection altitude versus sink rate and figure 3-6 for ejection altitude versus bank/dive angle. If rate of climb cannot be accomplished, level flight ejection should be accomplished immediately to avoid ejection with a sink rate. The automatic safety belt must not be opened before ejection, regardless of altitude. If the safety belt is opened manually, the automatic feature of the parachute is eliminated.

**WARNING**

- Improper routing of personal leads may cause inadvertent opening of the safety belt (aircraft equipped with MA-5) during ejection. Care must be taken to ensure that flight clothing, such as sleeves, will not catch and open the safety belt during ejection.

WARNING

- If the aircraft is not controllable, ejection must be accomplished at whatever speed exists, as this offers the only opportunity for survival. At sea level, wind blasts and deceleration will exert medium forces on the body up to approximately 450 KIAS (0.7 mach), severe forces causing flailing and skin injuries between 450 (0.7 mach) and 600 KIAS (0.9 mach), and excessive forces above 600 KIAS (0.9 mach). As altitude increases, the speed ranges of the injury-producing forces will be a function of the mach number.

The emergency MINIMUM ejection conditions, based on a level attitude, with no sink rate are as follows:

| | |
|---|---|
| BA-22 or BA-25 parachutes with 0.25 sec. delay opening or | Ground Level at 50 KIAS |
| BA-22 parachute with zero delay lanyard ATTACHED | |
| BA-22 parachute with zero delay lanyard NOT ATTACHED (1 second delay opening) | 100 feet AGL |

WARNING

The foregoing information is based on numerous rocket-sled tests using the ballistic rocket ejection catapult. No safety factor is provided for equipment malfunction. Since survival from an extremely low altitude ejection depends primarily on the aircraft attitude and altitude, the decision to eject must be left to the discretion of the pilot. Factors such as G-loads, high sink rates, and aircraft attitudes other than level or slightly nose high will decrease chances for survival. The emergency minimum ejection conditions (ground level and 50 KIAS) are provided only to show that zero altitude ejection can be accomplished in case of an emergency which would require immediate ejection. It must not be used as a basis for delaying ejection when above 2000 feet.

The emergency MAXIMUM ejection airspeeds from sea level through 14000 feet are as follows:

| | |
|---|---|
| BA-22 or BA-25 parachutes with 0.25 sec. delay opening | 500 KIAS |
| BA-22 parachute - zero delay lanyard ATTACHED | 400 KIAS |
| BA-22 parachute - zero-delay lanyard NOT ATTACHED (1-second delay opening) | 550 KIAS |

**BEFORE EJECTION.**

If time and conditions permit . . .

1. Notify crewmember of decision to eject.
2. Turn IFF to EMERGENCY, and if not in radio contact with appropriate agencies, turn radio to GUARD and transmit MAYDAY.
3. Turn aircraft toward uninhabited area.
4. Stow all loose equipment.
5. Survival Kit (AUTO/MANUAL)—As Required.
6. (High Altitude) Actuate emergency oxygen cylinder.
7. Disconnect oxygen hose and G-suit, tighten oxygen mask and chin strap securely (well beyond comfortable range), and lower and lock visor(s). Under controlled conditions above 2000 feet AGL, disconnect the zero-delay lanyard (if applicable).
8. Attain proper airspeed, altitude, and attitude.

NOTE

When the ejection seat is near the fully down position, the consoles extend above the triggers with the handgrips raised. Crewmembers should be aware of this reduced clearance between the triggers and the consoles to avoid interference when attempting to eject.

9. Assume proper position.

Refer to WARNINGS in figure 3-3.

EJECTION.

1. HANDGRIPS – RAISE.

   Raise handgrips until locked to expose triggers, using one or both hands.

2. TRIGGERS – SQUEEZE.

   Squeezing one or both triggers jettisons canopy and ejects seat. If the canopy fails to jettison, the seat will eject thru the canopy.

Figure 3-4. Deleted

## AFTER EJECTION.

Immediately after ejection the following procedures apply:

1. Safety Belt — Attempt To Open Manually.
   Attempt to manually open the safety belt as a precaution against the belt failing to open automatically.

**WARNING**

If the HBU safety belt fails to open, rotate and hold the release handle fully counterclockwise with the left hand and simultaneously twist, turn, jiggle or otherwise manipulate to gain latch release.

2. Safety Belt Released — Attempt To Separate from Seat.

   A determined effort must be made to separate from the seat to obtain full parachute deployment at maximum terrain clearance. This is extremely important for low altitude ejections.

3. If Safety Belt Opened Manually — Immediately Pull Parachute Arming Lanyard (Arming Ball) if Above 14,000 Feet or the Ripcord Handle if Below 14,000 Feet.

4. Survival Kit — Deploy after Parachute Opening.

**WARNING**

Refer to T.O. 14D1-2-1, Section III for various landing situations (i.e.: water, tree, power line, etc).

## STABILITY AUGMENTER MALFUNCTION.

A stability augmenter yaw system failure at high airspeeds may cause an abrupt yaw followed by a moderate rudder roll. If the yaw damper switch (pitch or yaw switch on Block 20 aircraft) is found in the OFF position during flight, the mission may be continued. Do not reengage the damper switch. Note the discrepancy on Form 781 after termination of the flight. If yaw or pitch oscillations are induced by the stability augmenter, the affected damper switch should be turned OFF.

1. Yaw Damper Switch — OFF.
2. Stability Augmenter Circuit Breakers—Pull (if applicable).

**NOTE**

If malfunction continues, reduce airspeed to minimize yaw and land as soon as conditions permit.

## RUDDER SYSTEM FAILURE.

If rudder system failure is experienced, the failure mode may be a full 30 degrees left hardover rudder. The pilot's only indications of this type failure will be uncontrolled yaw and roll movements. At airspeeds above 300 KIAS, aircraft control is probably impossible to retain. If uncontrollable conditions persist, eject at least 15,000 feet AGL whenever possible. At airspeeds below 300 KIAS, the aircraft will initially roll slightly to the right followed by a rapid roll to the left. If no pilot corrective action is taken within two seconds, the aircraft will stall. Aircraft control can be regained by the use of appropriate ailerons and stabilator if the pilot reacts within two seconds. Once under control, the aircraft cannot be landed safely due to high sideslip angles and lack of directional control on the ground. The aircraft should be flown to a safe bail-out altitude and area where a controlled ejection can be performed.

## FUEL QUANTITY INDICATOR AND LOW LEVEL CAUTION LIGHT SYSTEM MALFUNCTION.

When fuel quantity indicator failure is experienced, the fuel low-level caution light is unreliable. Failure of the fuel low-level caution system is indicated when remaining fuel supply in either system is less than 250 ± 25 pounds and the low-level light does not illuminate. With fuel quantity indicator failure or low-level caution system failure, closely monitor the fuel quantity and land as soon as conditions permit.

**WARNING**

Do not attempt crossfeed operation with a fuel quantity indicator inoperative as it will be impossible to monitor the fuel balance.

## LOW FUEL PRESSURE.

If a fuel pressure caution light comes on, proceed as follows:

1. Boost Pump Circuit Breakers — Check.

# EJECTION ALTITUDE VS SINK RATE

0.25 SECOND DELAY OPENING PARACHUTE OR ZERO DELAY LANYARD ATTACHED

*Note*

- ASSUMED REAR SEAT EJECTS FIRST FOLLOWED IN 0.75 SECOND BY FRONT SEAT.
- BOTH EJECTION SEATS HAVE THE SAME EJECTION CAPABILITY.

**WARNING**

- THE MINIMUM EJECTION ALTITUDES SHOW SEAT CAPABILITY (WITH 2-SECOND REACTION TIME) AS AFFECTED BY AIRCRAFT SINK RATE. THE MINIMUM ALTITUDES DO NOT PROVIDE ANY SAFETY FACTOR FOR EQUIPMENT MALFUNCTION, DELAY IN SEPARATING FROM THE SEAT, AND AIRCRAFT DIVE AND BANK ANGLES ABOVE 2000 FEET TERRAIN CLEARANCE.
- THE MINIMUM EJECTION ALTITUDES SHALL NOT BE USED AS THE BASIS FOR DELAYING EJECTION WHEN ABOVE 2000 FEET TERRAIN CLEARANCE.

CONDITIONS

AIRSPEED AT EJECTION — 150 KIAS
WINGS LEVEL — SLIGHT NOSE-UP ATTITUDE
2-SECOND REACTION TIME

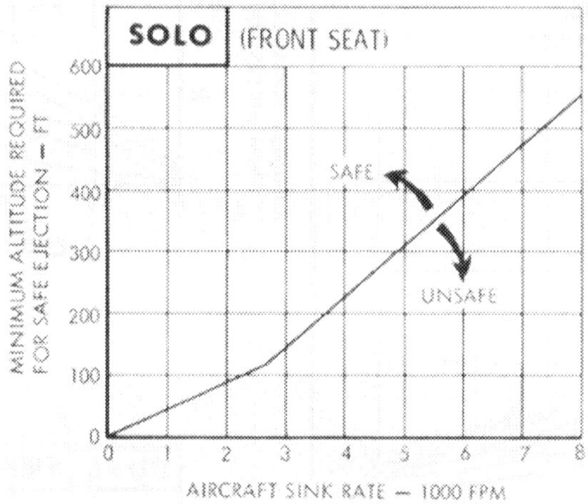


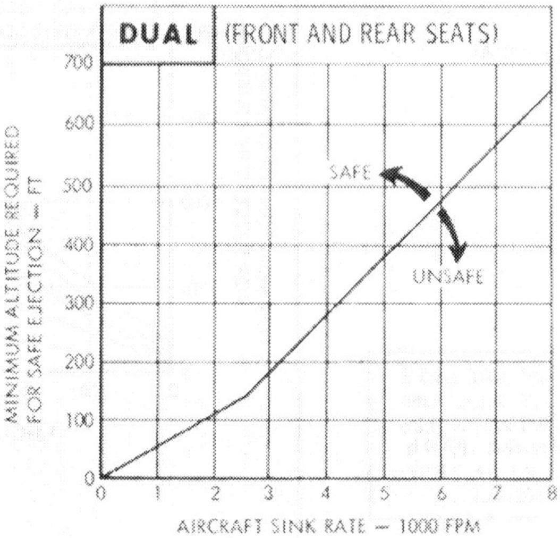


SINK RATE

F-5 1-140(1)

Figure 3-5.

# EJECTION ALTITUDE VS BANK/DIVE ANGLE

0.25 SECOND DELAY OPENING PARACHUTE OR WITH ZERO DELAY LANYARD ATTACHED.

**WARNING**

- EJECTIONS ABOVE EACH TIME REACTION, DIVE ANGLE, OR BANK ANGLE LINE ARE SAFE FOR GIVEN CONDITIONS.
- EJECTIONS BELOW EACH LINE ARE UNSAFE.

## BANK ANGLE

CONDITION
CONSTANT ALTITUDE

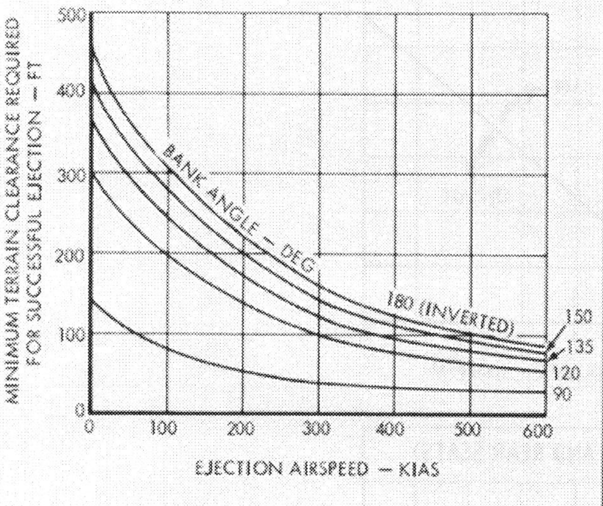


*Note*

BANK ANGLES UP TO 60° ARE SAFE FOR LEVEL FLIGHT AT ALL AIRSPEEDS UP TO 500 KIAS WITH 0.25 SECOND DELAY OPENING PARACHUTE OR 400 KIAS WITH ZERO DELAY LANYARD ATTACHED.

## DIVE ANGLE

CONDITIONS
WINGS LEVEL
2-SECOND REACTION TIME

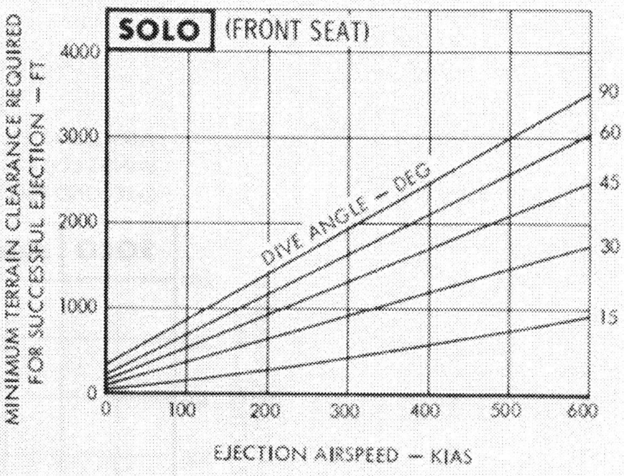


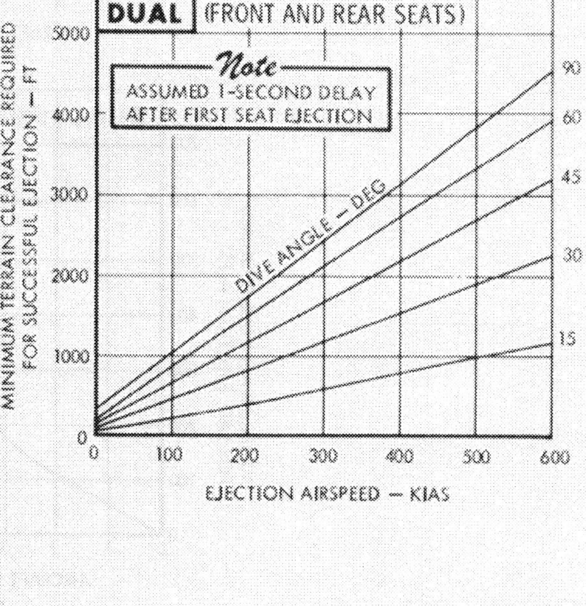


Figure 3-6.

# EJECTION SEQUENCE

**Note**

- TIME FROM TRIGGER SQUEEZE TO FULL 0.25 SECOND DELAY PARACHUTE DEPLOYMENT IS 3.5 SECONDS (APPROX) AT AN EJECTION AIRSPEED OF 150 KNOTS. VARIABLES SUCH AS LOWER AIRSPEEDS AND THE ATTITUDE OF THE PILOT AT TIME OF PACK OPENING CAN INCREASE PARACHUTE DEPLOYMENT TIME.
- TIME TO FULL PARACHUTE DEPLOYMENT IS SAME FOR BOTH SEATS.

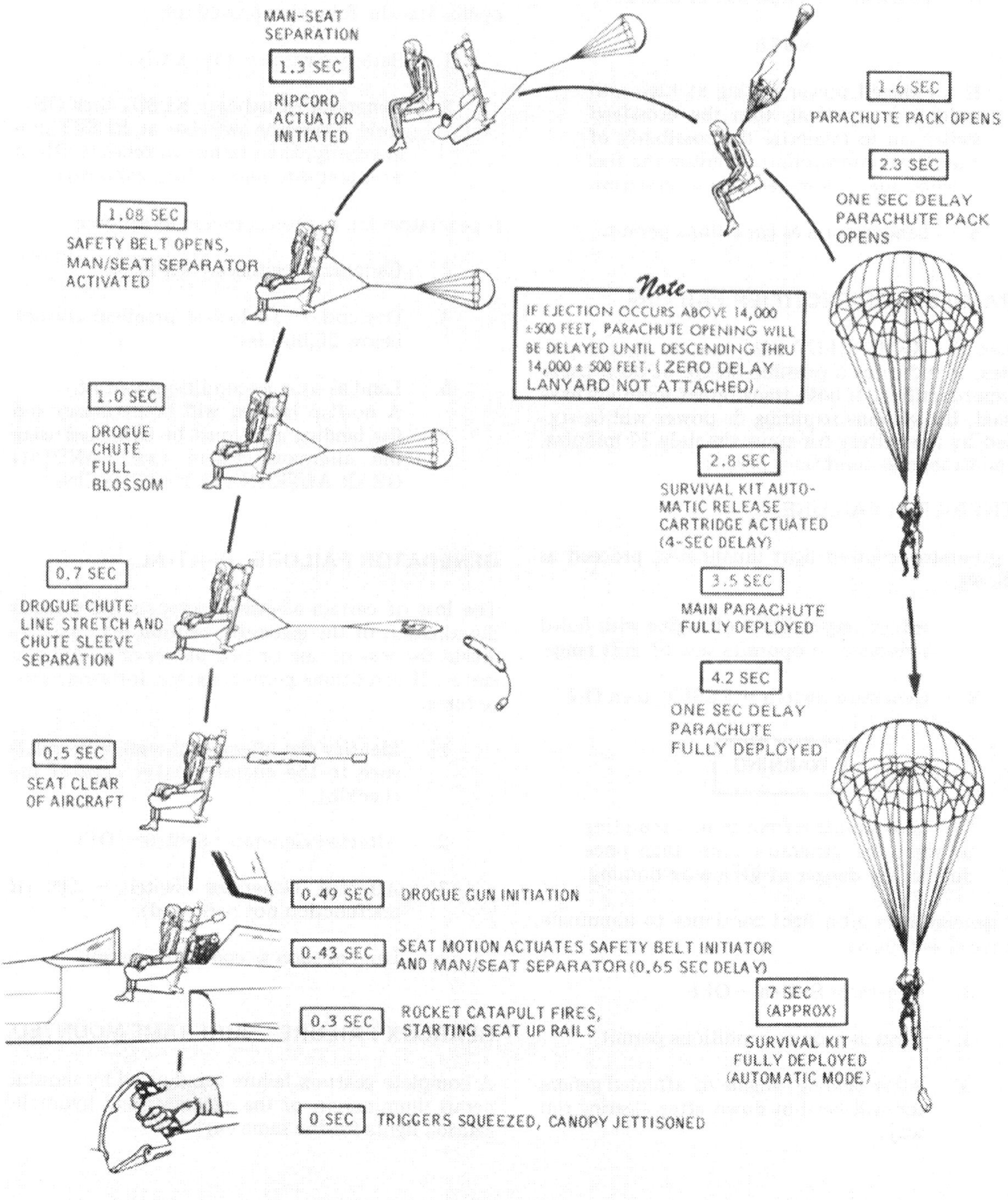


**Note**

IF EJECTION OCCURS ABOVE 14,000 ±500 FEET, PARACHUTE OPENING WILL BE DELAYED UNTIL DESCENDING THRU 14,000 ± 500 FEET. (ZERO DELAY LANYARD NOT ATTACHED).

Figure 3-7.

NOTE

If circuit breakers are found popped, turn off appropriate boost pump before resetting circuit breakers.

2. Power — Reduce.

3. Descend — 25,000 feet or below.

NOTE

If a reduced power setting at high altitude is impractical, turn the crossfeed switch on to minimize the possibility of fuel flow interruption. Monitor the fuel balance and descend as soon as practical.

4. Land as soon as conditions permit.

## TRANSFORMER-RECTIFIER FAILURE.

When the XFMR RECT OUT caution light illuminates, it indicates a possible failure of both transformer-rectifiers. If both transformer-rectifiers have failed, the systems requiring dc power will be supplied by the battery for approximately 17 minutes. Land as soon as conditions permit.

## GENERATOR FAILURE.

If generator caution light illuminates, proceed as follows:

1. Adjust engine RPM of engine with failed generator to opposite side of shift range.

2. Generator Switch — RESET, then ON.

WARNING

The pilot should refrain from attempting to reset the generator more than once due to the danger of generator burning.

If generator caution light continues to illuminate, proceed as follows:

3. Generator Switch — OFF.

4. Land as soon as conditions permit.

5. After landing, engine of affected generator will be shut down after clearing runway.

## ELECTRICAL FAILURE — COMPLETE.

With complete electrical failure, all warning systems, engine instruments (except engine tachometers), flight director, communication and navigation systems, speed brake, flaps, landing gear normal extension, landing gear indicators, nosewheel steering, fuel boost pumps, and engine ignition system are inoperative; and each engine anti-ice valve opens. Use the following procedures:

1. Battery Switch — Check ON.

2. Generator Switches — RESET then ON. Hold generator switches at RESET momentarily, then return switches to ON in an attempt to regain electrical power.

If generators fail to reset, proceed as follows:

3. Generator Switches — OFF.

4. Descend — To lowest practical altitude below 25,000 feet.

5. Land as soon as conditions permit. A no-flap landing will be necessary and the landing gear must be extended using the alternate system (see LANDING GEAR ALTERNATE EXTENSION).

## GENERATOR FAILURE - PARTIAL.

The loss of certain electrical components without illumination of the generator caution light may indicate the loss of one or two phases of an AC generator. If conditions permit, use the following procedures:

1. Identify the affected generator by reference to the circuit breaker page of the checklist.

2. Affected Generator Switch — OFF.

3. Affected Generator Switch — ON (if malfunction not corrected).

4. Land as soon as conditions permit.

## GEARBOX FAILURE — AIRFRAME-MOUNTED.

A complete gearbox failure is indicated by simultaneous illumination of the generator and hydraulic caution lights for the same engine.

If Gearbox Fails Completely:

1. Throttle (affected engine) — OFF, if excessive vibration exists.

A gearbox failure to shift is indicated when the LEFT or RIGHT GENERATOR caution light illuminates when accelerating or decelerating thru the shift range of 65% to 70% RPM.

If Gearbox Fails to Shift:

1. Engine RPM — Return to range where generator operation can be maintained.
2. Generator Switch — RESET then ON, if necessary.
3. Engine RPM — Leave in range of successful generator operation until on final approach; then use as necessary to complete landing.

## HYDRAULIC SYSTEMS MALFUNCTIONS.

Three different types of hydraulic system malfunctions may be encountered: hydraulic fluid overtemperature, low pressure, and high pressure. The hydraulic caution light will illuminate for either a fluid overtemperature or a low pressure condition. To determine the cause of a hydraulic caution light, check the indicators. Readings below 1500 psi indicate a low pressure situation. Momentary drops in pressure sufficient to cause illumination of the hydraulic caution light may be an indication of an unpressurized system. Normal or excessive pressure readings indicate a fluid overtemperature condition. The hydraulic indicators provide the only warning of high hydraulic pressure: a situation that can cause a hydraulic overtemperature condition. Although fluid overtemperature and high pressure usually occur together, it is possible to have one without the other. The corresponding engine should be shut down immediately whenever an overtemperature condition exists. If the pressure is high, but not accompanied by light or sluggish controls, the aircraft should be landed as soon as possible; be alert for the indications of overtemperature.

**WARNING**

- An ejection should be accomplished with dual hydraulic system failure since flight control is impossible.
- Hydraulic pressure provided solely by a windmilling engine is insufficient to control the aircraft for landing.

## HYDRAULIC MALFUNCTIONS (CAUTION LIGHT ILLUMINATED).

With a utility hydraulic system failure, the speed brake, nosewheel steering, normal landing gear system, and stability augmenter will be inoperative. If the UTILITY or FLIGHT HYDRAULIC caution light illuminates, use the following procedure:

1. Hydraulic Pressure Indicators — Check.

If hydraulic pressure is low:

1. Monitor both systems and avoid zero and negative G flight.
2. Land as soon as conditions permit.

**NOTE**

If utility hydraulic pressure is depleted, stop straight ahead and have gear pins installed prior to clearing runway.

If hydraulic pressure is normal or high (fluid overtemperature):

1. Shut down affected engine.

**NOTE**

If the Hydraulic Caution light goes out the engine may be restarted if necessary. However, the engine should be left shut down as long as possible to permit maximum cooling of hydraulic fluid.

2. Land as soon as conditions permit.

## EXCESSIVE HYDRAULIC PRESSURE (CAUTION LIGHT NOT ILLUMINATED).

A steady-state hydraulic pressure higher than 3200 psi in either system must be considered a system malfunction; proceed as follows:

If on the ground:

1. Shut down the affected engine.

If airborne:

1. Land as soon as conditions permit.
2. If accompanied by sluggish flight controls - Shut down the affected engine.
3. After landing and clear of runway — Shut down the affected engine (If step 2 not accomplished).

## STRUCTURAL DAMAGE OR CONTROLLABILITY CHECK.

If structural damage occurs or is suspected in flight, the pilot must make the decision whether to abandon the aircraft or attempt a landing. If it is determined the aircraft is controllable, proceed as follows:

1. Notify appropriate ground agency of intentions.
2. Climb to at least 15,000 feet above the terrain (if practical) at a controlled airspeed.
3. Simulate a landing approach (normally a straight-in full flap approach).

WARNING

Do not reposition flaps if damage to the flaps or flap actuating mechanism is known or suspected.

4. Determine airspeed at which aircraft becomes difficult to control (minimum controllable airspeed).

NOTE

In no case allow airspeed to decrease below touchdown speed.

5. Do not change aircraft configuration.
6. Maintain at least 20 KIAS above minimum controllable airspeed during descent and landing approach.
7. Fly a power-on straight-in approach requiring minimum flare. Plan to touchdown at normal speed or 10 knots above minimum control speed whichever is higher.

WARNING

Touchdowns as high as 200 knots are possible. High speed touchdown initially limits the effectiveness of aerodynamic and/or wheel braking.

## BRAKE SYSTEM MALFUNCTION.

Failure of certain components of the wheel brake master cylinders on brake lines located within the pressurized area of the cockpit may cause brake fluid overboarding through the brake fluid reservoir. If allowed to continue all the brake fluid could be forced overboard. If brake fluid overboarding is suspected or detected, proceed as follows:

1. Descend — 25,000 feet or below, if practical.
2. Cabin Pressure Switch — Ram dump, below 25,000 feet.
3. Land at lowest practical gross weight.

WARNING

If brake failure is encountered on landing roll, braking action may be regained by repeatedly pumping the brakes. The pedals should be released to the full UP position between strokes.

CAUTION

Do not pump the brakes in flight as this action could introduce air into the brake system which could result in complete brake failure.

# LANDING EMERGENCIES

## SINGLE-ENGINE LANDING.

A straight-in approach should be flown. See figure 3-8 or Section IX for single-engine approaches. The following procedure should be accomplished before landing:

1. Gear — Down and Check Down.

NOTE

- Power required under single-engine conditions may be in excess of that required to activate the landing gear warning system.
- If left engine is inoperative, normal windmilling rpm will provide adequate utility hydraulic pressure for a landing gear normal extension in a slightly longer extension time. If utility hydraulic system pressure is depleted, use the landing gear alternate extension system to extend the gear, and allow additional time for gear extension.

2. Wing Flaps — 60% (Set on Final Prior to Descent).
3. Wing Flaps — 100% when landing is assured only if stopping distance is critical.

Aerodynamic braking with less than 100% flaps is less effective and longer landing distances should be anticipated.

WARNING

Use maximum power, if necessary, to maintain landing pattern airspeeds. Refer to section VI and part 7 of the appendix for the effect of bank angle on vertical velocity.

## SINGLE-ENGINE GO-AROUND.

The available altitude and/or runway should be used to accelerate. The aircraft should be rotated at single engine takeoff speed plus 10 knots or in time to become airborne prior to the end of the runway, whichever comes first. Allow the aircraft to accelerate straight ahead, climbing only as necessary, until reaching 190 KIAS.

If, during the go-around, a touchdown occurs and take-off appears questionable, an abort may be warranted.

If go-around is continued:

1. THROTTLE(S) – MAX.
2. FLAPS – 60%.
3. (B) EXTERNAL STORE—JETTISON IF NECESSARY.
4. ATTAIN AIRSPEED ABOVE SETOS, 10 KNOTS DESIRED.
5. Gear — UP (As required above single-engine takeoff speed plus 10 knots).
6. Flaps — UP (As required above 190 KIAS).

WARNING

- A single-engine go-around should be attempted only at maximum power.
- With TOF above 4.5 (gross weight 11,800 lbs) or above 4.2 (gross weight 12,500 lbs) single engine takeoff is not considered possible.
- With other than 60% flaps, single-engine capability is impaired to such an extent that high takeoff factors coupled with heavy gross weights may make go-around impossible.
- It may be necessary to lower the nose to sacrifice altitude and perhaps allow the aircraft to settle to the runway to attain single engine takeoff speed. If the aircraft settles to the runway, lower the nose to facilitate acceleration.

NOTE

- If the left engine is inoperative but windmilling, generally gear retraction may be accomplished, but will require an extended time period; however, gear doors may not completely close. Gear retraction, when initiated between final approach speed and 190 KIAS, may require up to one minute.

NOTE

- If unable to retract the landing gear, best level flight/climb capability is obtained at 190 KIAS with 60 percent flaps or at 220 KIAS with the flaps up. At high gross weight flap retraction should not be initiated prior to 220 KIAS.

- After flaps are set at 60% the flap indicator should be checked to insure flaps are within 60% range.

## WING FLAP ASYMMETRY.

If lateral rolling and yawing is experienced during operation of the wing flaps, an asymmetric wing flap condition probably exists. If this occurs, use the following procedure:

1. Airspeed — 180 KIAS, minimum.

2. Wing Flap lever — actuate to eliminate or minimize the wing flap asymmetric condition.

3. Aux Flap Switch — Emergency.

NOTE

In the emergency position flap settings can be set to any intermediate position to eliminate the asymmetrical condition.

4. If asymmetry persists, land as soon as conditions permit, maintaining airspeeds 20 KIAS above the normal turn to final, final approach, and touchdown speeds. However, in no case touchdown below 160 KIAS.

NOTE

If the asymmetric condition cannot be corrected and conditions permit, land from a straight-in approach.

## WING FLAP-HORIZONTAL TAIL LINKAGE MALFUNCTION.

Aircraft response to horizontal tail interconnect failures depends on whether the self-locking device is part of the system, as well as the position of the wing flaps at the time of failure.

### ALL AIRCRAFT.

If the interconnect system fails with the flaps retracted, all aircraft will have the same characteristics. As flaps are lowered, a smooth but definite pitch-up will occur. This pitch-up can be counteracted, but requires heavy forward stick forces. As flaps approach the full down position, heavy stick forces will be required which cannot be completely trimmed and the stick will be very close to the forward stop. Although the aircraft may be flown in this configuration, very little nose down control authority is available. If such a condition is encountered, the flaps should be retracted and a no-flap landing made if at all possible. If conditions require the use of flaps to either reduce touch-down speed or improve aerodynamic braking, intermediate flap setting will increase the nose down control authority and reduce the stick forces required as compared to a full flap configuration.

### AIRCRAFT WITH LOCKING DEVICE.

If the interconnect system fails when flaps are extended, aircraft response and controllability will depend on whether the self-locking device is installed. Failure of the interconnect system will be manifested only when the flaps are moved from the position existing at the moment of failure. If flaps are raised, a pitch-down will result, accompanied by noticeable aft stick forces. Conversely, further flap deflection will cause a pitch-up, requiring progressively higher forward stick force to control the aircraft. An additional complication is introduced by the action of the locking device, however. Since the locking device maintains the status quo existing at the moment of failure, any additional horizontal tail authority and increased pitch control sensitivity previously added by the interconnect system will be preserved. Thus, the aircraft will exhibit normal handling qualities only when the flaps are positioned to that existing at the moment of failure. If the failure occurs at flap settings less than 45%, this effect will be slight. However, flight tests have verified that if interconnect cable fails when the flaps are down beyond 60 percent and flaps are later retracted, the control stick position necessary to maintain level flight at any airspeed will be aft of the horizontal tail trim limit. It will be necessary, therefore, to apply constant back stick force to maintain level flight. This apparent inability to trim out aft stick forces should not be confused with a horizontal tail trim malfunction. This situation can be identified by increased longitudinal control stick sensitivity and by the unusually far aft position of the control stick necessary to maintain level flight. If a PIO is encountered under these failure conditions, maintain back pressure on the stick while decreasing airspeed to 300 KIAS or below. For landing, position flaps as necessary to minimize stick forces.

WARNING

If interconnect cable failure is suspected, flight above 300 KIAS (or 0.9 IMN) should be avoided, due to increased stick sensitivity and the possibility of a pilot induced oscillation (PIO).

NOTE

- Do not release the stick in an untrimmed condition since an immediate negative "g" pitch-down will occur.
- Use of speed brakes should be avoided, if possible. Speed brakes produce a mild pitch-up which may compound the recovery problems.

### AIRCRAFT WITHOUT LOCKING DEVICE.

On aircraft without the interconnect cable locking device, if the interconnect cable fails with flaps lowered any amount, the horizontal tail will rapidly move to the position associated with a zero flap setting and as determined by the control stick position. This horizontal tail repositioning will always result in a pitch-up, with the severity being a function of airspeed and flap setting at the time of failure.

WARNING

If interconnect cable failure occurs on an aircraft without the locking device after flaps are down 60 percent or more, a sudden pitch-up will occur and the aircraft will stall instantaneously. Full forward stick will be necessary to arrest the rate of pitch-up and the pilot must take corrective action within three seconds to insure recovery without a loss of altitude.

## WING FLAP-HORIZONTAL TAIL LINKAGE MALFUNCTION WITHOUT LOCKING DEVICE.

If interconnect failure and pitch-up occur in the traffic pattern, proceed as follows:

1. Control stick — full forward to arrest pitch rate.
2. Throttles — MAX.
3. Flaps — 60 percent.
4. Landing gear — UP when continued flight is assured.
5. Flaps — UP when airspeed accelerates above no flap flying airspeed. (Be prepared to relax forward stick force as flaps are retracted.)
6. Land with wing flaps retracted.

NOTE

- Interconnect failure can occur even after flaps have stabilized in a given position.
- The amount of horizontal tail authority will be that available with zero flaps, regardless of the actual flap setting.
- The required stick position will be beyond the forward trim cutout limit.
- Moderate to heavy stick forces will be present until the flaps are retracted.

## LANDING WITH WING FLAPS RETRACTED.

If a landing is to be made with the wing flaps retracted, use the normal landing procedure modified as follows:

1. Downwind Leg — Extend.
2. Airspeed — Increase the final turn, final approach and touchdown airspeeds by 15 KIAS.

CAUTION

Extreme caution must be exercised when applying wheel brakes above 120 KIAS as locked wheels or tire skids are difficult to recognize. If a tire skid is detected, immediately release brakes and cautiously reapply.

NOTE

A no-flap full stop landing using aerobraking to just prior to loss of elevator authority and optimum braking thereafter may double the normal landing distance.

## LANDING GEAR EXTENSION FAILURE.

Unsafe cockpit gear indications should not be the only factor in the determination of an unsafe gear condition. Gear position should be determined by chase aircraft, if available, or other visual means. In the absense of visual confirmation of gear position any gear that indicates down in one or both cockpits is down and locked based upon the independent warning systems for each cockpit green light indicator. If all gear are fully down (verified by chase or other visual means) but one or more are indicating unsafe, stop straight ahead on the runway and have the gear safety pins installed.

Before attempting a landing with gear up, carefully consider whether to attempt a landing or to eject. The following table indicates that for a particular gear condition, a landing is considered feasible, or ejection is the best course of action.

| GEAR CONDITION * | | RECOMMENDED ACTION |
|---|---|---|
| NOSE | MAIN | |
| UP | BOTH DOWN | LAND |
| UP | BOTH UP | LAND EXCEPT (B) |
| UP | ONE DOWN | EJECT |
| DOWN | BOTH UP | |
| DOWN | ONE DOWN | |

*Actual landing gear position (not indication)

WARNING

- Landing in lieu of ejection for gear conditions recommending ejection is considered more hazardous.
- Recommendation to land presupposes that a favorable runway environment exists.
- (B) DO NOT ATTEMPT a landing with all gear up unless carrying an empty SUU-20 dispenser or an empty/soft or non-flammably loaded MXU-648. With other store configurations including pylon only, ejection is recommended.

Landing should always be made on a hard-surfaced runway. If time and conditions permit, request runway be foamed and expend excess fuel to reduce gross weight, minimize fire hazard, and provide a better sliding surface.

Use normal approach and touchdown speeds for all configurations. Minimize rate-of-sink at touchdown but maintain a normal landing attitude to avoid excessive "slam-down." The procedures to be used for landing with gear extension failure are contained in the following paragraphs.

## LANDING WITH NOSE GEAR UP OR UNSAFE.

1. Cabin Pressure Switch — RAM DUMP.
2. Shoulder Harness — LOCK.
3. Wing Flaps — FULL DOWN.
4. Landing Pattern — Normal.
5. Throttles — IDLE at touchdown.
6. Nose — Gently lower to runway.

NOTE

If nose gear is up, position throttles OFF when nose contacts runway.

7. Wheel Brakes — As Required.

NOTE

Do not use brakes if a safe stop can be made without them when the nose gear is down but indicating unsafe.

8. Battery Switch — OFF.

## LANDING WITH ALL GEAR UP.

This procedure should be used only under favorable conditions of the runway environment:

1. Gear — UP.
2. Cabin Pressure Switch — RAM DUMP.
3. Shoulder Harness — LOCK.
4. Wing Flaps — FULL DOWN.
5. Landing Pattern - Normal.
6. Throttles — OFF at touchdown.
7. Battery Switch — OFF.

## LANDING WITH BLOWN TIRE, LOCKED BRAKE, OR DIRECTIONAL CONTROL DIFFICULTY.

The aircraft may be safely landed with a blown tire, locked brake, or similar directional control difficulty. Plan to land at minimum gross weight unless landing sooner is necessitated. Go-around after touchdown on a blown tire or locked brake should be avoided as rubber or other debris may be ingested by the engines. When it has been determined that a main gear tire has blown or a brake is locked, land on the side of the runway away from the malfunction. Make maximum use of rudder and wheel braking to maintain directional control. Nosewheel steering should be engaged only as a final attempt to maintain or regain directional control.

WARNING

If one brake system fails or failure is suspected, plan to land in the center of the runway. Stop the aircraft by using aerodynamic braking followed by a combination of wheel brake and nosewheel steering. Rudder pedals should be neutralized prior to engaging the nosewheel steering to prevent violent swerving and possible loss of directional control.

## LANDING GEAR ALTERNATE EXTENSION.

If the landing gear normal extension procedure fails to extend the gear to a down and locked position, the landing gear alternate extension system may be utilized to extend the gear without hydraulic pressure or electrical power by using the following procedure:

1. Airspeed — 240 KIAS, or less.
2. Gear Door Switch — OPEN.
3. Landing Gear Lever — LG DOWN.
4. Landing Gear Alternate Release Handle — PULL approximately 10 inches and hold until gear unlocks; then stow handle.
5. Gear Position — Check.

WARNING

After lowering the landing gear with the alternate release handle, do not attempt to reset the switches by cycling the landing gear lever until the alternate release handle lanyard has been fully stowed.

CAUTION

Stop straight ahead on the runway, and have the landing gear safety pins installed prior to leaving runway.

NOTE

- If the main gear fails to extend fully, yawing the aircraft and applying negative or positive G forces may aid in extension.
- If the landing gear has been extended by use of the landing gear alternate release handle, nosewheel steering will not be available for taxiing.
- Once the three landing gear position indicators indicate that all three gears are down and locked, do not further activate landing gear controls.
- If the gear alternate extension system does not provide safe gear indication and utility hydraulic pressure is available, the landing gear system should be reset by recycling the landing gear lever to the LG UP position momentarily and returning it to the LG DOWN position to place utility hydraulic pressure on the down side on the landing gear system.

If the landing gear cannot be lowered by the normal or alternate procedures, it may be due to failure of the landing gear door selector valve or an electrical malfunction disrupting normal gear sequencing. In this case, the landing gear may not lower due to pressure in the utility hydraulic system or failure of the gear sequence relay to be activated. Dissipating this hydraulic pressure or bypassing the electrical circuit will allow the gear uplocks to release and the gear to extend. If the gear uplocks do not release the gear and when all the gears are in the full up position proceed as follows:

1. Gear Door Switch — Check OPEN.
2. Throttle (left engine) — OFF.
3. Control Stick — Rapid lateral stick movements to deplete utility hydraulic pressure.
4. Landing Gear Lever — LG DOWN.
5. Landing Gear Alternate Release Handle — PULL, approximately 10 inches, while pressure is depleted, and hold until gear unlocks; then stow handle.
6. Gear Position — Check; if indications are still unsafe, Landing Gear Lever LG UP, then LG DOWN.
7. Left Engine — Restart (see RESTART DURING FLIGHT).

NOTE

With gear extended, limiting airspeed of 240 KIAS may not provide sufficient airspeed for engine restart.

# SINGLE-ENGINE LANDING PATTERN (TYPICAL)

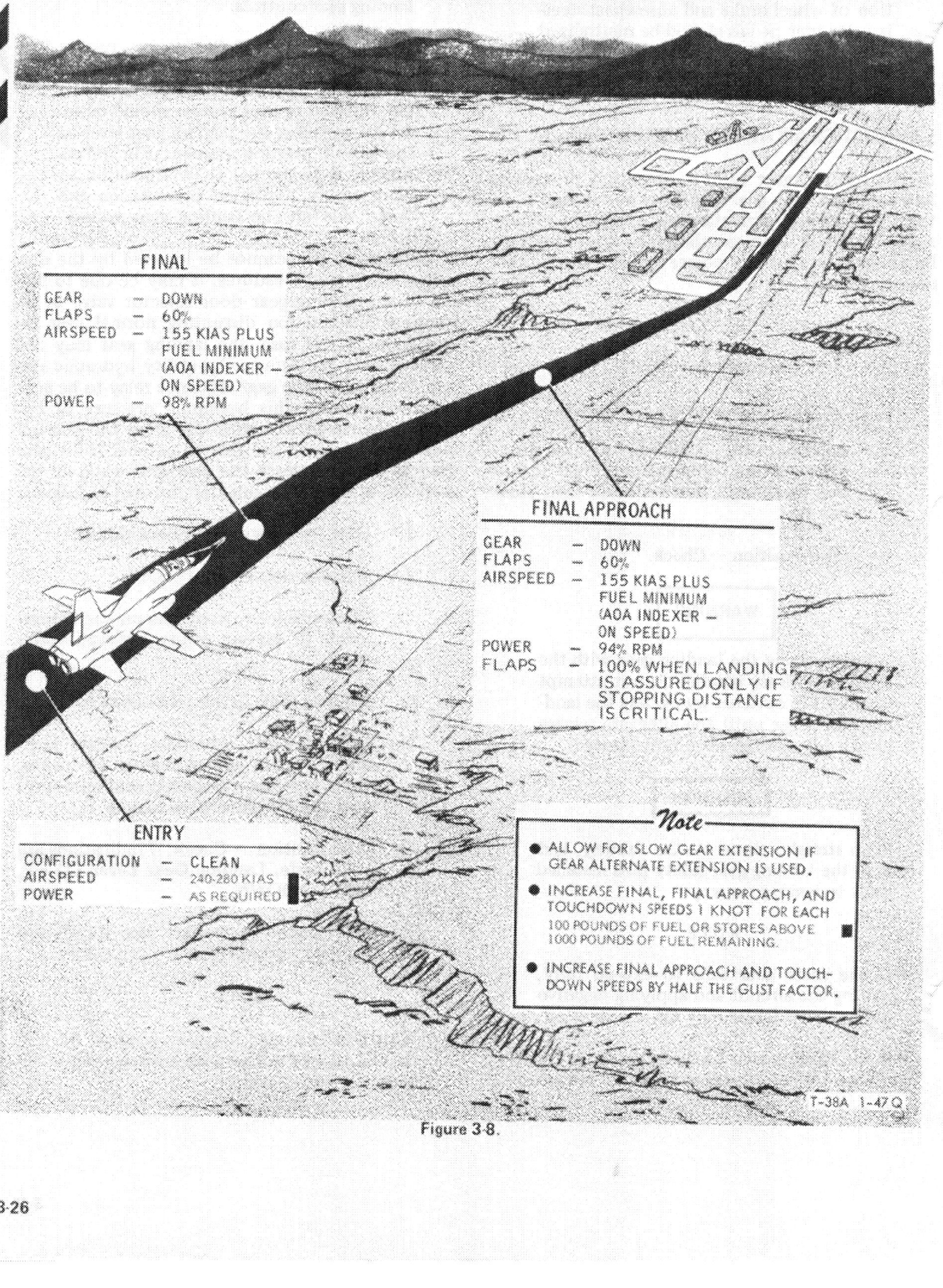


Figure 3-8.

# EMERGENCY ENTRANCE

## NORMAL ENTRANCE (LEFT SIDE OF FUSELAGE)

1 PUSH TWO LATCHES TO OPEN DOOR.
2 PULL HANDLE (OR HANDLES) OUT UNTIL ENGAGED.

**Note**

A MODERATE FORCE IS REQUIRED TO ROTATE HANDLES.

3 ROTATE HANDLE (OR HANDLES) FULLY CLOCKWISE TO UNLOCK AND RAISE CANOPY TO FULL OPEN.

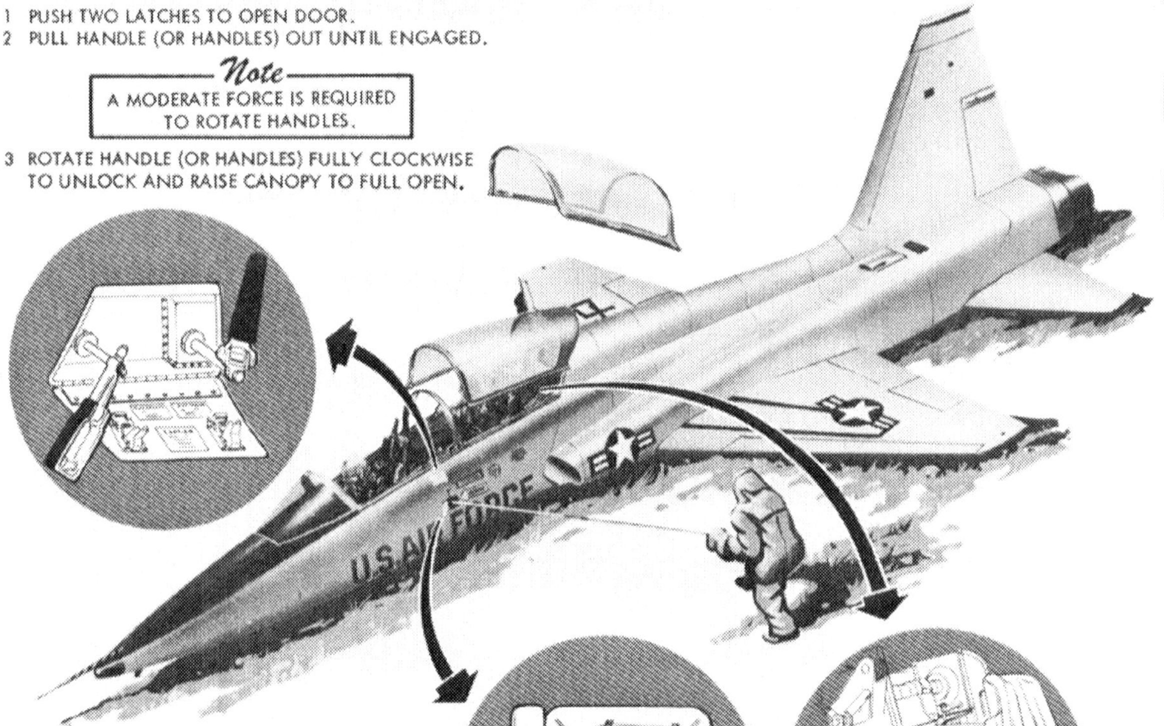


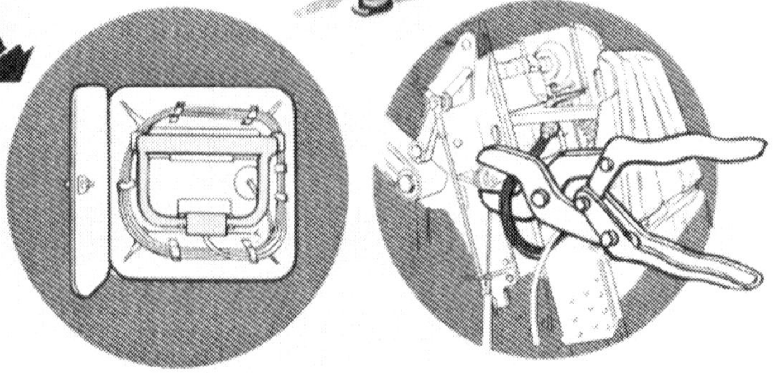

## CANOPY JETTISON ENTRANCE (EITHER SIDE OF FUSELAGE)

**WARNING**

Do not use this method when residual fuel is around cockpit area.

1 PUSH LATCH TO OPEN DOOR.
2 PULL D-HANDLE OUT TO FULL LENGTH (APPROXIMATELY 6 FEET).

**Note**

BOTH CANOPIES ARE JETTISONED WHEN EMERGENCY D-HANDLE IS PULLED.

## IF UNABLE TO OPEN CANOPY

BREAK CANOPY BEHIND PILOT/AIRCREW WITH AX OR SIMILAR IMPLEMENT.

**Note**

SPRAYING CANOPY WITH $CO_2$ WILL CAUSE THE GLASS TO BECOME BRITTLE AND EASY TO BREAK.

## AFTER ACCESS TO COCKPIT IS GAINED

**WARNING**

Inadvertent seat jettison is possible if handgrips are raised.

CUT CATAPULT HOSE, PAINTED YELLOW-ORANGE, USING WISS BULLDOG SHEARS NO. 5 OR BOLT CUTTER.

**Note**

IF HANDGRIPS HAVE NOT BEEN RAISED, INSERT SAFETY PIN IN RIGHT EJECTION SEAT LEGBRACE TO PREVENT INADVERTENT EJECTION.

T-38A 1-52 J

Figure 3-9.

# AUXILIARY EQUIPMENT

# SECTION IV

T-38A 1 103

TABLE OF CONTENTS



## CABIN AIR-CONDITIONING AND PRESSURIZATION SYSTEM.

### CABIN PRESSURE REGULATOR.

A cabin altimeter on the instrument panel of the front cockpit (figure 1-7) indicates the pressure altitude within the cabin. All controls in the air-conditioning and pressurization system, except the canopy defog, are electrically (ac) controlled. The canopy defog is pneumatically controlled and does not require ac power.

### CABIN PRESSURE SWITCH AND CABIN TEMPERATURE CONTROL KNOB.

A guarded cabin pressure switch (figure 4-1) is located on the right subpanel of the front cockpit. The switch controls cabin air-conditioning and pressurization. When the switch is placed at CABIN PRESS, both the cabin air-conditioning and pressurization systems are activated; the cabin temperature desired is then selected by rotating the cabin temperature control knob to the desired temperature. This is the automatic mode of operation. When the cabin pressure switch is placed at RAM DUMP, the anti-G suit, canopy defog, cabin pressurization and air-conditioning systems, and canopy seal are deactivated, and ram air enters the cabin for ventilating purposes. Placing the cabin pressure switch in RAM DUMP position does not deflate the canopy seal, but prevents air flow into the seal. The seal will remain inflated for an undetermined amount of time. Normal seal deflation is provided by an AC switch activated by opening the canopy locking lever, provided AC power is available.

NOTE

> To eliminate cabin conditioning duct "howl" with the rear cockpit cabin air inlet valve closed, adjust either the front cockpit cabin air inlet valve toward the closed position or adjust the rear cockpit cabin air inlet valve toward the open position.

### CABIN AIR TEMPERATURE SWITCH.

A cabin air temperature switch (figure 4-1) is located on the right subpanel of the front cockpit. The MAN HOT and MAN COLD positions provide for manual temperature control when the automa-temperature control system fails.

NOTE

> When controlling temperature manually, momentarily stop switch at the center position before going to desired position.

# CABIN AIR-CONDITIONING AND PRESSURIZATION SYSTEM

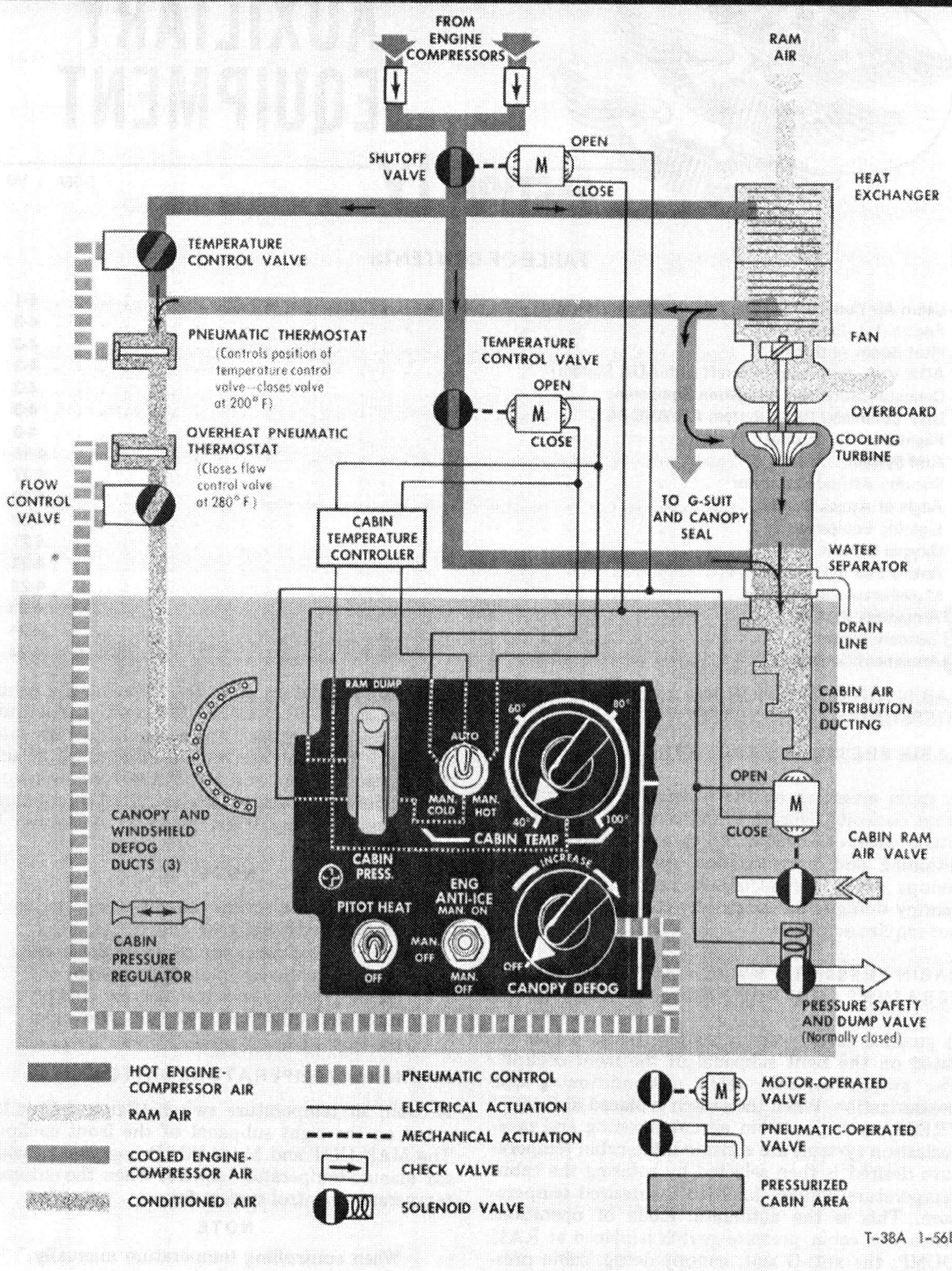


*Figure 4-1.*

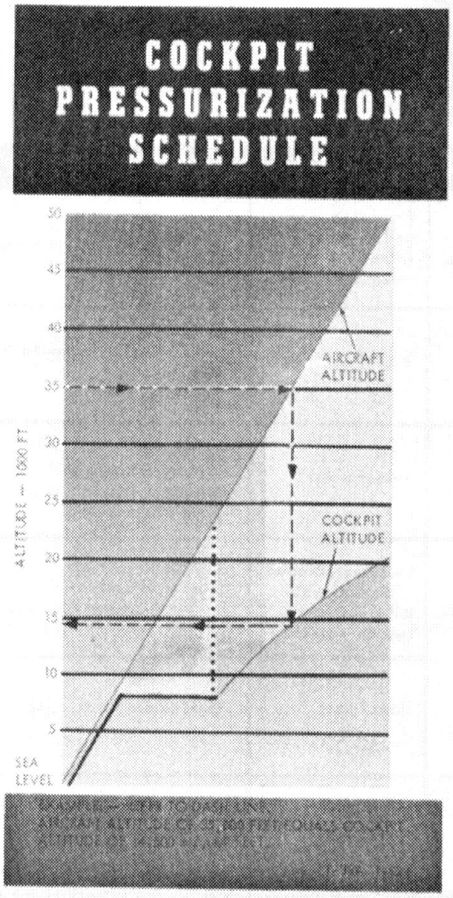


Figure 4-2.

## CANOPY DEFOG KNOB.

The flow of defog air to the windshield and both canopies is controlled by the canopy defog knob in the front cockpit (figure 4-1).

## ENGINE ANTI-ICE SYSTEM.

Engine anti-icing is accomplished by directing compressor eighth-stage air to the inlet guide vanes and bullet nose of the engine. A normally closed shutoff valve is controlled electrically by a three-position engine anti-ice switch (figure 4-1) on the right subpanel of the front cockpit. The switch positions are placarded MAN. ON in the up position and MAN. OFF in the center and down positions. Placing the switch at MAN. ON allows hot air to flow to the inlet guide vanes and bullet nose of the engine and causes the ENG ANTI-ICE ON light on the caution light panel and the MASTER CAUTION light in each cockpit to illuminate. The caution light alerts the crewmember that the switch is in the MAN. ON position but does not indicate that the system is operating. At engine speeds of 94% of 98% RPM, an increase in EGT of approximately 15°C is normal with the switch at MAN. ON. The engine anti-ice system fails to the on position with a complete loss of ac electrical power. Below 65% RPM, the anti-ice valve is always open, allowing hot air to flow to the inlet guide vanes and bullet nose of the engine, regardless of the position of the engine anti-ice switch. The switch should normally be at MAN. OFF. A 9% loss in MIL thrust and a 6.5% loss in MAX thrust can be expected with the engine anti-ice switch on.

## PITOT BOOM ANTI-ICING.

The pitot boom is de-iced by an electrical heating system. The heater is controlled by a pitot heat switch (figure 4-1) on the right subpanel in the front cockpit. Placing the switch to the up position (placarded PITOT HEAT) turns the pitot boom heat on.

## AOA VANE ANTI-ICING (AIRCRAFT WITH AOA SYSTEM).

The vane of the AOA transmitter is deiced by an electric heating element powered by the left ac bus and activated when the pitot heat switch is turned on.

## COMMUNICATION AND NAVIGATION EQUIPMENT.

Communication and navigation equipment installed in the aircraft is listed in figure 4-3. Refer to figure 1-15 for electrical power requirements to operate the communication and navigation equipment and the associated controls.

## UHF COMMAND RADIO SYSTEM AN/ARC-34.

The UHF command radio set provides line of sight voice or tone transmission and reception. Twenty of these frequencies can be preset prior to flight. An ARC-34 or ARC-34C radio control is located on the pedestal of each cockpit (figure 1-8). The two controls are similar with one exception; the ARC-34C has five manual frequency selector knobs. A restrainer prevents the use of the fifth digit selector knob. A four-position function control switch selects OFF, MAIN, BOTH, and ADF (inactive). A three-position switch selects MANUAL, PRESET,

# COMMUNICATION AND NAVIGATION EQUIPMENT

| TYPE | DESIGNATION | USE | OPERATOR | RANGE | CONTROL LOCATION |
|---|---|---|---|---|---|
| INTERPHONE | AN/AIC-18 | Crew intercommunication; flight crew and ground personnel intercommunication when aircraft is parked. | Both crewmembers. | Either cockpit and exterior when ground receptacle is used. | Left subpanel—both cockpits. |
| UHF COMMAND RADIO | AN/ARC-34<br>AN/ARC-164 | Air-to-air and air-to-ground communication. | Both crewmembers. | Line of sight. | Pedestal and left subpanel—both cockpits. |
| TACAN | AN/ARN-65<br>AN/ARN-84<br>AN/ARN-118 | Bearing and range information. Reception of coded identification signals. | Both crewmembers. | Line of sight. | Pedestal, left subpanel, and instrument panel—both cockpits. |
| ILS (LOCALIZER, GLIDE SLOPE, MARKER BEACON) | AN/ARN-58 | Reception of marker beacon signals and vertical and horizontal guidance during approach. | Both crewmembers. | Localizer—85 miles. Glide slope—35 miles. Marker beacon—vertical. | Pedestal, left subpanel, and instrument panel—both cockpits. |
| IFF/SIF | AN/APX-46 (A)<br>AN/APX-64 (A) | Automatic coded replies to ground interrogation for aircraft identification and air traffic control. | Front cockpit crewmember. | Line of sight. | Right console, front cockpit. |
| AIMS (A) | AN/APX-64 | Automatic coded replies to ground interrogation for aircraft identification, altitude reporting, and air traffic control. | Front cockpit crewmember. | Line of sight. | Right console, front cockpit |

(A) BLOCKS 20 THRU AF65-10363 AIRCRAFT (A) AF65-10364 THRU BLOCK 70 AIRCRAFT (A) BLOCK 75 AND LATER AND MODIFIED AIRCRAFT

T-38A 1-57 J

Figure 4-3.

and GUARD. MANUAL position permits use of the manual frequency selector knobs. With the switch at PRESET, the channel selector knob selects any of the 20 preset frequencies. When GUARD is selected, the main transmitter-receiver automatically channels to 243.0 megacycles. With the function switch at BOTH, an additional source of GUARD reception is available, which will operate simultaneously with and independently from the main transmitter-receiver. The tone button provides continuous tone transmission to aid ground stations in obtaining a directional fix on the aircraft.

### UHF COMMAND RADIO, AN/ARC-164(V) (After incorporation of T.O. 1T-38A-910). (Figure 4-4.)

The AN/ARC-164(V) UHF Command Radio set provides line-of-sight voice and tone transmission and reception. The control panel is located on the pedestal in each cockpit. A four-position function control switch (1) selects OFF, MAIN, BOTH, and ADF (inactive). The MAIN position of the switch permits normal operation on the selected main frequency. The BOTH position permits normal operation on the selected main frequency and simultaneous reception on guard channel. The ADF position is inoperative. A three-position mode control switch (2) selects MANUAL PRESET, and GUARD. The MANUAL position of the switch permits selection of any desired frequency within the range of the set to be manually selected by the manual frequency selector knobs (3). PRESET position permits selection of any of the 20 preset frequencies by use of the channel selector knob (4). GUARD position selects the fixed guard frequency (243.0) for the main receiver and transmitter. A button marked TONE (5) is adjacent to the function control switch and is used to provide a continuous wave (CW) for homing operations. The volume control knob (6) is inoperative. A PRESET button

(7) is located under the channel frequency card (8). New frequencies are preset by selecting PRESET on the mode control switch, placing the channel selector knob on the desired channel, setting the desired frequency with the manual frequency selector knobs and depressing the PRESET button under the channel frequency card. The 20 preset frequencies are normally standardized and set by communications personnel. The set is powered by the 28-volt DC bus. It can operate on as little as 18 volts and will operate on battery power in the event of electrical failure. In the event of AC electrical failure, the front cockpit radio is in control. The rear cockpit radio will transmit and receive only as determined by front cockpit control settings.

**NOTE**

Reception of weak signals may be aided by turning the squelch switch OFF.

## COMM ANTENNA SWITCH.

The aircraft is equipped with an upper and a lower UHF antenna. An antenna selector switch is located

UHF COMMAND RADIO, AN/ARC-164(V)

CONTROL PANEL

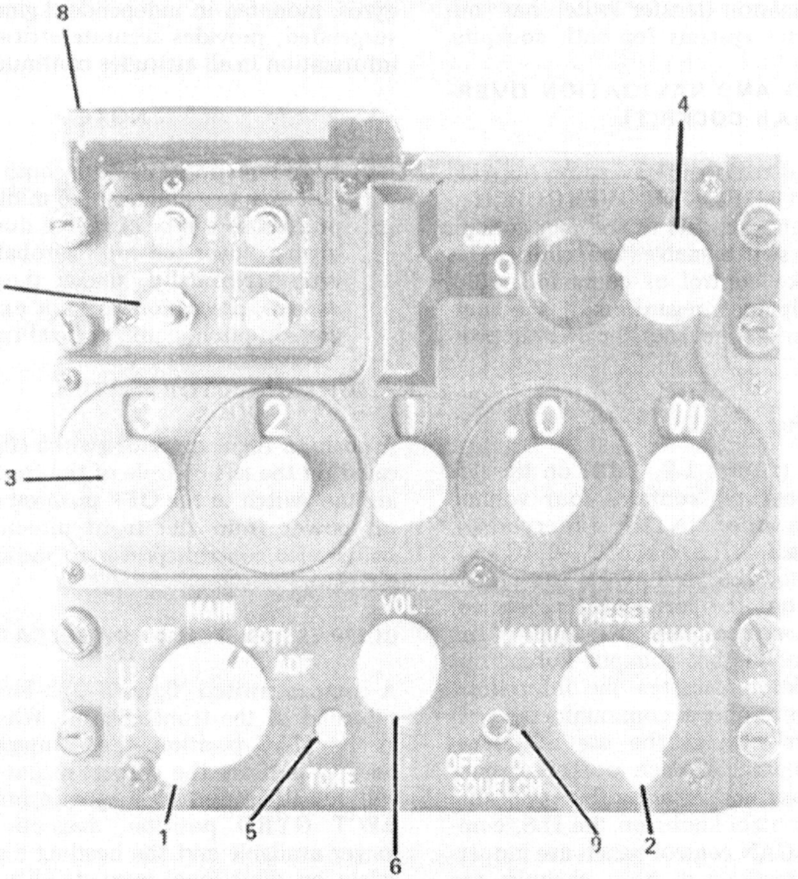


1. FUNCTION CONTROL SWITCH
2. MODE CONTROL SWITCH
3. MANUAL FREQUENCY SELECTOR KNOBS
4. CHANNEL SELECTOR KNOB
5. TONE BUTTON
6. VOLUME CONTROL KNOB
7. PRESET BUTTON
8. CHANNEL FREQUENCY CARD
9. SQUELCH SWITCH

**Figure 4-4.**

on the left subpanel of the front cockpit, placarded COMM ANTENNA (figure 1-9). Placing the switch at UPPER or LOWER permits reception and transmission thru the antenna manually selected. The switch is normally left at AUTO.

### COMMAND RADIO AND NAVIGATION TRANSFER SWITCHES (FRONT COCKPIT).

A command radio transfer switch and a navigation transfer switch are located on the left subpanel of the front cockpit, placarded RADIO TRANSFER (figure 1-9). The switches enable the front cockpit crewmember to transfer control of command radio and navigation equipment to either cockpit. The cockpit selected by the command radio transfer switch and the navigation transfer switch has control of the respective systems for both cockpits.

### COMMAND RADIO AND NAVIGATION OVERRIDE SWITCH (REAR COCKPIT).

A guarded command radio and navigation override switch, placarded COMM & NAV OVERRIDE (figure 1-10), on the left subpanel of the rear cockpit operates on ac. The switch enables the rear cockpit crewmember to take control of command radio and navigation equipment, regardless of the command radio and navigation transfer switch positions.

### INTERCOM PANEL.

An intercom panel (figures 1-9, 1-10) on the left subpanel of each cockpit contains four volume control knobs, placarded INTER (interphone), COMM (command radio), ILS, and NAV (TACAN). With command radio, ILS, or TACAN equipment turned on, pulling out the corresponding control knob permits headset reception of signals of the applicable equipment in that cockpit. Pulling out either interphone knob actuates the interphone system, providing interphone communication between crewmembers without the use of microphone switches. Volume for each cockpit is controlled by pulling out and rotating the applicable knob; the volume control knobs on the ILS, command radio, and TACAN control panels are inoperative. The signals received in both cockpits are those of the station selected in the cockpit designated by the command and navigation transfer switches or override switch.

## FLIGHT DIRECTOR SYSTEM.

The flight director system consists of an attitude director indicator and horizontal situation indicator (figure 4-5), a flight director switch (figure 1-11), a navigation mode switch (figure 1-7), a steering mode switch (figure 1-7), a compass switch (figure 1-9), a directional gyro indicator light (figure 1-10), and an attitude gyro control assembly. The instrument presentation is always identical in the two cockpits, with mode control in the cockpit selected by the navigation transfer switch. A button for fast erection of the ADI vertical gyro is located on the left subpanel in the front cockpit.

### ATTITUDE GYRO CONTROL ASSEMBLY.

The attitude gyro control assembly contains two gyros, which perform functions for both the compass system and the attitude director indicator. The combination of attitude (vertical) and directional gyros, mounted in independent gimbals but jointly suspended, provides accurate attitude and heading information in all attitudes continuously.

**NOTE**

> Gyro erection time for both ADI and HSI is approximately 3.5 minutes. Some precession can be expected during or following "over the top" aerobatic maneuvers. Normally, under these circumstances, precession will not exceed 4 degrees in pitch, bank, or heading.

### FLIGHT DIRECTOR SWITCH.

A guarded flight director switch (figure 1-11) is located on the left console of the front cockpit. Placing the switch to the OFF position removes electrical power from the flight director system. The switch also controls power to the standby attitude indicator.

### COMPASS SWITCH AND INDICATOR LIGHT.

A compass switch (figure 1-9) is located on the left subpanel of the front cockpit. When the switch is in the MAG position, the compass card will fast slave to indicate the correct magnetic heading and will remain slaved to magnetic north. In the DIRECT GYRO position, magnetic sensing is no longer available and the heading displayed is based solely on directional gyro stability. Returning the switch from DIRECT GYRO to MAG automatically fast slaves the system. Placing the switch momentarily at FAST SLAVE and returning it to MAG will also provide rapid correction of the system to magnetic north.

**NOTE**

> A 2-minute period should be allowed between FAST SLAVE cycle attempts.

# FLIGHT DIRECTOR SYSTEM DISPLAY

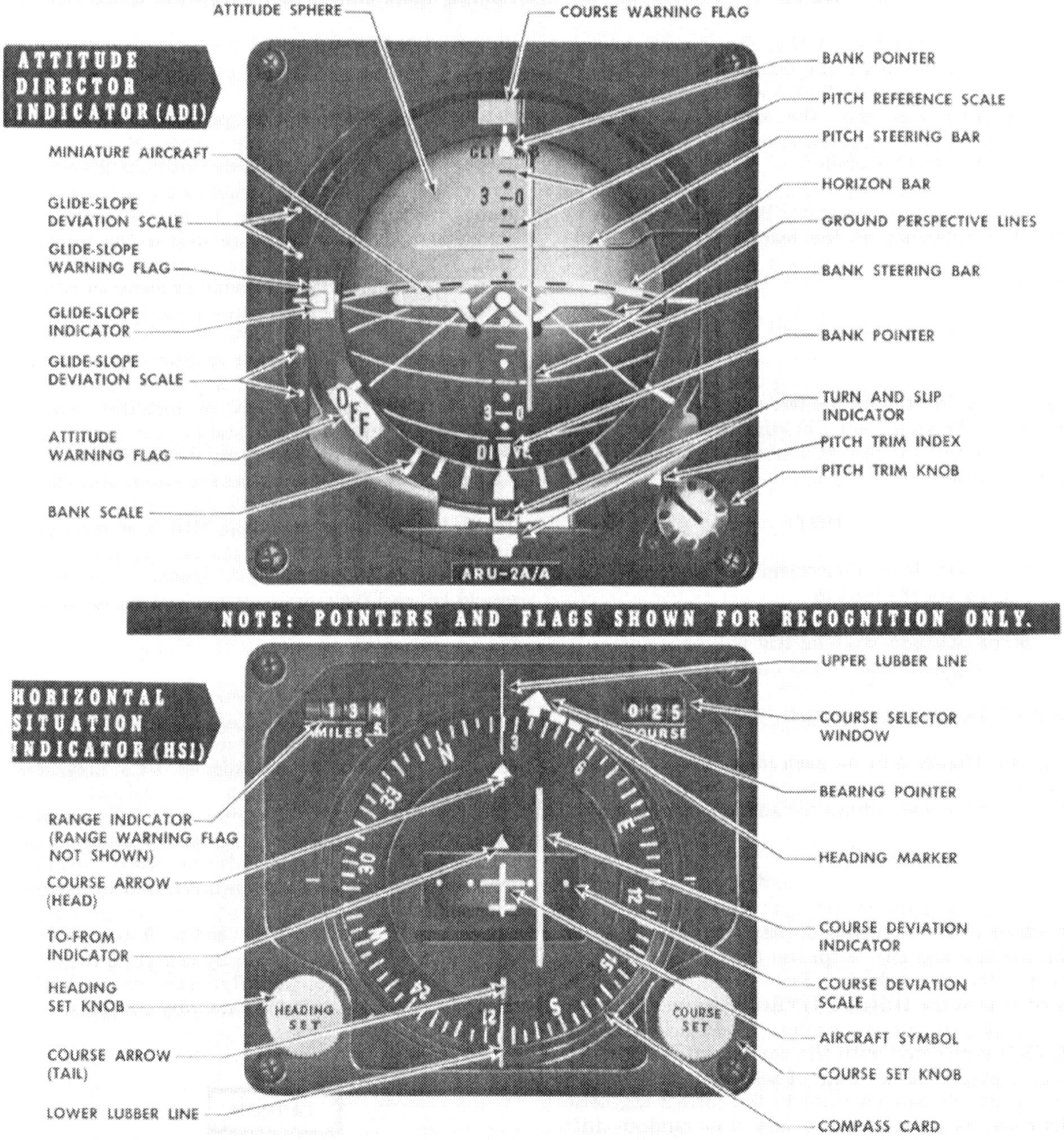


Figure 4-5.

When using FAST SLAVE or returning the system to MAG from DIRECT GYRO or after ac power interruption, the aircraft should remain in level unaccelerated flight for the 30-second FAST SLAVE cycle.

NOTE

It is recommended that the aircraft be stationary when the compass system is put into the FAST SLAVE cycle on the ground and that the aircraft not be moved until completion of the 30-second FAST SLAVE cycle.

A directional gyro indicator light (figure 1-10) on the left subpanel of the rear cockpit illuminates when the compass switch is placed at DIRECT GYRO.

## FAST ERECT BUTTON (ADI GYRO).

A button for rapidly erecting the vertical gyro (figure 1-9) is located on the left subpanel in the front cockpit. Pressing and holding the pushbutton erects the vertical gyro at a minimum rate of 15 degrees/minute.

NOTE

- Maintain level, unaccelerated flight while actuating the button.
- The attitude warning flag will be visible during actuation of the button.

## HORIZONTAL SITUATION INDICATOR (HSI).

An HSI (figure 4-5) on each instrument panel provides the pilot with a view of the navigation situation as if he were above the aircraft looking down.

### Heading Information.

When the compass switch is at MAG, the magnetic heading of the aircraft is displayed under the upper lubber line and the reciprocal heading is displayed under the lower lubber line. When the compass switch is in the DIRECT GYRO position, the heading displayed will be a random heading. If DIRECT GYRO is selected with the correct magnetic heading displayed at the time of selection, the heading will probably remain close to the correct magnetic heading, as the gyro has a very slow random drift rate. If DIRECT GYRO is selected when the compass card is not properly slaved to magnetic north, the compass card will be stabilized but will not indicate proper magnetic heading. In this case, the magnetic compass must be used for correct magnetic heading.

### Heading Marker and Heading Set Knob.

The heading marker may be positioned about the compass card by use of the heading set knob. Once positioned, the marker remains fixed relative to the card. Use of the heading marker is discussed under steering mode switch and navigation mode switch.

### Course Arrow, Course Set Knob, Course Selector Window, and Course Deviation Indicator.

The course arrow may be positioned about the compass card by use of the course set knob. The course set knob simultaneously positions the course arrow and course selector window so that they will always read the same course. Once positioned, the course arrow remains fixed relative to the compass card. When the course arrow is set, it will remain aligned (parallel) with the radial or localizer course selected, providing the compass card is slaved to magnetic north. The course deviation indicator, which consists of the center section of the course arrow, indicates lateral and angular displacement from the selected TACAN or localizer course. After tuning in a TACAN station and receiving a reliable signal, center the course deviation indicator (CDI) by rotating the course set knob, and check the reading of the course selector window. Rotate the course set knob until the CDI is at the outer dot, and check the course selector window for a change of 10 degrees ± 1.5. Radar, if available, should be used for any suspected HSI malfunction.

### Bearing Pointer.

The bearing pointer indicates correct magnetic bearing to a selected TACAN station when the compass card is functioning in the MAG mode. If the compass card is not aligned with magnetic north, which is possible when in the DIRECT GYRO mode, the bearing pointer will still indicate magnetic bearing to a selected TACAN station. The bearing pointer will not indicate proper relative bearing if the compass card is not slaved to magnetic north. With bearing pointer or compass malfunctions, the CDI may be used to find magnetic headings to a TACAN station by centering the CDI with a "to" indication, and flying the course in the course set window, using the standby compass.

CAUTION

With bearing pointer or compass malfunction, using the CDI to determine the magnetic course to a TACAN station should be attempted only as a last resort if unable to confirm position by radar.

#### To/From Indicator.

The to/from indicator functions only for TACAN. If the course deviation indicator is centered when the "to/from" reading is taken, it will immediately indicate whether the course selected, if intercepted and flown, will lead "to" or "from" the station. A "to" indication is presented when the "to/from" indicator appears on the same side of the instrument as the HEAD of the course arrow and conversely a "from" indication is presented when the indicator appears on the same side of the instrument as the TAIL of the course arrow.

#### Aircraft Symbol.

The aircraft symbol is presented at the center of the HSI and is fixed relative to the instrument. Comparison of the aircraft symbol with the compass card, course arrow, course deviation indicator, and heading marker will give a pictorial view of the angular relationship between the aircraft and the selected information.

#### Range Indicator.

The range indicator reads slant range in nautical miles to the selected TACAN station.

### ATTITUDE DIRECTOR INDICATOR (ADI).

An ADI (figure 4-5) is located on each instrument panel. For modes of operation of the ADI, refer to the steering mode switch and navigation mode switch discussion in this section.

#### Attitude Sphere, Pitch Trim Knob, and Miniature Aircraft.

The attitude sphere upper half is painted gray and the lower half is black. The gray area represents the sky and the black area, with etched perspective lines, represents the ground. At the junction of the gray and black is the horizon bar. General pitch attitude near level flight may be obtained by referencing the miniature aircraft against the sphere color. Specific pitch attitude may be obtained by referencing the miniature aircraft against the attitude sphere pitch markings. There are dots each 5 degrees of pitch, lines each 10 degrees of pitch, and numbered lines each 30 degrees of pitch. The pitch trim knob allows the attitude sphere to be adjusted to provide the desired pitch presentation relative to the miniature aircraft.

#### Bank Pointers.

A bank pointer is provided at the top and bottom of the instrument. The top pointer is without scale, but the bottom pointer is provided with a bank scale which is graduated in 10-degree increments up to 30 degrees and in 30-degree increments up to 90 degrees of bank. General bank information may be obtained by noting the angle between the miniature aircraft and numbered pitch lines. When the aircraft is erect, the legends on the attitude sphere will appear right side up.

NOTE

Since two bank pointers are provided, they cannot be used as a "sky pointer."

#### Attitude Warning Flag.

The attitude warning flag (OFF) will appear whenever electrical power to the system has failed or is interrupted. The flag will also appear during initial application of electrical power for approximately 1 minute. The instrument is unreliable until the flag disappears.

WARNING

- There is no warning of attitude sphere malfunctions other than power failure.
- The attitude warning flag will not appear with a slight electrical power reduction or failure of other components within the system. Failure of certain components can result in erroneous or complete loss of pitch and bank presentations without a visible flag. It is imperative that the attitude indicator be cross-checked with other flight instruments when under actual or simulated instrument conditions.

#### Turn and Slip Indicator.

One needle width deflection provides a 4-minute 360-degree turn.

#### Glide-Slope Indicator and Glide-Slope Warning Flag.

The glide-slope indicator indicates aircraft position relative to an ILS glide slope. The glide-slope warning flag retracts from view if the glide-slope signal strength is sufficient for satisfactory glide-slope information.

#### Course Warning Flag.

The course warning flag retracts from view if the localizer signal strength is sufficient for satisfactory localizer information. The course warning flag is at the top of the ADI, but serves as warning for localizer information displayed on the HSI course deviation indicator.

#### Bank Steering Bar.

The bank steering bar may be used in two ways: First, in the MANUAL mode, if the aircraft is flown in such a manner as to keep the bank steering bar centered, it will cause the aircraft to turn to a heading selected by the heading knob and displayed by the heading marker. Second, in the NORMAL mode, if the aircraft is flown in such a manner as to keep the bank steering bar centered, it will cause the aircraft to turn to and intercept a selected localizer beam in the direction of the approach course. In both of the above cases, the correct amount of bank is maintained during roll-in, turn, and roll-out by keeping the bank steering bar centered.

#### Pitch Steering Bar.

The pitch steering bar functions only to intercept and maintain a glide slope. If the aircraft is flown so as to keep the pitch steering bar centered, the aircraft will fly to and maintain a glide slope. The bar will center when (1) the pitch angle is correct to return to the glide slope, (2) the pitch angle is correct for leveling out on the glide slope, and (3) the pitch angle is correct for remaining on the glide slope.

**NOTE**

> Although the course and glide-slope warning flags are positioned on the ADI near the pitch and bank steering bars, they do not warn of pitch and bank steering malfunctions. If the pitch and bank steering bars are being used for an ILS approach, the warning flags must be out of view. The steering bars may malfunction without warning, so the glide-slope indicator and the course deviation indicator must be monitored during an ILS approach to ensure that desired aircraft positioning is being obtained using the steering bars.

### STEERING MODE SWITCH AND NAVIGATION MODE SWITCH.

A steering mode switch and a navigation mode switch (figure 1-7) are located on each instrument panel. The following discussion assumes that desired navigation facilities are tuned in.

#### Steering Mode Switch.

The steering mode switch has two positions, (1) MANUAL and (2) NORMAL. In the MANUAL position, the bank steering bar is displayed on the ADI. If the aircraft is flown in such a manner as to center the bank steering bar, the aircraft will roll in, turn to, roll out, and maintain the heading selected by the heading set knob and displayed by the heading marker. This is the sole function of the MANUAL position and it will operate in this manner regardless of the position of the navigation mode switch. Operation of the system with the switch in the NORMAL position will be discussed under Navigation Mode Switch.

#### Navigation Mode Switch.

The navigation mode switch has three positions: (1) TACAN, (2) LOCALIZER, and (3) INSTRUMENT LANDING SYSTEM (ILS). The following discussion of switch selections assumes that the steering mode switch is in the NORMAL position.

**TACAN Selected.** When TACAN is selected, the bearing pointer indicates magnetic bearing to the TACAN station. The course arrow and course window, which are set simultaneously with the course set knob, indicate the TACAN course selected. The course deviation indicator indicates the aircraft position relative to the selected TACAN course, and the range indicator indicates range to the TACAN station in nautical miles. The "to/from" indicator indicates whether the course selected, if intercepted and flown, will lead the aircraft "to" or "from" the station. No steering bars are in view.

**LOCALIZER Selected.** When LOCALIZER is selected, the course arrow and course window should be set with the published localizer front course. The course deviation indicator will then show aircraft position relative to the localizer course. If within the area of the glide-slope reception, the glide-slope indicator will provide indications of the aircraft position relative to the glide-slope. The bank steering bar will be in view.

**ILS Selected.** When ILS is selected, the operation is the same as in LOCALIZER, except that the bank required to center the bank steering bar is reduced from a maximum of 35 degrees to 15 degrees. The pitch steering bar is in view to provide pitch steering relative to the glide-slope. Crosswind correction is also provided in this mode.

### FLIGHT DIRECTOR OPERATION.

#### Manual Heading Mode.

1. Navigation Mode Switch — TACAN.

2. Steering Mode Switch — MANUAL.

3. Heading Marker — Set to desired heading.

4. Bank Steering Bar — Centered.

NOTE

The maximum bank angle commanded by the bank steering bar in the manual mode is 35 degrees.

**TACAN Course Interceptions.**

Refer to AFM 51-37, INSTRUMENT FLYING, for course interceptions using the flight director system. Select TACAN on navigation mode switch when making TACAN course interceptions.

**ILS Approach.**

1. ILS Receiver — Tune, identify, and monitor.

2. Course Arrow and Course Window — Set localizer front course.

3. Navigation Mode Switch — LOCALIZER.

NOTE

With the localizer front course selected, the aircraft symbol is always directional in relation to the course deviation indicator (CDI).

4. Steering Mode Switch — NORMAL.

5. Bank Steering Bar — Centered.
   The bank steering bar may be used when the aircraft heading is within 90 degrees of the localizer front course. The flight director directs an intercept angle up to 45 degrees to the localizer. A maximum bank angle of 35 degrees is required to center the bank steering bar.

WARNING

- The bank steering bar may be used only for a front course approach.
- If the published front course has not been set in the course selector window, the bank steering bar will be unreliable.

6. Navigation Mode Switch — ILS when on the localizer.
   Keeping the bank steering bar centered will maintain the aircraft on or correct it to the localizer course. Wind drift corrections are accomplished automatically.

NOTE

The bank steering bar will command excessive or erroneous steering indications if the aircraft is not on or near the localizer course when ILS is selected.

7. Pitch Steering Bar — Centered.
   As the glide-slope indicator (GSI) approaches midscale, adjust the pitch to center the pitch steering bar. Keeping the pitch steering bar centered will maintain the aircraft on or correct it to the glide slope.

8. CDI and GSI — Cross-check throughout the approach.
   The navigation mode switch must be at LOCALIZER or ILS to obtain localizer or glide-slope indications from the CDI and GSI. The course and glide-slope warning flags function only in LOCALIZER and ILS and are out of view in TACAN. TACAN bearing and range are available in the LOCALIZER or ILS positions.

## TACAN.

### AN/ARN-65(V)

**Channel Selector Switch.**

Any desired operating TACAN channel from 01 to 126 may be selected by actuating the channel selector switch on the pedestal in the cockpit selected by the position of the navigational transfer switch.

**Function Switch.**

A TACAN function switch on each pedestal has positions marked T/R, REC, and OFF. When the switch is placed at T/R, the set is energized to receive both bearing and distance signals. With the switch at REC, only bearing data is received.

WARNING

Errors may exist in the TACAN system which will appear as a false bearing indication in multiples of 40 degrees. TACAN bearing information should be verified with ground radar when possible. If a false lock-on is suspected, change to another frequency momentarily, then return to the original frequency to regain a positive lock-on.

### AIRCRAFT MODIFIED WITH AN/ARN-84(V) TACAN. (Figure 4-6.)

### CHANNEL SELECTOR SWITCH.

A mechanical indicator, controlled by CHAN and X/Y switches, indicates what channel has been selected. The X/Y switch provides 126X channels and 126Y channels for a total of 252 channels of operation (Y channel mode of operation provides future expansion capabilities. Ground beacons now in existence provide for X channel operation only.)

#### Function Selector Switch.

A TACAN function switch on each pedestal has positions marked OFF, REC, T/R, and A/A. When the switch is placed at REC only, bearing data to a ground station is provided. With the switch at T/R, the set is engaged to receive both bearing and distance signals to a ground station. When A/A (air-to-air) is selected, the set operates similar to the T/R mode except that no bearing information is provided and the range in nautical miles is to another TACAN (A/A) equipped aircraft. To use this feature, A/A should be selected by both aircraft with a 63 channel frequency separation on the X or Y band between aircraft sets.

**NOTE**

When applying power, the set should not be considered reliable until after a two-minute warm-up period.

A TEST switch is provided on the control head to allow navigational set self test. To initiate self test, press and release the Test switch. (Other control switches can be in any position, except OFF.) If the system is functioning correctly, after 10 seconds and for a period of approximately 10 seconds, the bearing pointer should indicate 180 ± 2 and the DME window should indicate 000 NM.

AN/ARN-84 (V) TACAN CONTROL HEAD

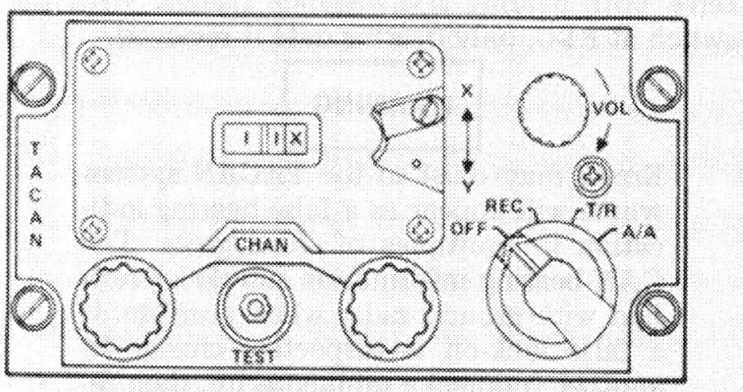


Figure 4-6.

### AN/ARN-118(V) (AIRCRAFT MODIFIED BY T.O. 38A-934) (FIGURE 4-7).

The AN/ARN-118(V) TACAN system provides range and bearing navigation information for air-to-air operation with a suitably equipped cooperating aircraft. A suitably equipped cooperating aircraft is defined as an aircraft equipped with TACAN bearing and/or distance transmitting equipment using prearranged 63-channel separation. The AN/ARN-118(V) will act as a distance transmitter but will not transmit bearing information. It eliminates 40-degree lock-on error and long search cycles. In case of co-channel interference in T/R mode, the interfering identifier is garbled. A warning flag appears when the desired station signal is invalid. When a temporary loss of signal occurs, memories keep range and bearing indications tracking for 15 seconds and 3 seconds respectively until the signal is regained. It automatically self tests after any temporary signal loss and displays its status on the control head. Operational range is up to 390 NM depending on aircraft altitude. The set requires ac and dc power.

### AN/ARN-118(V) TACAN CONTROLS.

Function Selector Switch (3)

1. OFF — Turns TACAN equipment off.
2. REC — Receives only. Indicators display bearing to ground station; no distance is displayed.

AN/ARN-118 (V) TACAN CONTROL HEAD

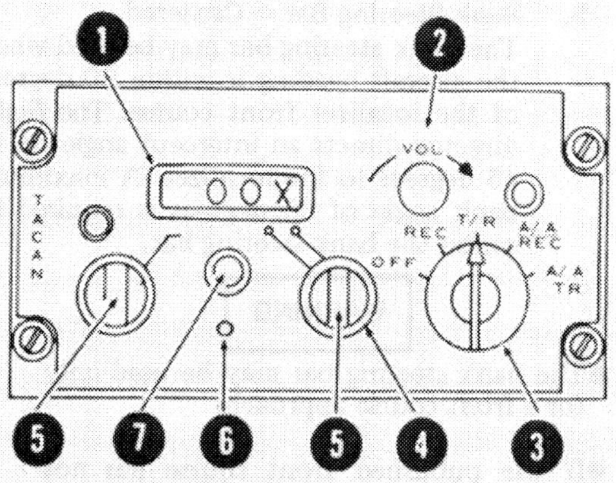


1. CHANNEL DIGITAL DISPLAY
2. VOLUME CONTROL KNOB
3. FUNCTION SELECTOR SWITCH
4. CHANNEL MODE SELECTRO SWITCH (X, Y)
5. CHANNEL SELECTOR SWITCHES
6. PRESS - TO - TEST BURRON
7. TEST INDICATOR LIGHT

Figure 4-7.

3. T/R — Indicators display bearing and distance to ground station. This is the normal position.

4. A/A REC — Indicators display bearing to other suitably equipped aircraft. No distance is displayed.

5. A/A T/R — Indicators display bearing and distance to other cooperating suitably equipped aircraft.

6. Channel Selector Switches (5).

   The Channel selector switches select TACAN channels 1 through 126.

7. Channel Mode Selector Switch (4).

   The Channel Mode Selector Switch selects either X or Y modes making a total of 252 TACAN channels available.

8. Volume Control Knob (2).

   VOL — Not operational.

9. Press-to-Test Button/Test Indicator Light (6 and 7).

TEST — Self-test is initiated by pressing the button. The test indicator light will flash to confirm lamp operation. If the test indicator light illuminates at a subsequent time, a receiver-transmitter fault has been detected. For maximum test capability, allow 2 minutes of operation before initiation. The function selector switch should be set to T/R or A/A T/R for self-test. A test cycle may be terminated at any time by rotating either a channel or mode selector switch. Only receiver-transmitter faults are displayed by the test indicator light.

## AN/ARN-118(V) TACAN OPERATION.

Preflight Check.

1. Function Selector Switch — T/R.

2. HSI Course — Set to 180 degrees.

**NOTE**

Allow 90-second warm-up.

3. Press-to-Test Button — Depress and release. Observe HSI and test indicator light:

   a. Test indicator light flashes momentarily.

   b. DME warning flag comes into view.

   c. Bearing pointer may slew to 270 degrees for approximately 7 seconds.

**NOTE**

The following indications last approximately 15 seconds.

   d. DME warning flag goes out of view.

   e. Range Indication — 000.0 ± 0.5 NM.

   f. Bearing Indication — 180 ± 3 degrees.

   g. CDI — Centered ± 1/2 dot.

   h. To-From arrow — To.

   i. DME warning flag comes into view until a usable signal is received.

Normal Operation for Air to Ground Navigation.

**NOTE**

Allow 90-second warm-up.

1. Function Selector Switch — T/R.

2. Channel Mode and Selector Switches — Set frequency, adjust volume, and identify station.

3. HSI — Check (DME, CDI, To-From arrow and bearing pointer information displayed).

4. HSI Course Set — Adjust to proper course. Fly normal intercept procedures.

Normal Operation for Air to Air Navigation.

**NOTE**

For A/A operation, use preassigned channels or contact a cooperating aircraft. The channel of the receiving aircraft must be either 63 channels above or below the cooperating aircraft and within the 1 through 126 channel X or Y range. Y mode is preferred to preclude interference since no Y mode ground stations are presently operating. Interference shows mainly as reduced bearing and distance lock-on range.

NOTE

- To prevent IFF/TACAN interference, avoid channels 1 through 11, 58 through 74, and 121 through 126.
- As many as five aircraft can lock-on to a "parent" aircraft in A/A T/R mode. The radius of operation for all aircraft involved will be limited to a distance equal to four times the distance between the "parent" aircraft and the nearest other aircraft unless the system is otherwise set by ground maintenance personnel. If system is otherwise set, bearing information received from a "parent" aircraft may be erratic due to resultant noise pickup when an automatic gain control is disabled.

1. Function Selector Switch — A/A T/R.
2. Channel Mode and Selector Switches — Set to desired frequency. If the cooperating aircraft is equipped to transmit bearing signals, CDI, distance, to-from, and bearing information will be displayed. If not, DME only will be displayed.

NOTE

- Automatic Self-Test. If the TACAN signal is lost, an automatic self-test is initiated. This is indicated by the bearing pointer slewing to 270 degrees for about 7 seconds. If the test indicator light illuminates, a system malfunction has occurred, and a press-to-test should be accomplished. Changing channel or mode will not initiate a self-test.
- Press-to-Test. If the test indicator light illuminates during flight, perform a preflight test. If the light remains on, repeat the test in REC or A/A REC function. If the light goes out, the malfunction is probably in the transmitter and bearing information is valid. If the light is illuminated in both T/R and REC functions, all information in considered invalid.

## AIM SYSTEM.

The AIM system provides for identification (IFF/SIF), altitude reporting, and a corrected display of the aircraft altitude on an AAU-19/A counter-drum-pointer altimeter. The IFF/SIF function and altitude reporting are accomplished thru the receiver-transmitter (AN/APX-64), which enables the aircraft to identify itself automatically and report the aircraft altitude when challenged by surface or airborne radar equipment capable of interrogation. The set can also identify the aircraft in which it is installed as a firendly aircraft within a group of specific friendly aircraft. The modes of operation have the following significance: Mode 1 — Friend Identity, Mode 2 — Personal Identity, Mode 3/A — Air Traffic Control, Mode 4 — Not Used, and Mode C — Altitude Reporting. The receiver is sensitive to all interrogation signals within operating frequency; however, only those signals meeting the complete predetermined requirements of the mode being used will be recognized and answered. Mode 2 code settings are set into the receiver-transmitter on the ground and thus are fixed for any one flight. Mode 1 and 3/A codes are set up at the control panel and all modes can be turned on or off. The corrected altitude function is accomplished by an altitude computer and the counter-drum-pointer altimeters. The altitude computer (CPU-46/A) provides digital altitude information, corrected for static pressure effect, to the receiver-transmitter, and as an electrical input to the counter-drum-pointer altimeters (AAU-19/A). An airborne test set is a component of the system to self-interrogate or monitor the replies to external interrogation. When the test set is not installed, all self-interrogation and external checks are inoperative. The system is powered by the left ac bus except for the test set and AAU-19/A altimeter vibrator, which are powered by the 28-volt dc bus.

### AIM SYSTEM CONTROL PANEL.

The AIM system control panel (figure 4-8) is located on the right console of the front cockpit.

#### Master Control Knob.

The master control knob has five positions, placarded OFF, STBY, LOW, NORM, and EMER. When the master control knob is positioned to STBY (standby), the system is inoperative but ready for use after the initial 3-minute warmup period. In the LOW position, the system operates at reduced sensitivity and replies only in the area of strong interrogations. In the NORM (normal) position, the system operates at full sensitivity, which provides maximum performance. To select the EMER (emergency) position, the master control knob must be pulled out and rotated. When the knob is positioned at EMER, modes 1, 2, and 3A are automatically enabled regardless of the position of the mode select switch or code selector wheel. Mode 3 code 7700 is transmitted each time the set is interrogated by ground radar. On some control units, the STBY position has a detent stop; when returning from any selected position to OFF, the master control knob must be pulled out and rotated to return to the OFF position.

Radiation Test/Monitor Switch.

The radiation test/monitor switch has three positions, placarded RAD TEST, OUT, and MON. The switch is spring-loaded for momentary contact in the RAD TEST (radiation test) position and will return to the OUT position when released. The RAD TEST position is used by the ground crew to preflight the system. With the switch at the MON (monitor) position, illumination of the TEST light indicates a normal operating condition for the signal response to external interrogations for the mode switches that are ON. With the radiation test/monitor switch in the OUT position, the TEST light will not illuminate in response to external interrogations. The MON position is inoperative when the airborne test set is not installed.

Mode Select/Test Switch.

Four mode select/test switches are placarded TEST, ON, and OUT. The switches grouped under the TEST heading are labeled M-1, M-2, M-3/A, and M-C. The OUT position for each switch deactivates the mode selected. If more than one switch is placed at ON, the receiver-transmitter will reply to interrogations for all modes selected. With the M-C mode switch at ON, the aircraft altitude is reported in increments of 100 feet referenced to a barometric pressure of 29.92 inches of Hg to an altitude of 80,000 feet when interrogated. The switches are spring-loaded to the ON position from the TEST position. With the radiation test/monitor switch at the OUT position and a mode select switch held in the TEST position, the selected mode can be self-interrogated; illumination of the TEST light indicates a normal operating condition. The TEST function is inoperative when the airborne test set is not installed.

Code Selector Wheels.

Two sets of code selector wheels are provided to set Mode 1 and Mode 3/A. A set of two wheels placarded Mode 1 will select 32 different codes. A set of four wheels placarded Mode 3/A will select 4096 codes. Each wheel is placarded with digits 0 thru 7, which can be seen thru the recessed windows on the face of the control panel.

Identification of Position (I/P) Switch.

The identification of position (I/P) switch has three positions, placarded IDENT, OUT, and MIC. When the switch is momentarily held in the spring-loaded IDENT (identification) position, the I/P timer is energized for approximately 15 to 30 seconds. The receiver-transmitter will transmit an identification-of-position pulse group during the period if a Mode 1, 2, or 3/A interrogation is recognized. When the switch is placed in the MIC (microphone) position, the system will function in an identical manner as it did in the IDENT position except the system will not be activated until the microphone button on the right throttle in either cockpit is pressed. Placing the switch to the OUT position prevents transmission of identification-of-position pulse groups.

Mode 4.

Mode 4 is not used and all controls and lights are inoperative.

COUNTER-DRUM-POINTER ALTIMETER.

A servo/pneumatic counter-drum-pointer altimeter (AAU-19/A) on the instrument panel in each cockpit (figure 4-9) consists of a precision pressure altimeter combined with a servomechanism. The altimeter has two modes of operation; primary (servoed) mode and standby (pneumatic) mode. In the primary mode of operation, the altimeter is controlled by signal inputs from the altitude computer. Direct readout of the altitude is accomplished by the numbers on the 10,000-foot counter, 1000-foot counter, and the 100-foot drum on the face of the instrument. A single pointer indicates hundreds of feet around the fixed circular scale. The 100-foot pointer serves as a precise readout of values less than 100 feet. Below an altitude of 10,000 feet, a diagonal warning symbol will appear on the 10,000-foot counter. A barometric pressure set knob is provided to insert the desired altimeter setting in inches of Hg. Rapid rotation of the barometric pressure set knob or use of abnormal force to overcome binding of the knob may cause internal gear disengagement or gear failure, resulting in excessive altitude indication errors in both the primary and standby modes. In case of an electrical power interruption longer than 3 seconds or a system failure in the altimeter or altitude computer, a warning flag placarded STBY (standby) will appear in the upper left portion of the instrument face, indicating that the altimeter has automatically reverted to standby mode of operation (pressure altimeter) and uncorrected altitude is displayed. Simultaneously, a dc operated vibrator is activated in the altimeter. A function switch, placarded STBY (standby) and RESET, is a spring-loaded self-centering switch used to select the primary or standby mode of operation. To select the primary mode of operation, momentarily place the function switch to RESET after ac electrical power is available. The standby mode of operation may be selected while the altimeter is in the primary mode of operation by momentarily placing the function switch to STBY. Each altimeter can be operated independent

# AIM SYSTEM CONTROL PANEL

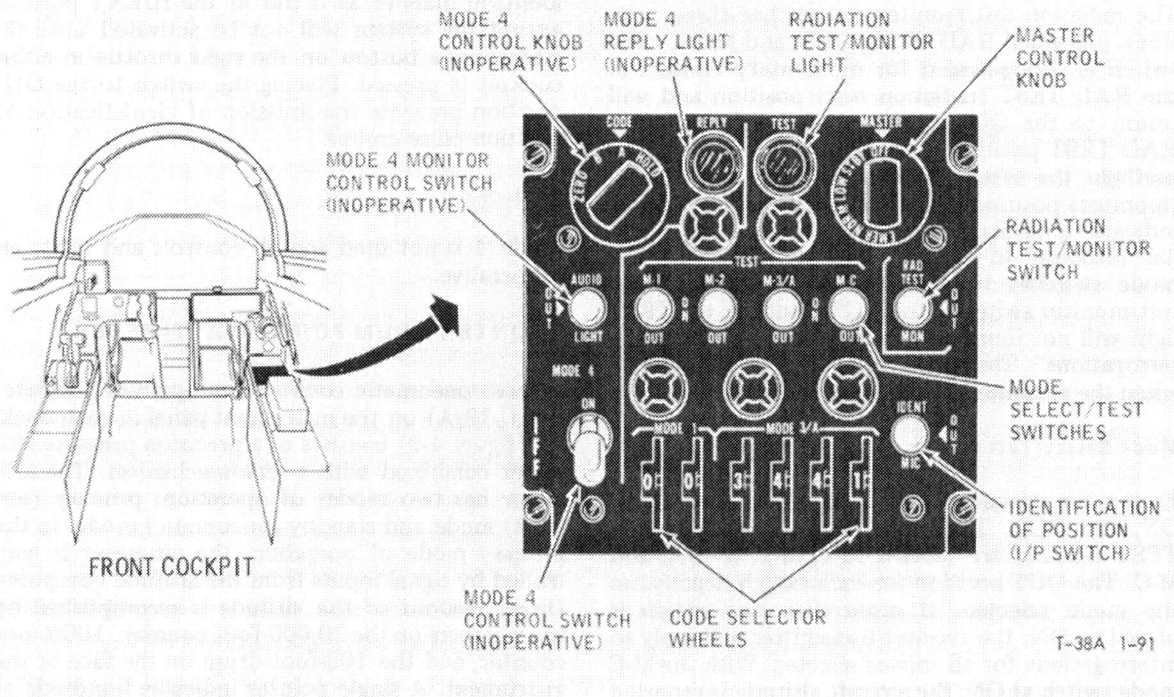


Figure 4-8.

# COUNTER-DRUM-POINTER ALTIMETER (AAU-19/A)

Figure 4-9.

of the other in the primary mode or in the standby mode.

#### Primary (Servoed) Mode of Operation.

In the primary mode of operation, corrected pressure altitude (installation error correction) synchro signals are sent from the altitude computer to the receiver-transmitter and to a servomechanism in the altimeter. These signals are computed only for a barometric pressure of 29.92 inches of Hg. To correct the altimeter indicated altitude for other than 29.92 inches of Hg, set the current altimeter setting in the altimeter barometric scale. When the system is interrogated for altitude reporting (mode C), the receiver-transmitter will automatically report the aircraft altitude to the nearest 100 feet for a barometric pressure of 29.92 inches of Hg, regardless of the setting in the altimeter barometric scale.

#### Standby (Pneumatic) Mode of Operation.

In the standby mode of operation, the altimeter receives static air pressure directly from the pitot-static system and operates in exactly the same manner as the standard pressure altimeter (AAU-7). Altimeter installation error corrections from the appendix must be used to correct the aircraft altitude. Mode C altitude reporting is available if the standby mode of operation has been selected by the crewmember and is not the result of the system automatically reverting to standby operation due to an altitude computer failure. In the standby mode of operation, the vibrator is automatically energized to remove the friction from the counter-drum-pointer mechanism, decreasing the lag in the altimeter indications.

**WARNING**

> If the AAU-19/A altimeter internal vibrator is inoperative, due either to internal failure or dc power failure, the 100-foot pointer may momentarily hang up when passing thru 0 (12-o'clock position). If the vibrator has failed, the 100-foot pointer "hangup" can be minimized by tapping the case of the altimeter.

### ALTIMETER ERROR CROSS-CHECK.

Operational checks of the altimeter should be performed routinely as prescribed in Section II and anytime a malfunction is suspected. When changing from RESET to STBY to RESET positions, a change in readings will be observed due to the difference between corrected altitude (RESET position) and uncorrected altitude (STBY position). At sea level static conditions, the change may be up to 75 feet. At airspeeds up to 0.9 mach, below 10,000 feet, the indicated change may be as much as 150 feet and above 10,000 feet as great as 250 feet. This difference may be observed between the modes of one altimeter or the modes of both front and rear altimeters (front in RESET, rear in STBY, etc). The allowable difference between the primary mode (RESET) readings of both altimeters is 75 feet during preflight and at all altitudes and airspeeds throughout the operating range.

**NOTE**

> If the difference in indicated altitude between RESET and STBY modes of one altimeter or between altimeters exceeds allowable limits, continue the mission in the STBY mode(s).

## STANDBY ATTITUDE INDICATOR.

A standby attitude indicator (figure 1-7) is located on the instrument panel to provide an attitude indicating system if the flight director system malfunctions. The indicator is remotely operated by signals from an MD-1 vertical gyro, which is separate from the flight director system and located in the dorsal section of the fuselage. Complete erection requires 5 minutes after ac power is applied. The MD-1 gyro senses pitch-and-bank angles and incorporates a pitch-and-bank erection system. The aircraft attitude is shown accurately thru 360 degrees of roll and plus or minus 82 degrees of pitch. The pitch-and-bank erection system reduces turning errors to a minimum. Acceleration and deceleration cause slight errors in pitch indications, which are most noticeable on takeoff. Pitch and roll attitudes are shown by the circular motion of a universally mounted sphere displayed as the background for a miniature reference aircraft. The miniature reference aircraft is always in proper physical relationship to the simulated earth, horizon, and sky areas of the background sphere. On the sphere, the horizon is represented by a solid fluorescent line, the sky by a light gray area, and the earth by a dull black area. Horizontal markings on the face of the sphere show accurate aircraft attitudes up to 82 degrees of climb or dive. The pitch trim knob on the lower right side of the instrument electrically rotates the sphere to the proper position in relation to the fixed miniature reference aircraft to correct for pitch attitude changes. This adjustment is necessary, since the level-flight attitude of the aircraft varies with weight and speed.

### ATTITUDE WARNING FLAG.

The attitude warning flag (OFF) will appear whenever electrical power to the system has failed or is interrupted. The flag will also appear during initial application of electrical power for approximately 1 minute. The instrument is unreliable until the flag disappears.

WARNING

- There is no warning of attitude sphere malfunctions other than power failure.
- The attitude warning flag will not appear with a slight electrical power reduction or failure of other components within the system. Failure of certain components can result in erroneous or complete loss of pitch and bank presentations without a visible flag. It is imperative that the attitude indicator be cross-checked with other flight instruments when under actual or simulated instrument conditions.

NOTE

During high G maneuvering the warning flag may appear without system malfunction.

## ANGLE-OF-ATTACK SYSTEM.

The angle-of-attack (AOA) system (figure 4-10) senses aircraft angle of attack and displays this information to both crewmembers. The AOA system consists of an AOA vane transmitter, AOA CPU-115/A computer, and in each cockpit, an AOA indicator, AOA indexer, and indexer lights dimmer control. The system provides compensation for various wing flap and landing gear configurations. The AOA system presents the following displays in each cockpit:

a. Optimum AOA for final approach.

b. AOA when buffet and stall will occur.

c. Approximate AOA for maximum range and maximum endurance.

The vane of the AOA transmitter is located on the forward right side of the fuselage. The vane is electrically heated for anti-ice and is activated when the pitot heat switch is turned on. The AOA computer, which is powered by the left ac bus, receives signals from the AOA vane transmitter, wing flap position synchro-transmitter, and nose gear downlock indicating system. The computer automatically computes and sends the appropriate signals to the AOA indicator and AOA indexer in each cockpit.

### AOA INDICATOR.

The ARU-26/A AOA dial indicator on the instrument panel operates during all phases of flight and indicates AOA information. The indicator presents AOA as a percentage of maximum lift AOA. The dial is calibrated in units of .1 counterclockwise from 0 to 1.1. Each unit represents approximately 10% of aircraft lift, from 0% at 0 indication to 100% at 1.0 indication. Three preset fixed indices and two colored arcs on the dial indicate the following:

.18 White Index — Maximum Range (1-G flight)

.3 White Index — Maximum Endurance (1-G flight)

.6 White Index — Optimum Final Approach at 3-o'clock Position (1-G flight)

.9 to 1.0 Yellow Arc — Buffet Warning

1.0 to 1.1 Red Arc — Stall Warning

The red OFF flag will appear on the face of the dial when electrical power is removed from the AOA system or when the system has failed. The AOA indicator is powered by the left ac bus.

WARNING

The airspeed indicator should be cross-checked frequently when using AOA information; some system malfunctions may not necessarily trigger the OFF flag or be repeated in the other cockpit.

### AOA INDEXER.

The ARU-27/A AOA indexer above the instrument panel is controlled by the AOA computer and provides an illuminated heads-up display of the AOA information in the form of low-speed, on-speed, and high-speed indexer lights. (See figure 4-10). The three indexer lights are powered by the dc bus. The lights are operative in the landing configuration with the wing flaps up or down, or when the landing gear is up and the wing flaps are extended 5 percent or more. With the landing gear and wing flaps up, the high-speed indexer light is inoperative to eliminate continuous illumination during cruise flight conditions. See figure 4-11 for allowable On-Speed Band for AOA indexer. AOA system failure is indicated when all three symbols of the indexer are illuminated. The three indexer lights can be tested by placing the warning test switch on the right console at TEST.

# AOA SYSTEM AND DISPLAYS

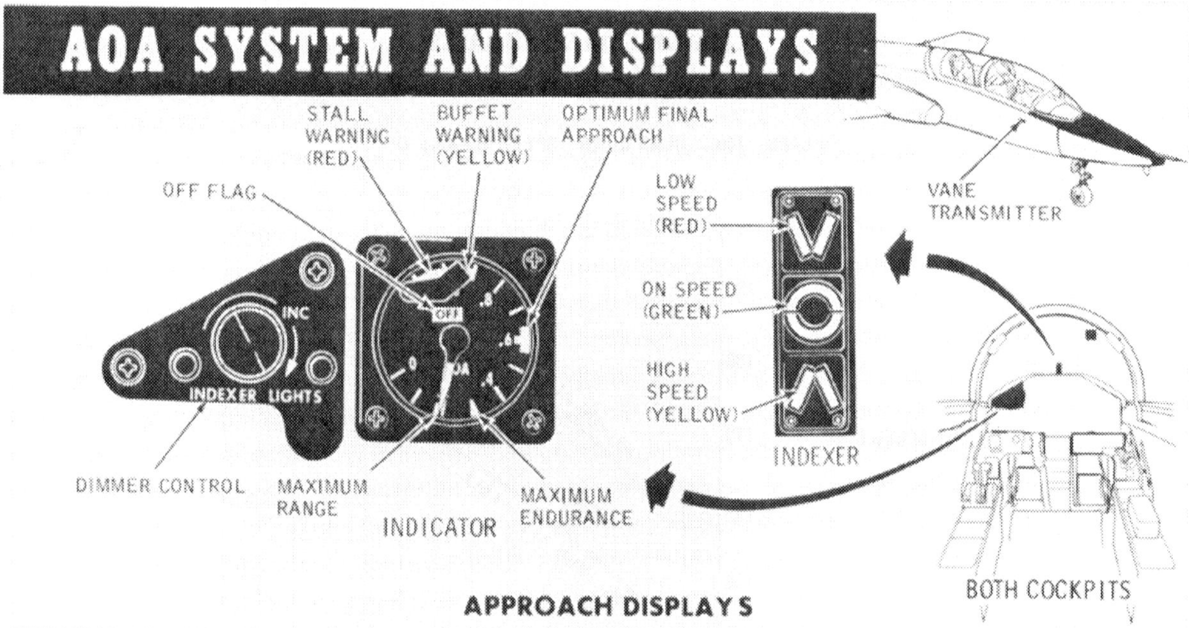


## APPROACH DISPLAYS

| INDICATOR | INDEXER | AIRSPEED | ATTITUDE |
|---|---|---|---|
|  |  | SLOW | VERY HIGH AOA  |
|  | 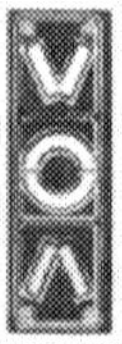 | SLIGHTLY SLOW | HIGH AOA 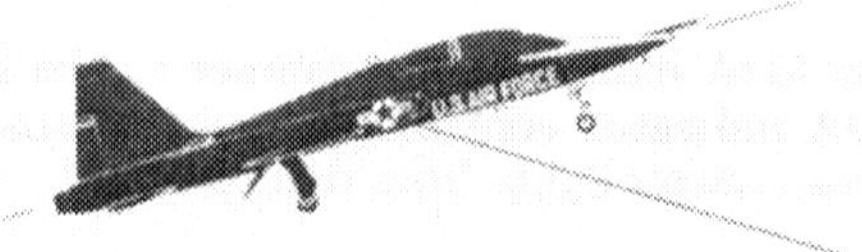 |
| 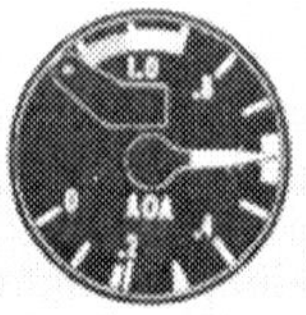 |  | ON SPEED | OPTIMUM AOA 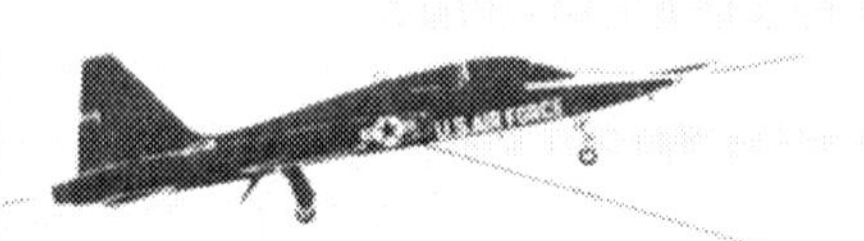 |
|  |  | SLIGHTLY FAST | LOW AOA 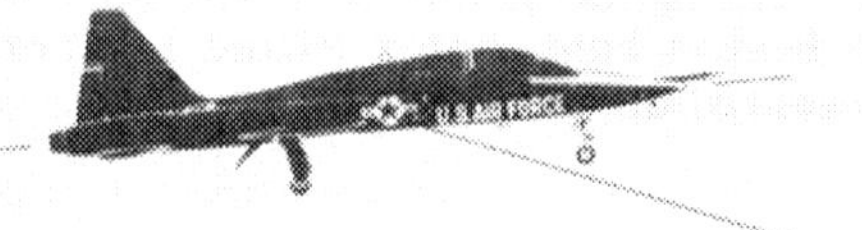 |
| 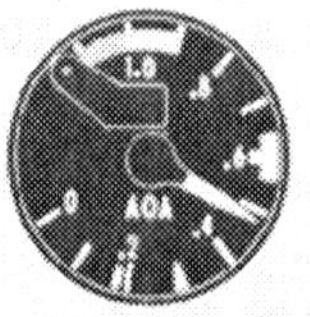 |  | FAST | VERY LOW AOA  |

T-38A 1-98A

Figure 4-10.

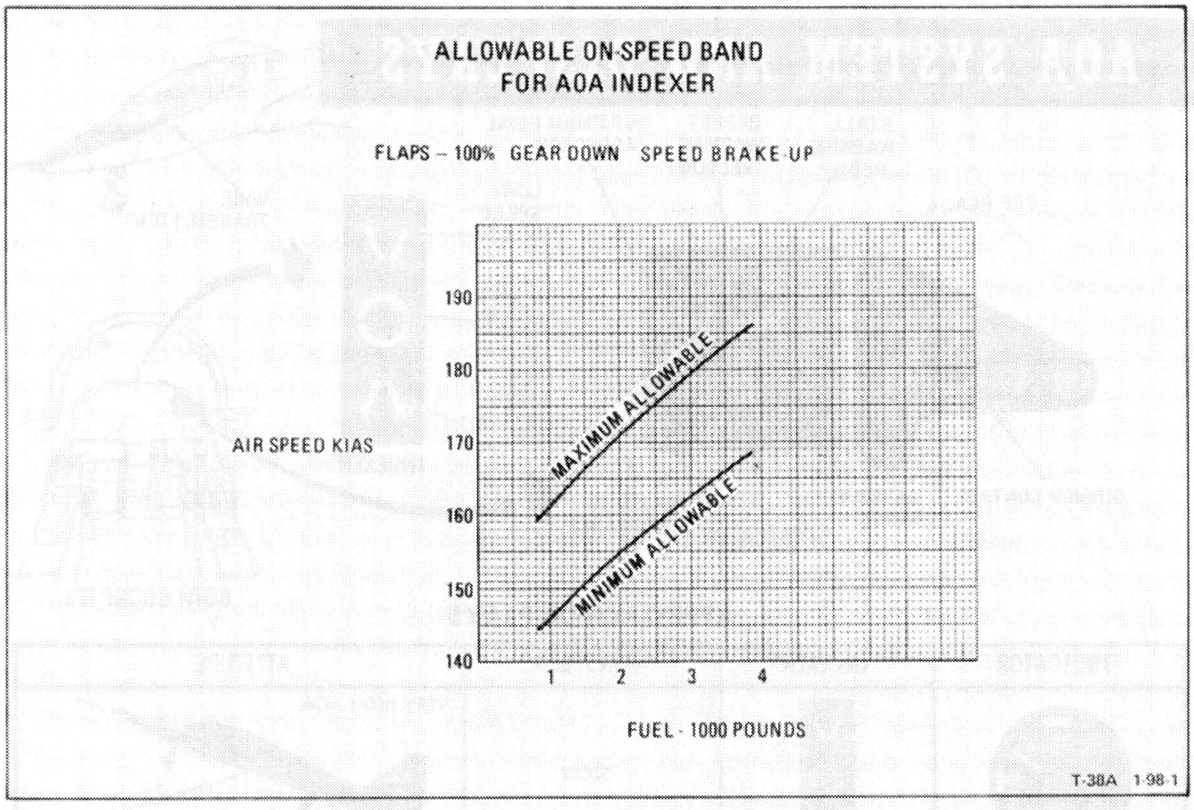


Figure 4-11.

### AOA INDEXER LIGHTS DIMMER.

The AOA indexer lights dimmer to the left of the AOA indicator controls the light intensity of the three indexer lights from dim to bright.

## LIGHTING EQUIPMENT.

### EXTERIOR LIGHTING.

#### Rotating Beacon Lights and Switch.

One rotating beacon light is located near the top of the vertical stabilizer and one on the lower fuselage. The lights operate on ac and are controlled by the beacon light switch (figure 1-11) on the right console of the front cockpit.

#### Position Lights and Switch.

The position lights, which operate on 6-volt ac from a transformer off the left ac bus, are individually located in each wingtip, in the vertical stabilizer, and in the lower fuselage. The position lights are controlled by a bright/dim switch (figure 1-11) on the right console of the front cockpit.

#### Formation Lights and Switch.

Formation lights, operating on dc bus power, are individually located on each side of the forward nose section. Formation lights are controlled by a switch (figure 1-11) on the right console of the front cockpit.

#### Landing-Taxi Light.

A single retractable landing-taxi light with dual filaments is installed. When the position lights are turned on, and the gear is extended, the light also extends. The landing-taxi light switch (figure 1-9) on the left subpanel of the front cockpit controls only the filament power. When the weight of the aircraft is off the main gear and the landing-taxi light switch is at ON, both filaments are burning. When the weight of the aircraft is on the main gear, the light moves to the taxi position and one filament is extinguished. Turning off the position lights retracts the landing light in about 10 seconds.

### INTERIOR LIGHTING.

The instrument lights operate on ac power. A knob (figures 1-11, 1-12) on the right console of each cockpit controls operation and intensity of the instrument lights. White floodlights, operating on ac, aid in illuminating the instrument panel, console panels, and the cockpit area. The floodlights are controlled by a knob (figures 1-11, 1-12) on the right console of each cockpit. The two floodlights over each cockpit instrument panel (figure 1-7) automatically switch from ac to dc if the ac power supply fails, provided the floodlight control knob is not at the OFF position. These floodlights serve as an alternate lighting source under this condition and cannot be dimmed when operating on dc power. The integral console, subpanel, and pedestal lights operate on ac. Operation and intensity of these lights are controlled by rotating the console lights knob (figures 1-11, 1-12) on each right console.

NOTE

If the left generator and bus transfer relay fail instrument and console lights will not be operational. Floodlights which are powered by the right AC bus will not be automatically available, and the floodlight rheostat must be adjusted to obtain cockpit lighting.

## UTILITY LIGHTS.

Two removable utility lights, one in each cockpit, are normally mounted on the right console aft of the map case (figure 1-5) in the front cockpit and on the right canopy frame (figure 1-6) in the rear cockpit. Each light is controlled by a rheostat, which is an integral part of the light. Each light can be removed from the mounting bracket and is equipped with a spring extension cord, enabling use anywhere in the cockpit, or it can be placed in various other mounting brackets in the cockpit. The lights operate on ac power.

WARNING

Stow after use to prevent interference with ejection seat and man-seat separator system.

## OXYGEN SYSTEM.

The aircraft uses a liquid oxygen system to supply breathing oxygen to crewmembers. The oxygen regulators (automatic diluter demand) control the flow and pressure of the oxygen and distribute it in the proper proportions to the masks. One of two types of oxygen regulators (figures 1-11, 1-12) on the right console of each cockpit contains a pressure gage, a blinker type flow indicator, emergency flow lever, diluter lever, and a supply lever. The later type regulator differs from the earlier type regulator in that when the supply lever is at OFF, the flow of oxygen and cockpit air to the oxygen mask are both cut off. On earlier type regulators, cockpit air only will flow to the mask with the supply lever at OFF.

## OXYGEN QUANTITY INDICATOR.

An oxygen quantity indicator, operating on ac and located on the right subpanel of each cockpit (figures 1-9, 1-10), indicates converter liquid oxygen quantity in liters. The indicator is provided with an OFF flag, which will appear in case of electrical power failure.

## OXYGEN LOW-LEVEL CAUTION LIGHT.

An oxygen low-level caution light (figures 1-11, 1-12) on the right console of each cockpit illuminates when the oxygen indicator reads 1 liter or less of liquid oxygen. The light may blink, due to oxygen sloshing, if the system contains less than 3 liters.

## OXYGEN SYSTEM PREFLIGHT CHECK (PRICE).

P — PRESSURE.

The pressure gage should read 50 to 120 psi (figure 5-1) and should agree with the pressure gage in the other cockpit.

R — REGULATOR.

Check regulator supply lever at ON. Hook up mask and perform a pressure check. Place the emergency flow lever at EMERGENCY position, take a deep breath and hold it. If mask leaks, readjust mask and check pressure. The oxygen should stop flowing if the mask is properly fitted; if the oxygen continues to flow, the regulator, the hose, or the valve is not holding pressure, and the cause of the leak should be corrected. Return the emergency lever to NORMAL. If you cannot exhale, the valve has malfunctioned and the discrepancy should be corrected.

WARNING

It is possible for the supply lever to stop in an intermediate position between OFF and ON. Care should be taken to push the supply lever full ON and visually check the flow indicator blinker for proper functioning.

# OXYGEN DURATION HOURS TABLE

## CREWMEMBER DURATION IN HOURS

| | COCKPIT ALTITUDE FEET | DILUTER LEVER | 10 | 9 | 8 | 7 | 6 | 5 | 4 | 3 | 2 | 1 | BELOW 1 |
|---|---|---|---|---|---|---|---|---|---|---|---|---|---|
| ONE CREWMEMBER | 40,000 & ABOVE | 100% OXYGEN | 56 | 50 | 45 | 39 | 33 | 28 | 22 | 16 | 11 | 5.6 | EMERGENCY DESCEND TO ALTITUDE NOT REQUIRING OXYGEN. |
| | | NORMAL OXYGEN | 56 | 50 | 45 | 39 | 33 | 28 | 22 | 16 | 11 | 5.6 | |
| | 35,000 | 100% OXYGEN | 56 | 50 | 45 | 39 | 33 | 28 | 22 | 16 | 11 | 5.6 | |
| | | NORMAL OXYGEN | 56 | 50 | 45 | 39 | 33 | 28 | 22 | 16 | 11 | 5.6 | |
| | 30,000 | 100% OXYGEN | 40 | 36 | 32 | 28 | 24 | 20 | 16 | 12 | 8.1 | 4.0 | |
| | | NORMAL OXYGEN | 41 | 37 | 32 | 29 | 25 | 20 | 16 | 12 | 8.3 | 4.1 | |
| | 25,000 | 100% OXYGEN | 31 | 28 | 25 | 21 | 18 | 15 | 12 | 9.4 | 6.2 | 3.1 | |
| | | NORMAL OXYGEN | 39 | 35 | 31 | 27 | 23 | 19 | 15 | 11 | 7.8 | 3.9 | |
| | 20,000 | 100% OXYGEN | 23 | 21 | 19 | 16 | 14 | 11 | 9.5 | 7.1 | 4.7 | 2.3 | |
| | | NORMAL OXYGEN | 44 | 40 | 35 | 31 | 26 | 22 | 17 | 13 | 8.9 | 4.4 | |
| | 15,000 | 100% OXYGEN | 19 | 17 | 15 | 13 | 11 | 9.5 | 7.6 | 5.7 | 3.8 | 1.9 | |
| | | NORMAL OXYGEN | 54 | 48 | 43 | 37 | 32 | 27 | 21 | 16 | 10 | 5.4 | |
| | 10,000 | 100% OXYGEN | 15 | 13 | 12 | 10 | 9.2 | 7.6 | 6.1 | 4.6 | 3.0 | 1.5 | |
| | | NORMAL OXYGEN | 54 | 48 | 43 | 37 | 32 | 27 | 21 | 16 | 10 | 5.4 | |
| TWO CREWMEMBERS | 40,000 & ABOVE | 100% OXYGEN | 28 | 25 | 22 | 19 | 16 | 14 | 11 | 8.4 | 5.6 | 2.8 | EMERGENCY DESCEND TO ALTITUDE NOT REQUIRING OXYGEN. |
| | | NORMAL OXYGEN | 28 | 25 | 22 | 19 | 16 | 14 | 11 | 8.4 | 5.6 | 2.8 | |
| | 35,000 | 100% OXYGEN | 28 | 25 | 22 | 19 | 16 | 14 | 11 | 8.4 | 5.6 | 2.8 | |
| | | NORMAL OXYGEN | 28 | 25 | 22 | 19 | 16 | 14 | 11 | 8.4 | 5.6 | 2.8 | |
| | 30,000 | 100% OXYGEN | 20 | 18 | 16 | 14 | 12 | 10 | 8.1 | 6.1 | 4.0 | 2.0 | |
| | | NORMAL OXYGEN | 20 | 18 | 16 | 14 | 12 | 10 | 8.3 | 6.2 | 4.1 | 2.0 | |
| | 25,000 | 100% OXYGEN | 15 | 14 | 12 | 11 | 9.4 | 7.8 | 6.2 | 4.7 | 3.1 | 1.5 | |
| | | NORMAL OXYGEN | 19 | 17 | 15 | 13 | 11 | 9.8 | 7.8 | 5.9 | 3.9 | 1.9 | |
| | 20,000 | 100% OXYGEN | 11 | 10 | 9.5 | 8.3 | 7.1 | 5.9 | 4.7 | 3.5 | 2.3 | 1.1 | |
| | | NORMAL OXYGEN | 22 | 20 | 17 | 15 | 13 | 11 | 8.9 | 6.6 | 4.4 | 2.2 | |
| | 15,000 | 100% OXYGEN | 9.5 | 8.6 | 7.6 | 6.6 | 5.7 | 4.7 | 3.8 | 2.8 | 1.9 | 0.9 | |
| | | NORMAL OXYGEN | 27 | 24 | 21 | 18 | 16 | 13 | 10 | 8.1 | 5.4 | 2.7 | |
| | 10,000 | 100% OXYGEN | 7.6 | 6.9 | 6.1 | 5.3 | 4.6 | 3.8 | 3.0 | 2.3 | 1.5 | 0.7 | |
| | | NORMAL OXYGEN | 27 | 24 | 21 | 18 | 16 | 13 | 10 | 8.1 | 5.4 | 2.7 | |
| | LIQUID CONTENTS LITERS | | 10 | 9 | 8 | 7 | 6 | 5 | 4 | 3 | 2 | 1 | BELOW 1 |

- TOP FIGURES INDICATE DILUTER LEVER "100% OXYGEN".
- BOTTOM FIGURES INDICATE DILUTER LEVER "NORMAL OXYGEN".

T38A-1-59A

Figure 4-12.

I — INDICATOR.

With the diluter lever in 100% OXYGEN position, check blinker for proper operation.

C — CONNECTIONS.

Check connection secure at the seat. Check regulator hose for kinks, cuts, or cover fraying. Check that male part of the disconnect is not warped and rubber gasket is in place. A 10- to 20-pound pull should be required to separate the two parts. Check mask hose properly installed to connector.

E — EMERGENCY.

Check emergency oxygen cylinder properly connected and a minimum pressure of 1800 psi. (Pressure gage must be checked during parachute preflight.)

## ANTI-G SUIT SYSTEM.

Air pressure from the air-conditioning system is used to inflate the anti-G suit in each cockpit to offset the effects of high load factor.

### ANTI-G SUIT TEST BUTTON.

An anti-G suit press-to-test button in the top of each regulator (figures 1-11, 1-12) is located on the left console of each cockpit. The button is used to manually test operation of the anti-G suit valve; the further the button is pressed, the greater is the anti-G suit pressure available.

## MISCELLANEOUS EQUIPMENT.

Additional items provided include:

a. Instrument hood.

b. Rearview mirrors.

c. Map data case.

d. Weapon System Support Pod (WSSP).

(1) Some aircraft are equipped to carry a WSSP which mounts under the center section of the fuselage. The nose section of the pod is attached to a tray which slides out for loading and, when stowed, is secured in place by a metal over-center latch type strap on each side. Each latch strap is covered by a stream-lined fairing which is secured by a wing nut dzus-type fastener. The pod is approximately 84 inches long, 24 inches wide, and 16 inches deep. The nose and tail sections are faired. Normal load capacity is approximately 140 pounds. Maximum loaded weight of the pod (with attached pylon) is 250 pounds.

(B) e. MXU-648 Aircraft Baggage/Cargo Pod.

(1) The MXU-648 baggage/cargo pod is a modification of a fire bomb container. Each pod has a hinged access door on the left side. Some pods have a removable tail cone for loading a variety of cargo in size and length. The cargo compartment contains a metal floor and a cargo tie-down system which consists of straps and/or netting secured to permanent hooks installed in the floor. Refer to Section V for external stores limitation.

## (B) ARMAMENT SYSTEM.

The aircraft is fitted with a centerline pylon containing an MA-4 bomb rack. The rack can carry stores weighing up to 1000 pounds with 14 inch suspensions. The pylon is bolted to the airplane and is not jettisonable. Practice ordnance is released from the pylon through the store use of the armament release system. Emergency jettison of a store from the pylon is accomplished with the emergency jettison system. Depressing the jettison button opens the pylon hooks allowing the store to freefall. Forced ejection is not provided. A fixed reticle sight is provided to assist in weapons delivery, and a 16mm gunsight camera is installed.

### (B) AUTHORIZED CONFIGURATIONS.

Refer to Section V for the external weapon configurations authorized to be carried on this aircraft.

### (B) OPTICAL SIGHT.

The CA-513 gunsight (figure 1-5) provides a collimated reticle image which is reflected from the combining glass to produce the reticle image focused at infinity. The combining glass angle is adjustable by the pilot, producing a reticle image depression from 0 to 270 milliradians from the 0 mil reference line. The sight has an etched reticle containing an outer 100 mil dia circle, tick marks at 10 mil intervals along the major axes, plus a 2 mil center aim dot. The reticle is illuminated by two lamps which produce a reticle image. Reticle intensity is controlled by the reticle intensity knob in the front cockpit on the left hand console. The combining glass angle is manually adjusted by the pilot by means of a knob with an affixed graduated dial. The dial is marked in 5 mil increments, and rotation of the knob is in discrete 2.5 mil increments. The sight mirror is provided with a removable dust cover which should be removed before flight and replaced after flight to prevent dust accumulation and possible damage to the sight and its wiring from concentrated sunlight.

NOTE

The reticle intensity knob should be rotated to OFF when the sight is not in use to prevent wiring damage and blown fuzes from overheating.

## B GUN CAMERA SYSTEM

A KB-26A, 16mm motion picture camera is mounted on the optical sight. Before takeoff, the desired frame speed, over-run time, and lens aperture setting should be set. The camera has frames per second speeds of 24 and 48 fps and f-stop settings from 2.8 to 22. The system can be actuated by the front cockpit bomb button or trigger, the rear cockpit CAMERA START button, or a prerun switch on the camera itself. When power is on the aircraft and the camera circuit breaker is in, the camera can be run by simply depressing either the CAMERA START switch in the rear cockpit or the prerun button on the camera. The camera will not run when the aircraft has weight on the right main gear "squat" switch. The trigger in the front cockpit can be used to run the camera if the MSTR switch is in the CAM ONLY position and the MODE switch is in any position or if the MSTR switch is in the STO/CAM position and the MODE switch is out of SAFE. With the above conditions met, depressing the trigger to the first detent will start the camera, and depressing the trigger fully will run the camera and light the EVENT marker in the camera. The bomb button can be used to run the camera if the MSTR switch is in the STO/CAM position and the MODE switch is out of SAFE. With these conditions met, depressing the bomb button will cause the camera to run and illuminate the EVENT marker. The EVENT marker is used to identify, on film, the point at which an armament fire signal was generated.

CAUTION

The gun camera must have the camera dust cover (dummy film magazine) or a fully loaded film magazine installed to ensure proper balance of the gunsight assembly and to provide lens protection. Applying power to the camera system without a magazine or dust cover installed can result in system damage.

## B GUNNERY SYSTEM.

One SUU-11, 7.62 mm, gun pod can be mounted on the centerline pylon. The gun pod offers accurate and concentrated firepower, minimum weight and high reliability. The pod carries 1500 rounds of ammunition which is 15 seconds of firepower at the high rate of fire and 30 seconds at the low rate. Expended ammunition cases are ejected from the bottom of the pod.

## B ARMAMENT CONTROLS.

Armament controls provide normal and emergency release, firing, or jettison of all externally carried weapons. There are no armament controls, as such, in the rear cockpit; there is, however a camera start switch and stores ARMT HOT light on the left hand vertical panel.

### B MASTER ARMAMENT SWITCH.

The MSTR ARMT switch, located on the front cockpit left hand vertical panel, is a three position toggle switch with a STO/CAM, SAFE, and CAM ONLY position. With this switch in the SAFE position, the release system is deactivated and the gun camera can be run only via the camera mounted prerun switch or the rear cockpit camera start button. When the MSTR ARMT switch is in the CAM ONLY position, the camera can be operated using the trigger on front cockpit stick grip. When the MSTR ARMT switch is in the STO/CAM position, the camera can be operated via the front cockpit trigger or bomb-rocket button; additionally, the release system is activated.

### B MODE SWITCH.

The MODE switch is located on the front cockpit left hand vertical panel just below the MSTR ARMT switch. This is a three position toggle switch with RKT/GUN, SAFE, and BOMB positions.

### B JETTISON BUTTON.

There is a ring guarded push button type jettison button located on the front cockpit left hand subpanel just outboard of the Landing Gear Alternate Release Handle. Pushing this button when 28 volt DC power is on the airplane will open the hooks of the pylon bomb rack and any loaded store will fall away.

CAUTION

Once the jettison circuit has been activated, a store will drop from the pylon if and when the rack safety pin is removed (the pin will be tight).

# OPERATING LIMITATIONS

SECTION V

T-38A 1-104

TABLE OF CONTENTS



## INTRODUCTION.

Cognizance must be taken of instrument markings in figure 5-1, since they represent limitations that are not necessarily repeated in the text.

## MINIMUM CREW REQUIREMENT.

The minimum crew requirement for this aircraft is one pilot. Solo flights must be made with the pilot flying the aircraft from the front cockpit.

## THROTTLE SETTING THRUST DEFINITIONS.

### NORMAL THRUST.

Normal (maximum continuous) thrust is the thrust obtained at 98.5% RPM or 630°C EGT, whichever occurs first.

### MILITARY THRUST.

MIL (military) thrust is the thrust obtained at 100% RPM without afterburner operation.

### MAXIMUM THRUST.

MAX (maximum) thrust is the thrust obtained at 100% RPM with the afterburner operating. Afterburner range extends from minimum afterburner of approximately 5 percent augmentation above MIL thrust to maximum afterburner, which is approximately 40 percent augmentation above MIL thrust.

## AIRSPEED LIMITATIONS.

### WING FLAPS.

Do not exceed the following airspeeds for the wing flap deflections:

1% to 45% . . . . . . . . 300 KIAS
46% to 60% . . . . . . . . 240 KIAS
Over 60% . . . . . . . . 220 KIAS

### LANDING GEAR.

Do not exceed 240 KIAS with the landing gear extended and/or landing gear doors open.

**CAUTION**

Extension/retraction of landing gear at bank angles greater than 45 degrees or at load factors greater than 1.5 G's can result in overstress failure of the main landing gear sidebrace trunnion.

### WEAPON SYSTEM SUPPORT POD (WSSP).

Do not exceed the following airspeed when the WSSP is installed:

350 KIAS in severe turbulence or with speed brake open.
400 KIAS under all other conditions.

**CAUTION**

Avoid abrupt control movements at airspeeds greater than 240 KIAS.

# INSTRUMENT MARKINGS

## BASED ON FUEL GRADE JP-4

EHU-31A/A

TEMPERATURE EXHAUST GAS

WARNING

WITH EHU-31A/A INDICATORS, IT IS POSSIBLE TO EXPERIENCE AN ENGINE START OR FLAMEOUT UNRECOGNIZED BY THE PILOT. OTHER ENGINE INSTRUMENTS MUST BE REFERENCED TO CONFIRM AN ENGINE START OR FLAMEOUT.

| | |
|---|---|
| 140°C | MINIMUM |
| 140°C TO 630°C | CONTINUOUS OPERATION |
| 645°C | MAXIMUM STEADY STATE |
| 925°C | MAXIMUM DURING START OR ACCELERATION, MOMENTARY |
| 630°C TO 645°C | MIL AND MAX THRUST RANGE |

OIL PRESSURE

- 5 PSI MINIMUM (IDLE)
- 20 TO 55 PSI NORMAL OPERATING RANGE
- 55 PSI MAXIMUM

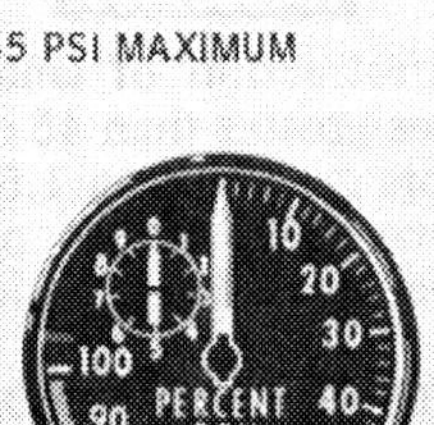

ENGINE TACHOMETER

- 83% TO 98.5% RPM CONTINUOUS
- 104% RPM MIL AND MAX THRUST

(99.0% TO 104% RPM– MIL AND MAX THRUST RANGE)

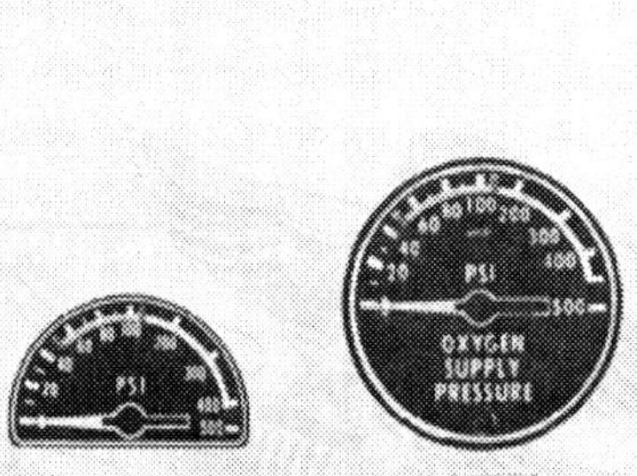

(EARLIER) (LATER)

OXYGEN PRESSURE

- 50 TO 120 PSI NORMAL RANGE

HYDRAULIC PRESSURE

- 1500 PSI MINIMUM
- 2850 TO 3200 PSI NORMAL RANGE
- 3200 PSI MAXIMUM

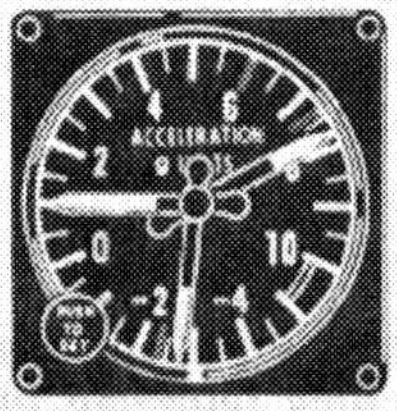

ACCELEROMETER

- −2.3 G'S FULLY FUELED
- +5.6 G'S FULLY FUELED
- +7.33 G'S WITH 900 POUNDS OR LESS OF FUEL REMAINING

AIRSPEED-MACH NO. INDICATOR

220 KNOTS IAS MAXIMUM ALLOWABLE AIRSPEED WITH FLAPS EXTENDED OVER 60%.

*Note*

A RED POINTER ON THE INSTRUMENT IS SET TO INDICATE A MAXIMUM ALLOWABLE AIRSPEED OF 710 KNOTS EAS.

T-38A 1-60 S

SOLID BLACK PLATE (+3 COLORS)

Figure 5-1

# ENGINE OPERATING LIMITATIONS

| CONDITION | EGT °C | RPM % | NOZZLE POSITION % | FUEL FLOW LB/HR | OIL PRESSURE PSI | TIME DURATION (MINUTES) |
|---|---|---|---|---|---|---|
| **GROUND STEADY STATE** | | | | | | |
| START | 925 (MAX) * 845 | --- | --- | 360 (MAX) | INDICATION | --- |
| IDLE | --- | 46.5-49.5 | 77-92 | 400-600 (STD DAY) | 5-20 | --- |
| MILITARY | 630-645 | 99.0-100.5 | 0-20 | 2100-2500 (SEA LEVEL) | 20-55 | 30 |
| (MAX) AFTERBURNER | 630-645 | 99.0-100.5 | 50-85 | --- | 20-55 | 5 |
| **FLIGHT STEADY STATE** | | | | | | |
| START | 925 (MAX) * 845 | --- | --- | 360 (MAX) | INDICATION | --- |
| IDLE | 140 (MIN) | --- | --- | 200 (MIN) (STD DAY) | 5 (MIN) | --- |
| MILITARY | 630-645 | 99.0-104 | 0-20 | --- | 20-55 | 30 |
| (MAX) AFTERBURNER | 630-645 | 99.0-104 | 50-85 | --- | 20-55 | 15 |
| **FLUCTUATION LIMITS** | | | | | | |
| IDLE (GROUND) | --- | 46.5-49.5 | NONE ALLOWED | ±25 | ±2 | --- |
| MILITARY AND AB (GROUND) | ** | 99.0-100.5 | ±3 | ±50 | ±2 | --- |
| MILITARY AND AB (FLIGHT) | ** | ±1% WITHIN STEADY STATE LIMITS | ±3 | ±50 | ±2 | --- |

## OTHER LIMITATIONS

EGT:

* 1. ABORT START IF EGT REACHES 845°C TO PRECLUDE EXCEEDING TEMPERATURE LIMITS.

2. ABORT AIRCRAFT DURING GROUND START IF EGT EXCEEDS 925°C MOMENTARILY.

** 3. TOTAL FLUCTUATIONS IN EGT OF 15°C (±7.5°C) ARE ACCEPTABLE IF THE AVERAGE EGT IS BETWEEN 630°C AND 645°C.

4. AT LOW COMPRESSOR INLET TEMPERATURES, MILITARY AND AFTERBURNER EGT AND RPM MAY BE BELOW NORMAL OPERATING LIMITS. (SEE SECTION VII.)

RPM:

1. MAXIMUM ALLOWABLE TRANSIENT RPM IS 107%.

NOZZLE POSITION:

1. FOLLOWING RAPID THROTTLE MOVEMENTS, NOZZLE POSITION SHOULD STABILIZE WITHIN PERMISSIBLE FLUCTUATION RANGE WITHIN 5 SECONDS ON GROUND AND 10 SECONDS IN FLIGHT.
2. NOZZLE POSITION MAY BE LESS THAN 50% WHEN OPERATING THE AFTERBURNER AT LESS THAN MAX AB.

OIL PRESSURE:

1. DURING COLD WEATHER STARTS, OIL PRESSURE USUALLY EXCEEDS 55 PSI. TO EXPEDITE OIL WARM-UP, ENGINE MAY BE OPERATED AT MILITARY POWER OR BELOW. IF OIL PRESSURE DOES NOT RETURN TO OPERATING LIMITS WITHIN 6 MINUTES AFTER ENGINE START, SHUT DOWN ENGINE.
2. IF A SUDDEN CHANGE OF 10 PSI OR GREATER IN OIL PRESSURE INDICATION OCCURS AT ANY STABILIZED RPM, FOLLOW ENGINE OIL SYSTEM MALFUNCTION PROCEDURES IN SECTION III.

T-38A 1-93R

Figure 5-2.

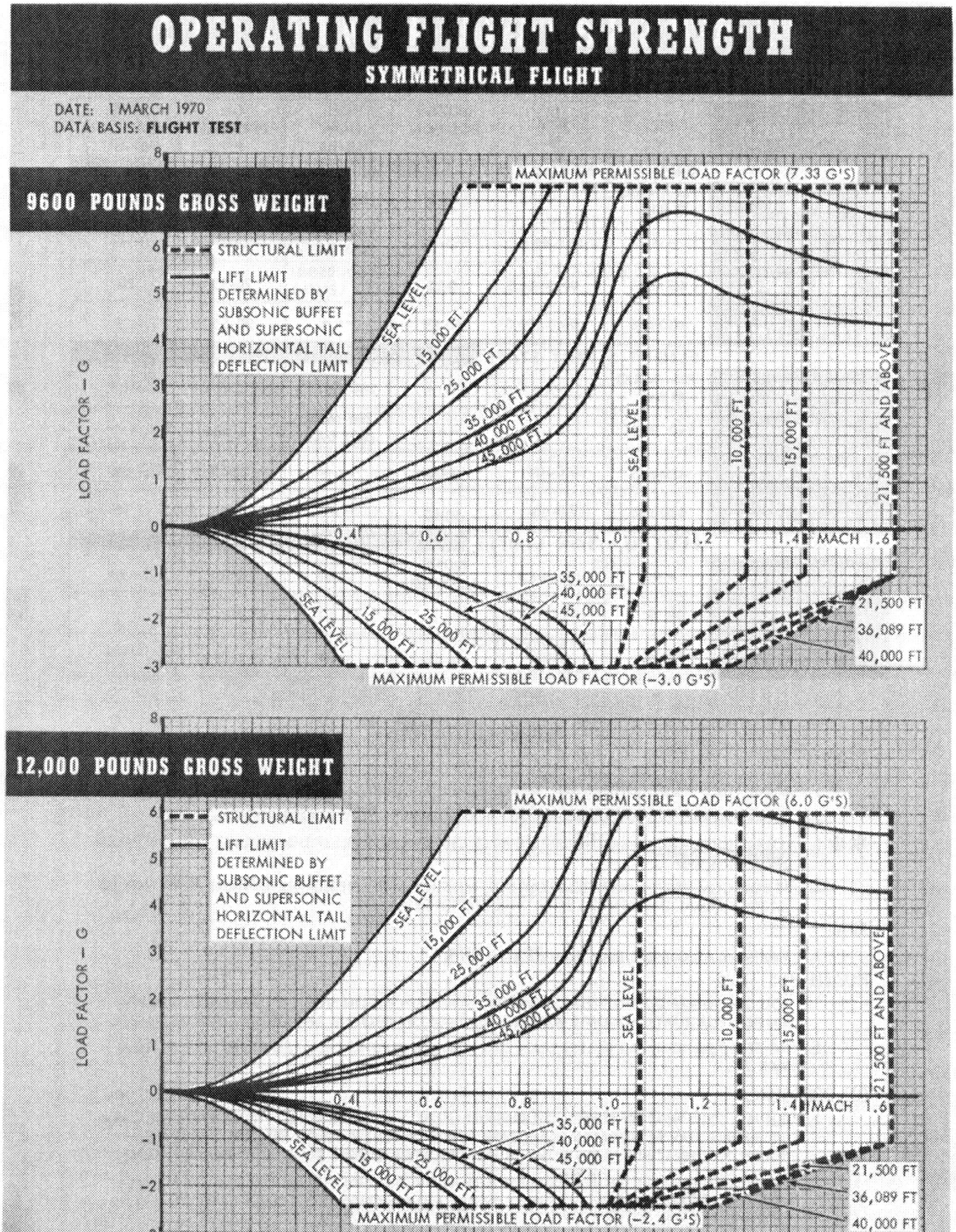
OPERATING FLIGHT STRENGTH
SYMMETRICAL FLIGHT
DATE: 1 MARCH 1970
DATA BASIS: FLIGHT TEST
9600 POUNDS GROSS WEIGHT
MAXIMUM PERMISSIBLE LOAD FACTOR (7.33 G'S)
STRUCTURAL LIMIT
LIFT LIMIT DETERMINED BY SUBSONIC BUFFET AND SUPERSONIC HORIZONTAL TAIL DEFLECTION LIMIT
LOAD FACTOR — G
SEA LEVEL
15,000 FT
25,000 FT
35,000 FT
40,000 FT
45,000 FT
10,000 FT
21,500 FT AND ABOVE
MACH
21,500 FT
36,089 FT
MAXIMUM PERMISSIBLE LOAD FACTOR (−3.0 G'S)
12,000 POUNDS GROSS WEIGHT
MAXIMUM PERMISSIBLE LOAD FACTOR (6.0 G'S)
MAXIMUM PERMISSIBLE LOAD FACTOR (−2.4 G'S)
T-38A 1-61A

Figure 5-3.

# MAXIMUM PERMISSIBLE LOAD FACTOR

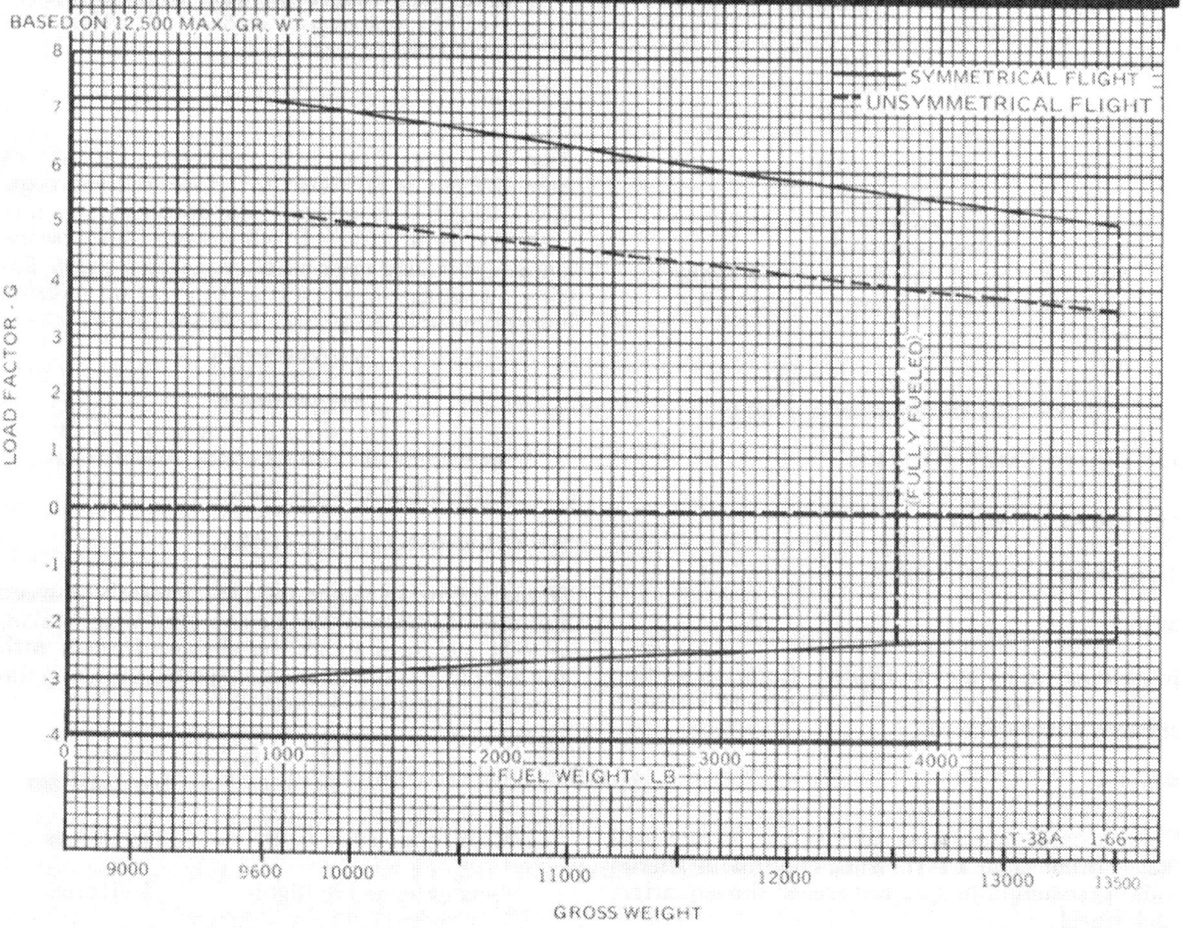


Figure 5-4.

**NOSEWHEEL STEERING.**

Do not exceed 65 KIAS with nosewheel steering engaged.

**CANOPY.**

Do not exceed 50 KIAS while taxiing with a canopy open.

## LOAD FACTOR LIMITATIONS.

Do not exceed the following (see figures 5-3, 5-4).

**SYMMETRICAL FLIGHT.**

| Load Factor (G's) | Weight of Fuel Remaining (Pounds) |
|---|---|
| -2.5 to +6.0 | 3000 |
| -2.7 to +6.6 | 2000 |
| -2.9 to +7.2 | 1000 |

### UNSYMMETRICAL FLIGHT

| Load Factor (G's) | Weight of Fuel Remaining (Pounds) |
|---|---|
| 0 to +4.3 | 3000 |
| 0 to +4.7 | 2000 |
| 0 to +5.1 | 1000 |

**CAUTION**

Do not exceed the following G limits when the Weapon System Support Pod is installed:

0 to +4.0G Symmetrical Flight
0 to +3.0G Unsymmetrical Flight

## PROHIBITED MANEUVERS.

### VERTICAL STALLS.

Vertical stalls are prohibited.

### SPINS.

Intentional spins are prohibited. Refer to section VI for spin recovery procedure in case an inadvertent spin is experienced.

### ROLLS.

Do not enter continuous aileron rolls at any load factor other than 1.0 G. When continuous aileron rolls are accomplished, do not exceed three-quarters stick travel.

## MISCELLANEOUS LIMITATIONS.

### FUEL SYSTEM.

To prevent fuel starvation and subsequent engine flameout, do not exceed the following:

1. Maximum thrust dives with less than 650 pounds of fuel in either fuel supply system.

2. Maximum thrust power in zero G flight or at negative load factors exceeding 10 seconds at 10,000 feet or 30 seconds at 30,000 feet. With less than 650 pounds of fuel in either supply system, time for successful engine operation is further reduced.

**NOTE**

Lower power settings will result in proportionally longer operating times; however, do not exceed engine oil system supply limitations.

### ENGINE OIL SYSTEM.

Due to engine oil supply and pressure requirements zero-G flight is restricted to 10 seconds and negative-G flight (any attitude) to 60 seconds. A momentary drop or loss of oil pressure may be experienced during negative-G or inverted flight. Engine oil venting overboard and/or low oil pressure may occur until positive-G loads are applied.

**CAUTION**

If oil pressure does not recover within approximately 10 seconds, return to normal flight conditions.

### WHEEL BRAKES AND TIRES.

If the following minimum time intervals between full stop landings cannot be complied with, brakes, wheels, and tires should be allowed to cool with the aircraft parked in an uncongested area, and the condition reported in Form 781.

| | Minimum Time Interval Between Full Stop Landings |
|---|---|
| Gear retracted in flight . . . | 45 minutes |
| Gear extended in flight . . . | 15 minutes |

### LANDING RATE OF DESCENT.

Landing should be made with as low a sink rate as practicable. Do not exceed the following sink rates at touchdown:

590 feet per minute normal landing, 395 feet per minute crab landing, with less than 1700 pounds of fuel.
340 feet per minute normal landing, 200 feet per minute crab landing, with full fuel.

### WEIGHT AND CENTER OF GRAVITY LIMITATIONS.

The weight and balance limitations cannot be exceeded by normal operating or loading conditions; however it is possible to attain an aft center of gravity when the right fuel system contains more fuel than the left fuel system. To avoid exceeding the aft center of gravity limit during solo flight, do not allow the right (aft) fuel system quantity to equal more than twice the left (forward) fuel system quantity. If this should occur, longitudinal static stability is reduced and caution should be exercised to prevent overcontrolling during high speed subsonic flight or landings.

## HYDRAULIC PRESSURE.

Hydraulic pressure readings outside the normal range with no demand on the respective system are indicative of a malfunction within the system. High pressures pose the greater danger because of possible fluid overtemperatures. However, operating hydraulically powered equipment (e.g., making rapid flight control movements) will cause pressure fluctuations well outside the static limit. These fluctuations are not considered a malfunction.

## B AUTHORIZED STORES

| STORE |
| --- |
| SUU-11 A/A, B/A<br>with the GAU-2B/A GUN |
| SUU-20/A(M), A/A, B/A<br>with BDU-33(/)<br>and 2.75 FFAR |
| MXU-648 Baggage/Cargo Pod |

Figure 5-5.

## (B) EXTERNAL STORES LIMITATIONS

| STORE | AIRSPEED LIMITATIONS | | | ACCELERATION – G | | | | MAX DIVE FOR DEL | STORES CONFIGURATION WEIGHT LBS. | TOTAL DRAG INDEX | REMARKS |
|---|---|---|---|---|---|---|---|---|---|---|---|
| | CARRIAGE | EMPLOY-MENT | JETTISON | CARRIAGE | | EMPLOY-MENT | JETTISON | | | | |
| | | | | SYM | UNSYM | | | | | | |
| PYLON | 600<br>1.2 | NA | NA | BAL | BAL | NA | NA | NA | 29 | 5 | |
| B-37K-1 | 500<br>1.0 | 300<br>to<br>*500<br>or 0.9 | 300/<br>0.7 * | +6.0<br>-1.0 | +4.0<br>-0.0 | +4.0<br>+0.7 | 1.0<br>LEVEL<br>FLT | 45° | F186<br>E86 | F 45<br>E 40<br>** | *** |
| SUU-20(/) with BDU-33(/) 2.75 FFAR | 500<br>1.0 | BOMBS<br>300<br>to<br>*500<br>or 0.9 | 300/<br>0.7 * | +6.0<br>-1.0 | +4.0<br>-0.0 | +4.0<br>+0.7 | 1.0<br>LEVEL<br>FLT | 45° | F485<br>E270 | F 60<br>E 63<br>** | *** |
| | | RX<br>300<br>to<br>*450<br>or 0.9 | | -1.0 | -0.0 | +4.0<br>+0.5 | | 60° | | | |
| SUU-11(/) | 500<br>1.0 | 200<br>to<br>450<br>0.9 | 300/<br>0.7 * | +6.0<br>-1.0 | +4.0<br>-0.0 | +5.0<br>+0.5 | 1.0<br>LEVEL<br>FLT | 30° | F325<br>E245<br>with 250<br>Rounds,<br>260 lbs. | 34<br>** | *** |
| MXU-648 | 500<br>1.0 | NA | NA | +3.0<br>0.0 | +2.4<br>+0.5 | NA | NA | NA | Remov-able<br>Tail<br>E130<br>F430<br>Fixed<br>Tail<br>E98<br>F398 | 25<br>** | ***<br>MAX CARGO LOAD - 300 LBS |

NA - Not Applicable/Not Authorized
* Whichever is less
** Store drag includes pylon
*** If gross weight exceeds 11,700 lbs., refer to Figure 5-4

**CAUTION**

FFAR and the SUU-11 shall not be fired with the speed brake open.

Figure 5-6.

# FLIGHT CHARACTERISTICS

## SECTION VI

T-38A 1-105

TABLE OF CONTENTS



### WAKE TURBULENCE

Avoid wake turbulence. The aircraft because of the short wingspan, is particularly susceptible to wake turbulence upset. The vortex-produced rolling moment can exceed aileron authority in the takeoff and/or landing configuration. The rapid changes in lift can result in a stall without sufficient altitude to recover.

### STALLS.

The stall is characterized by airframe buffet and a high sink rate rather than by a clean nose-down pitch motion. As angle of attack is increased, there is a corresponding increase in buffet intensity. The buffet is most severe with flaps fully extended. The stall condition is immediately preceded by heavy low-speed buffet and moderate wing rock. The wing rock can be controlled with rudder. The actual stall is normally not accompanied by any abrupt aircraft motion, but is indicated only by the very high sink rate. However, if the stall condition is aggravated by abrupt control inputs, unusual aircraft attitudes may result.

#### STALL RECOVERIES.

Stalls can be terminated by relaxing back stick pressure, rolling wings level, and moving throttles to MAX simultaneously. If in the landing configuration, raise gear and speed brake, allowing flaps to remain extended until stall recovery has been accomplished. While it is normally not necessary to allow the nose to pitch down, relaxation of back pressure is critical in breaking the stall and allowing the aircraft to accelerate, reducing the buffet, eliminating wing rock, and maintaining adequate aileron control. Reducing the bank angle will lower the stall speed and decrease the sink rate (see figures 6-1, 6-2, and the Effect of Bank Angle on Vertical Velocity charts in Part 7 of appendix). Since timely identification of an actual stall is difficult, stall recovery should be initiated at the first indication of increasing buffet or rate of sink. Recovery from a stalled condition can be accomplished with a minimum loss of altitude using the above stall recovery technique.

**WARNING**

If a high sink rate condition is allowed to develop, excessive altitude loss will occur and recovery may not be possible at traffic pattern altitudes.

**NOTE**

See section VII for engine operating instructions during stall.

#### SUBSONIC ACCELERATED STALLS.

Accelerated stalls are similar to 1-G stalls.

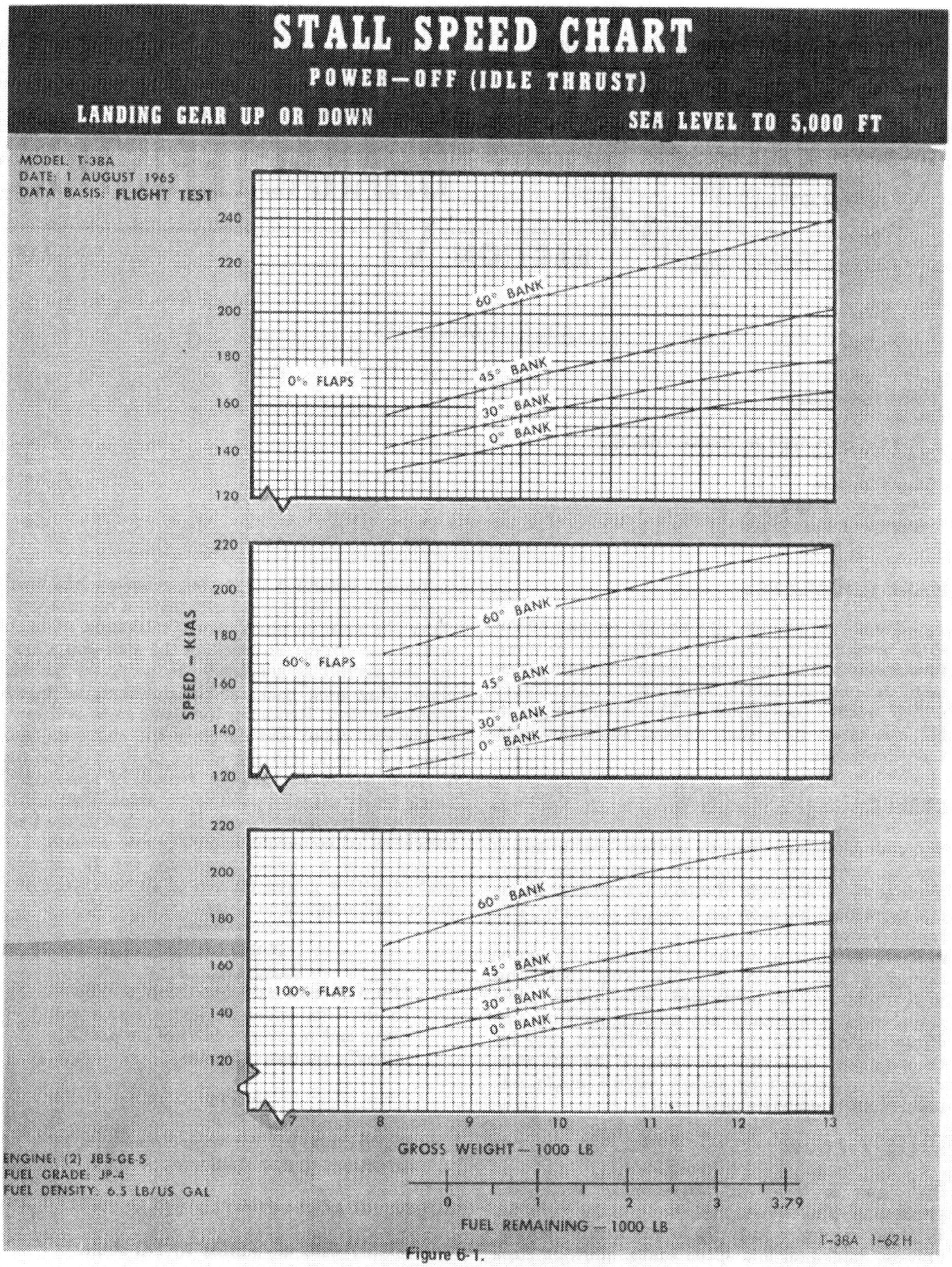
STALL SPEED CHART
POWER—OFF (IDLE THRUST)
LANDING GEAR UP OR DOWN
SEA LEVEL TO 5,000 FT
MODEL: T-38A
DATE: 1 AUGUST 1965
DATA BASIS: FLIGHT TEST
SPEED — KIAS
0% FLAPS
60% FLAPS
100% FLAPS
60° BANK
45° BANK
30° BANK
0° BANK
GROSS WEIGHT — 1000 LB
FUEL REMAINING — 1000 LB
ENGINE: (2) J85-GE-5
FUEL GRADE: JP-4
FUEL DENSITY: 6.5 LB/US GAL
T-38A 1-62H

Figure 6-1.

# EFFECT OF BANK ANGLE ON VERTICAL VELOCITY

## SEA LEVEL STANDARD DAY

60% FLAPS AND GEAR DOWN

DATE: 1 APRIL 1969
DATA BASIS: **FLIGHT TEST**

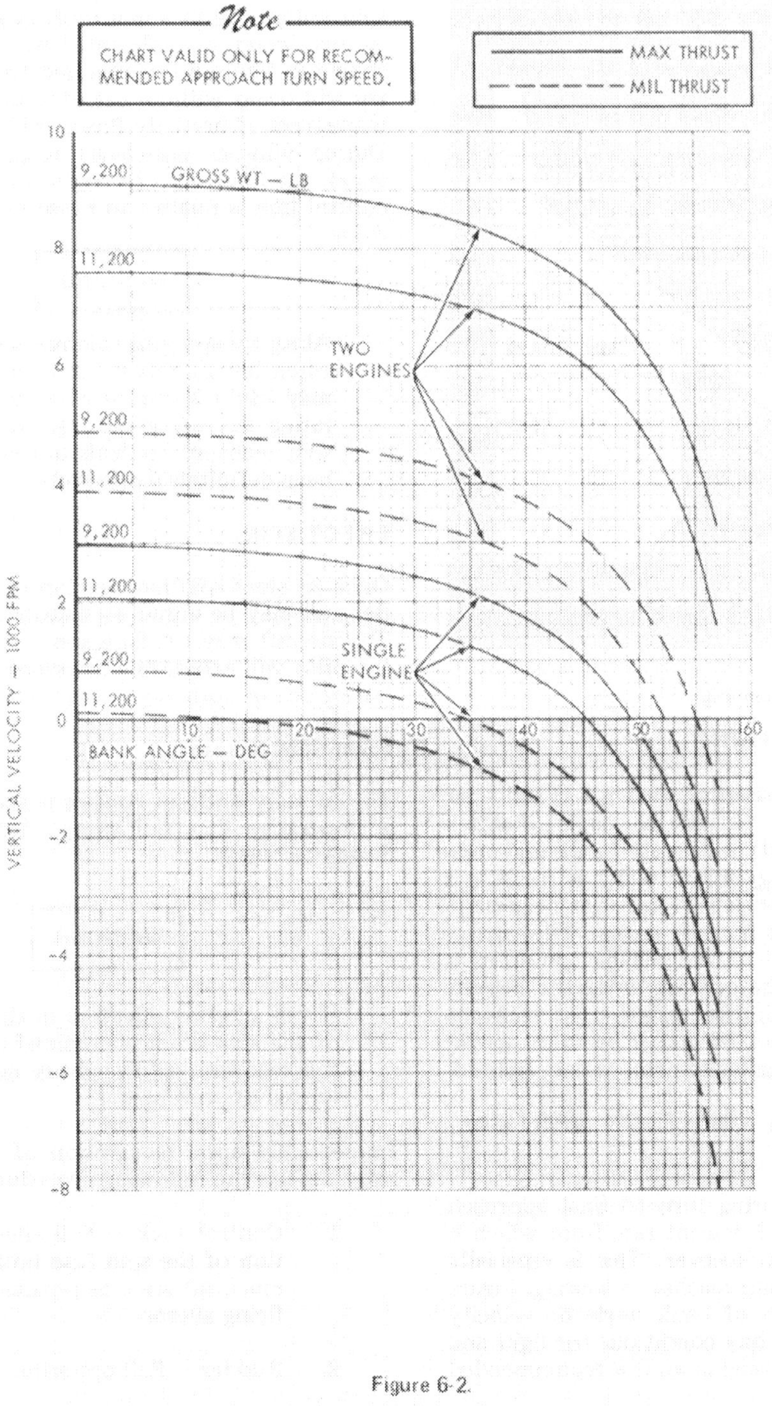


Figure 6-2.

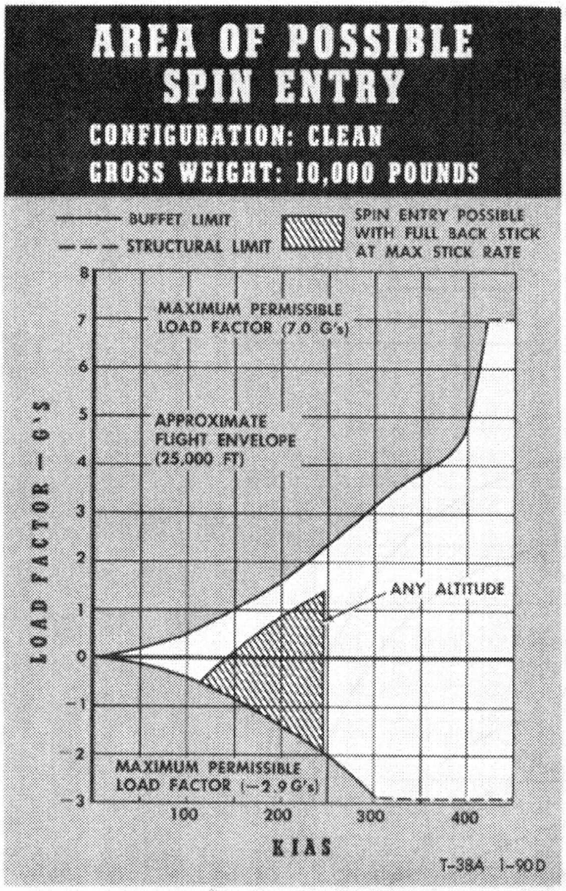


Figure 6-3.

## POST STALL GYRATIONS.

Gyrations can be experienced during 1-G stalls, inverted stalls (negative G, negative angle of attack and stick held forward), accelerated stalls and cross control stalls. These gyrations will not result in a spin (abrupt full aft stick movement at near maximum rate is required for spin entry). The corrective procedure for all unrecognizable gyrations is to smoothly neutralize controls until the aircraft settles into a recognizable maneuver or recovers. Expect a short period of erratic motion and/or negative load factors after controls are neutralized.

## EFFECT OF BANK ANGLE ON VERTICAL VELOCITY.

Steep bank angles during turn to final approach can cause a very rapid descent rate from which it may be impossible to recover. This is especially true for single-engine approaches to landing. Figure 6-2 shows the effects of bank angle on velocity for sea level standard day conditions for light and heavy aircraft gross weights at the recommended final turn speed. Single-engine landing patterns should be planned so that steep bank angles are not required. A complete set of charts showing the effects of bank angle on vertical velocity for various conditions can be found in part 7 of the Appendix.

## SPINS.

The aircraft exhibits a high degree of resistance to spin entry; abrupt application of aft stick at close to maximum possible rates within the envelope shown in figure 6-3 is required to enter a spin. Entry will occur without use of rudder. Normal flight maneuvers, if properly flown, will not cause a spin. During unusual maneuvers (e.g., collision avoidance), the pilot must be aware of his airspeed and control inputs relative to those required for a spin entry.

**WARNING**

Abruptly applying spin recovery controls when the aircraft is not in an actual spin may cause a spin or extremely disorientating aircraft gyrations. Do not apply spin recovery controls unless a spin has been definitely diagnosed.

## ERECT SPIN.

Once an erect spin has developed, the spin will be flat and may be either oscillatory or very smooth. The aircraft may oscillate about all three axes, and the pilot will experience transverse G-loads. Flameout of one or both engines can be expected.

### Erect Spin Recovery.

The primary antispin control is the aileron, and it is imperative that full aileron deflection be held during recovery.

**WARNING**

If full aileron deflection in the direction of the spin is not maintained throughout the recovery, spin recovery may be prolonged or prevented.

Immediately upon recognition of the direction of rotation, use the following procedure.

1. Control stick — Full aileron in the direction of the spin (use both hands) and as much aft stick as possible without sacrificing aileron.

2. Rudder — Full opposite.

3. Do not change gear, flaps and speed brake positions during recovery.

4. Neutralize controls after recovery.

NOTE

Recovery from the spin is normally abrupt and may be followed by some spiraling during the resultant dive.

### INVERTED SPIN.

An inverted spin is very oscillatory about all axes and is easily recoverable.

#### Inverted Spin Recovery.

Immediately upon experiencing an inverted spin, use the following procedure:

1. All flight controls — Neutralize.

WARNING

- Maintain controls in neutral position throughout the spin recovery. Any aileron or rudder deflection can induce a transition to an erect spin.
- Ejection from either an erect or inverted spin is to be accomplished if a spin recovery is not completed by 15000 feet above the terrain, or if transverse G-loads preclude maintaining antispin controls, whichever occurs first.

## FLIGHT CONTROLS.

### STABILITY AUGMENTATION.

The stability augmenter system positions the rudder control surfaces to automatically damp out yaw short period oscillations. On Block 20 aircraft, the stability augmenter system additionally damps out pitch oscillations. The aircraft may be flown safely throughout the flight envelope without the stability augmenter system engaged.

### G-OVERSHOOT.

The horizontal tail control system incorporates a bob-weight to increase stick forces under G-loads. Since the pilot does not feel the effect of the bob-weight until the aircraft responds to the stick movement, G-overshoots may occur if the stick is deflected too abruptly.

CAUTION

Abrupt forward or aft deflection or "pulsing" of the stick in the mach range from 0.80 to 0.95 may result in overshoot of the limit load factor.

### LATERAL CONTROL.

Aileron deflection does not increase proportionally with stick travel. The first 4-1/2 inches of stick travel provide one-half aileron deflection, while the remaining 1-1/2 inches of stick travel provide full aileron deflection.

## MANEUVERING FLIGHT.

NOTE

Maneuvering and handling qualities are degraded at lower airspeeds; therefore, a minimum of 300 KIAS should be maintained except for instrument approaches, maximum range descents, landings, and tactical maneuvering. The objective for establishing a minimum airspeed is to maintain a satisfactory energy state (i.e. "G" available that will provide desired recovery response if an undesirable flight parameter is encountered below 15,000 feet AGL).

### STICK FORCES.

Minimum stick forces per G occur at approximately mach 0.9. Be careful not to overcontrol when maneuvering near this airspeed so that the allowable load factor is not exceeded.

### PILOT INDUCED OSCILLATIONS.

The relationship between pilot response and aircraft pitch response in high subsonic-low altitude flight is such that overcontrolling may lead to severe pilot induced oscillations. This oscillation is characterized by a sudden and violent divergence in pitch attitude resulting in very large positive and negative load factors, which are actually made larger by the pilot attempting to control the oscillation. Because the basic aircraft is stable, the pilot should immediately release the stick so that the aircraft can damp itself or if at very low altitude or close to another aircraft, the pilot should attempt to apply and rigidly hold back-pressure on the stick. In addition to the above, a reduction in airspeed will aid in recovery. It should be noted that if the pilot is not securely strapped into his seat,

# FLIGHT ENVELOPE
## STANDARD DAY

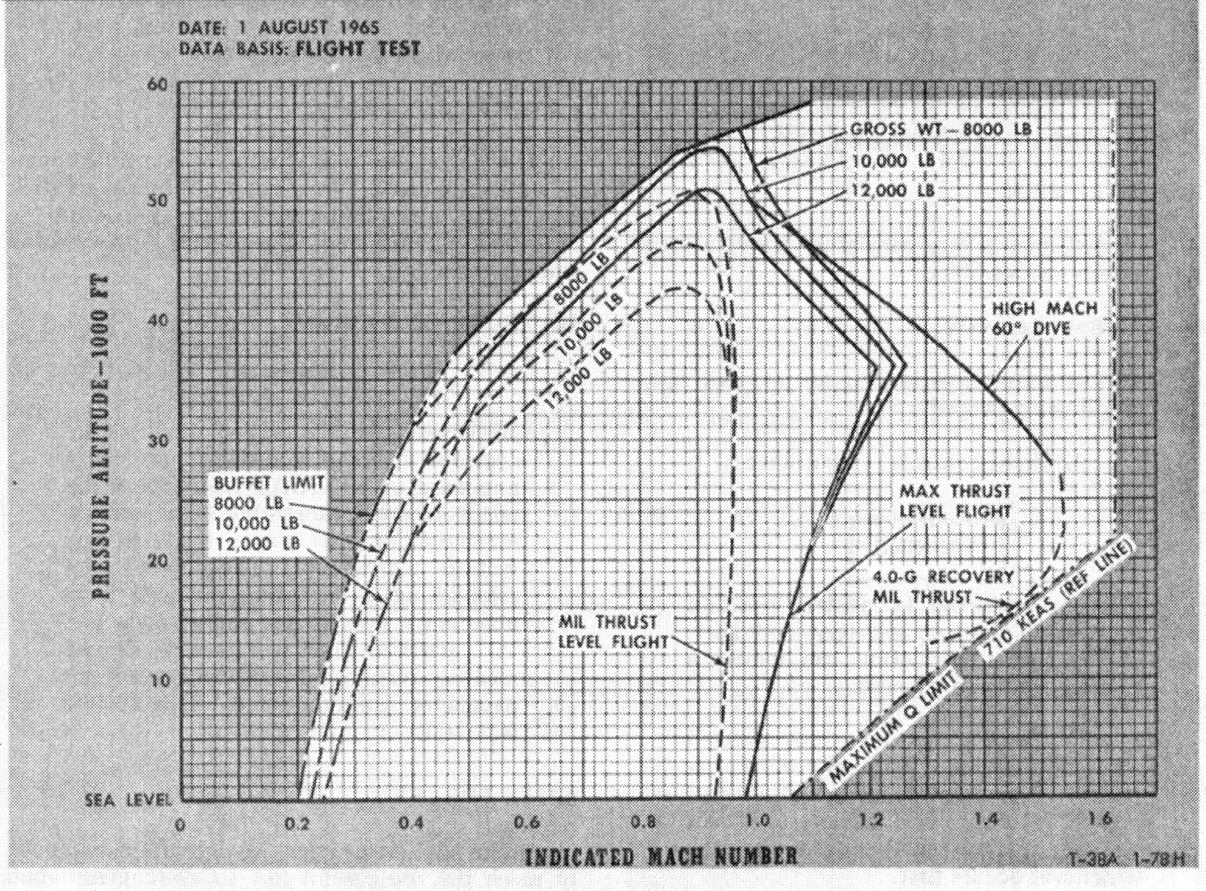


Figure 6-4.

the above recovery procedures may be difficult to accomplish.

### ROLLS.

Roll rates obtainable in this aircraft with full aileron deflection are extremely high and could cause the pilot to become disoriented. Caution should be exercised when using rudder in conjunction with aileron application during rapid roll or turn entry. Rapid input of both rudder and half (or more) aileron, can cause large load factor excursions during the maneuver.

### UNSYMMETRICAL-G.

Unsymmetrical-G forces occur anytime the aircraft has a roll rate. A phenomenon known as roll coupling can also super-impose an additional G increment during rolling maneuvers. In steady state banked coordinated flight (roll rate = 0), G forces are symmetrical.

## HIGH SPEED DIVE RECOVERY.

To recover from a high speed dive, simultaneously retard throttles to IDLE, open the speed brake, level the wings, and pull out with sufficient G-forces for a safe recovery.

# HIGH MACH 60° DIVE

## STANDARD DAY GROSS WEIGHT 10,000 LB

DATE: 15 JANUARY 1965
DATA BASIS: **FLIGHT TEST**

1 BEGIN 1.5 G's MAX THRUST ENTRY.

2 ATTAIN 60° DIVE ANGLE.

3 REDUCE THRUST TO MIL AND BEGIN 4-G DIVE RECOVERY:
AT 26,400 FT FOR ENTRY AT 45,000 FT.
AT 27,800 FT FOR ENTRY AT 50,000 FT.

4 END DIVE RECOVERY AT LEVEL FLIGHT ATTITUDE.

*Note*

**Altitude loss shown includes a 4-second time allowance for load factor to build up to 4 G's.**

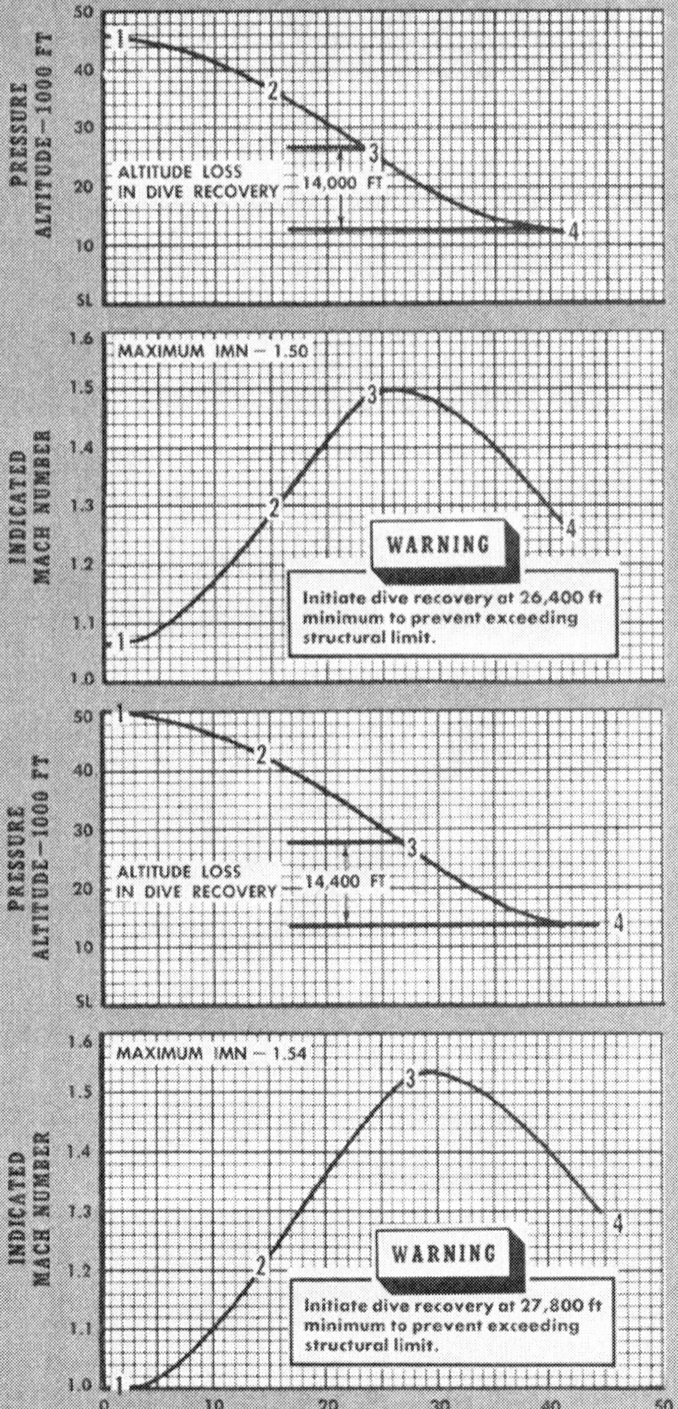


T-38A 1-65E

Figure 6-5.

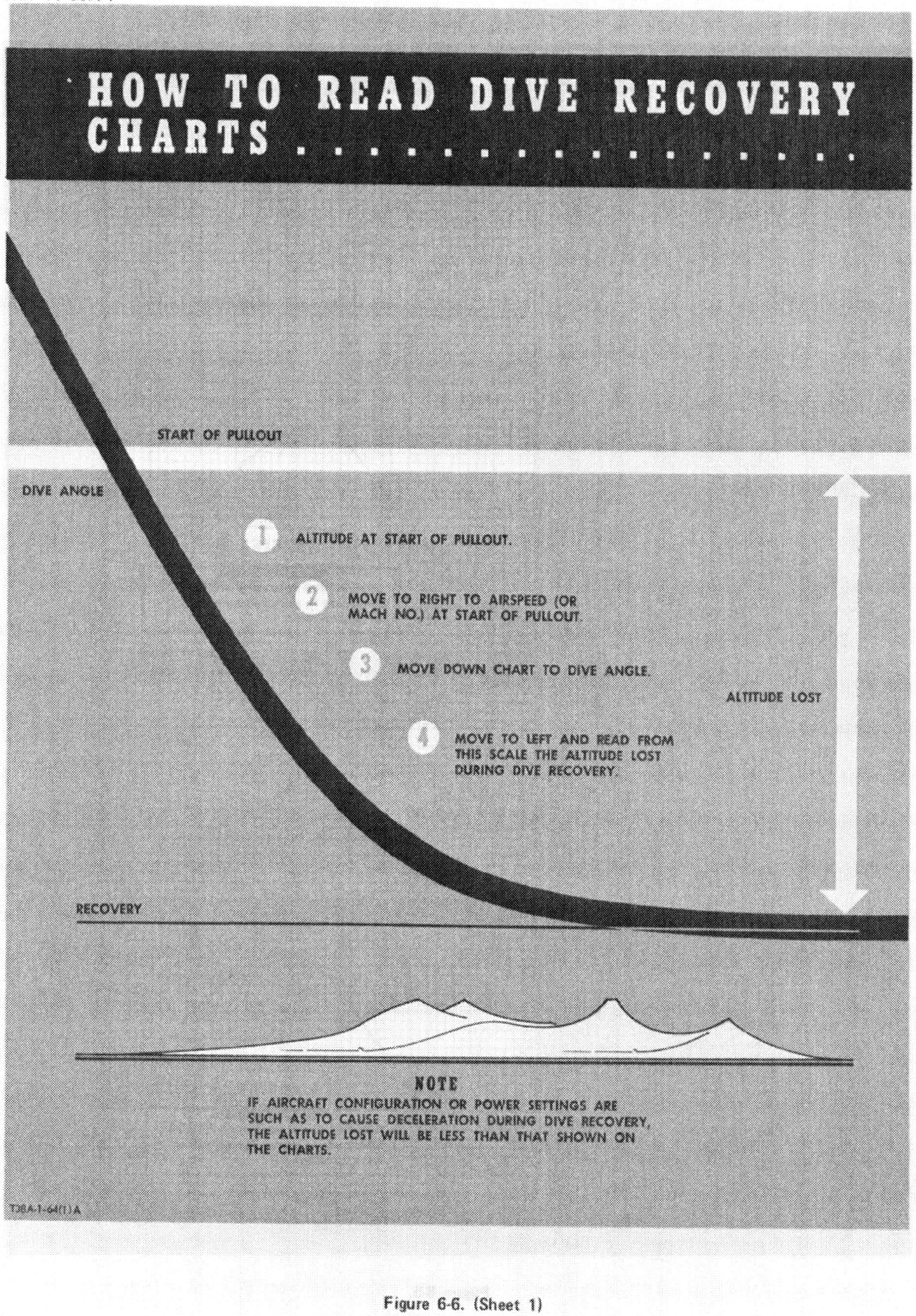


Figure 6-6. (Sheet 1)

# ALTITUDE LOST DURING DIVE RECOVERY

**CONSTANT 4.0 G ACCELERATION**

DATE: 15 JANUARY 1965
DATA BASIS: **FLIGHT TEST**

- SUBSONIC LIFT LIMIT IS DETERMINED BY BUFFET.
- SUPERSONIC LIFT LIMIT IS DETERMINED BY HORIZONTAL TAIL DEFLECTION LIMIT.
- THE DASHED LINES (LIFT LIMITS) ON THE LEFT OF THE CHART SHOW THE AIRSPEED AT WHICH THE AIRCRAFT WILL ENTER AN ACCELERATED STALL AT THE G'S INDICATED.

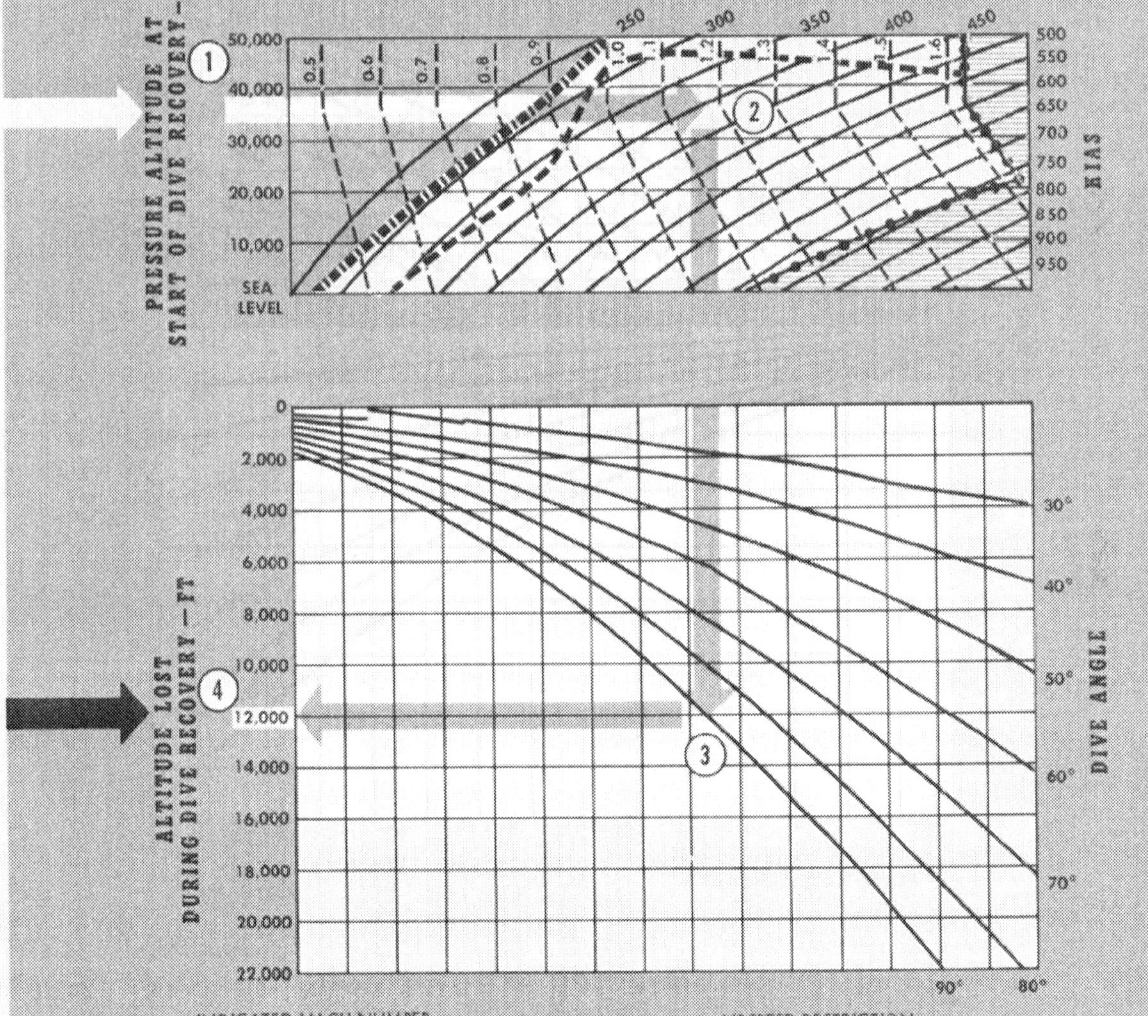


– – – – INDICATED MACH NUMBER
▮▮▮▮▮ LIFT LIMIT FOR 8,000 LB GROSS WEIGHT
–●–●–●– AIRSPEED RESTRICTION
▬ ▬ ▬ LIFT LIMIT FOR 12,000 LB GROSS WEIGHT

T-38A 1-64(3)F

Figure 6-6. (Sheet 2)

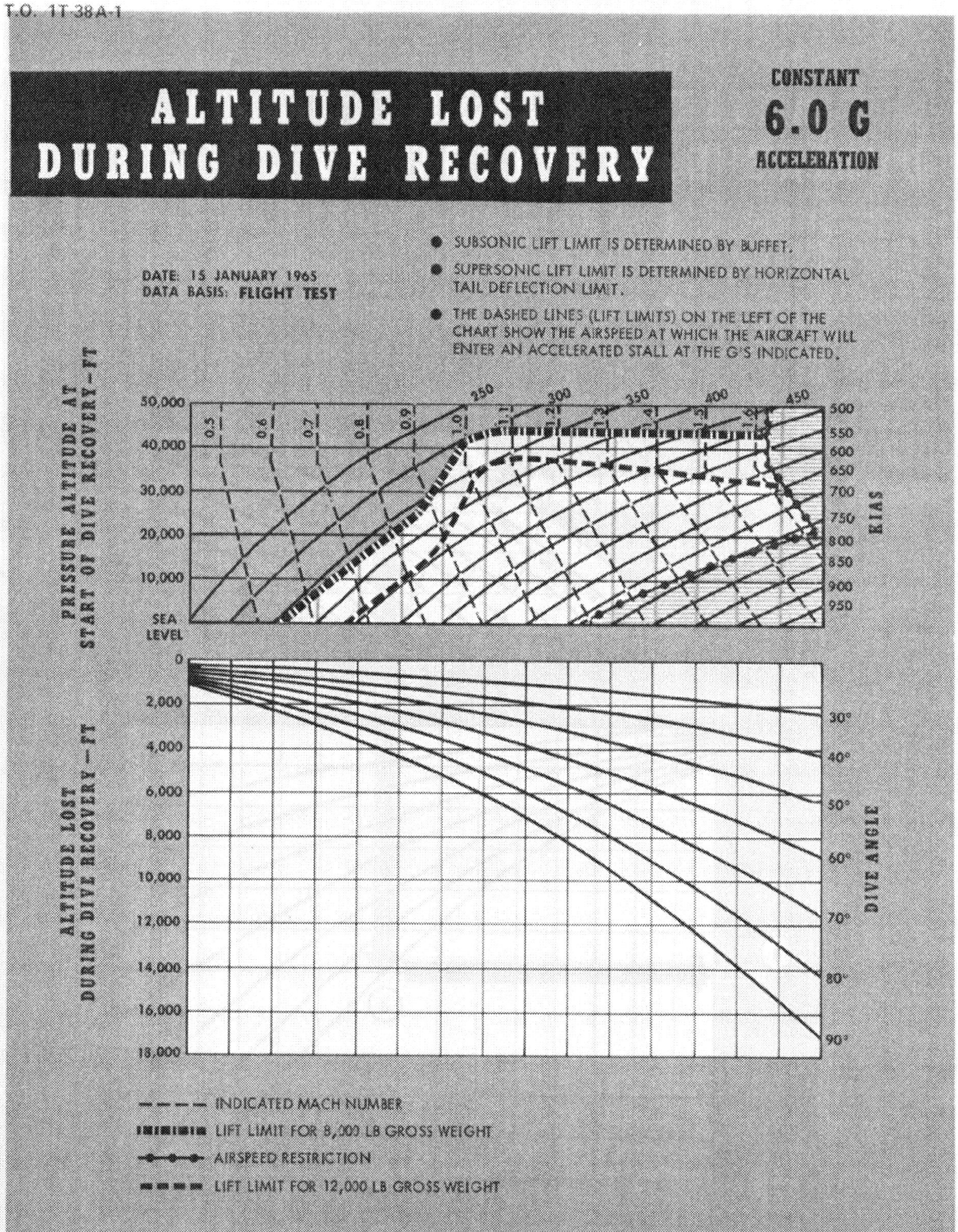


Figure 6-6. (Sheet 3)

# ALTITUDE LOST DURING DIVE RECOVERY

CONSTANT
**7.33 G**
ACCELERATION

DATE: 15 JANUARY 1965
DATA BASIS: **FLIGHT TEST**

- SUBSONIC LIFT LIMIT IS DETERMINED BY BUFFET.
- SUPERSONIC LIFT LIMIT IS DETERMINED BY HORIZONTAL TAIL DEFLECTION LIMIT.
- THE DASHED LINES (LIFT LIMITS) ON THE LEFT OF THE CHART SHOW THE AIRSPEED AT WHICH THE AIRCRAFT WILL ENTER AN ACCELERATED STALL AT THE G'S INDICATED.

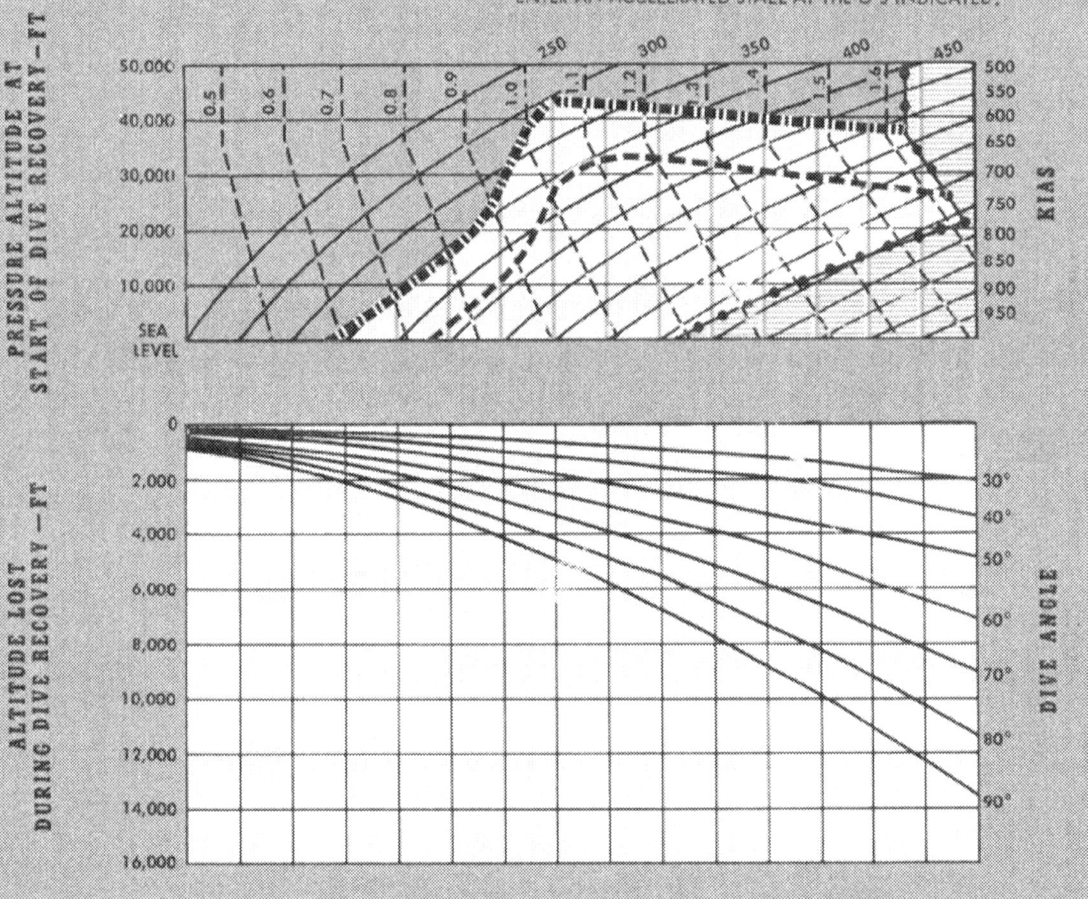


- - - - INDICATED MACH NUMBER
▮▮▮▮ LIFT LIMIT FOR 8,000 LB GROSS WEIGHT
-●-●-●- AIRSPEED RESTRICTION
▬ ▬ ▬ LIFT LIMIT FOR 12,000 LB GROSS WEIGHT

T-38A 1-64(4)E

Figure 6-6. (Sheet 4)

# SYSTEMS OPERATION

## SECTION VII

T-38A 1-106

TABLE OF CONTENTS



### FUEL MANAGEMENT.

The fuel systems function automatically to supply fuel to the engines once the throttles have been moved from the OFF position and the fuel boost pumps turned on. The fuel quantity indicators should be monitored to maintain the two systems within 200 pounds of each other to ensure the aircraft center-of-gravity (CG) is maintained within limits. Maintaining the two systems within the 200 pound parameter is accomplished by using the fuel balancing (crossfeed) procedures in this section. Crossfeeding is not recommended during low fuel conditions or while at low altitudes, instead use differential power settings to obtain proper balance. With the fuel low caution light illuminated, a slightly nose up flight attitude should be maintained to assure maximum usable fuel from both systems. Maintaining this attitude is necessary to preclude uncovering the fuel boost pump inlets, allowing air to enter the fuel supply lines, causing engine flameout.

NOTE

During low fuel state descents do not maintain a nose down attitude for extended periods. Occasionally transit to a positive pitch attitude to refill the boost pump sump.

### FUEL BALANCING (CROSSFEED).

Crossfeeding is recommended when fuel differences exceed 200 pounds. Attempt to enter traffic pattern in a fuel-balanced condition. Differential power settings should be used to balance fuel to avoid use of crossfeed operation during low fuel conditions.

NOTE

If a fuel unbalance is experienced for no apparent reason, perform gage test prior to crossfeeding.

1. Fuel Quantity — Check.
2. Crossfeed Switch — ON.
3. Boost Pump Switch (on side of lower fuel quantity) — OFF.

WARNING

With the crossfeed switch ON and either both boost pumps ON — or both boost pumps OFF — a rapid fuel imbalance can occur. Fuel will feed predominantly from the left system.

**WARNING**

If crossfeed operation is continued until the active system runs dry, dual engine flameout will occur.

4. Boost Pump Switches — Both ON When Quantities Are Equal.

5. Crossfeed Switch — OFF.

## COMPRESSOR STALL.

A compressor stall is an aerodynamic interruption of airflow thru the compressor section. Factors that can increase the stall sensitivity and decrease the compressor stall margin are: Foreign object damage, high aircraft angles of attack at low airspeeds, low compressor inlet temperatures (CIT), maneuvering flights, unusual flight attitudes, atmospheric variations, jet wash, temperature and pressure distortion, ice formation on inlet ducts and engine inlet guide vanes or a combination of the above. Compressor stalls can be caused by various other factors such as: Engine component malfunction, incorrect engine rigging, incorrect RPM and fuel flow trim, throttle burst to military or maximum power at high altitude and low airspeed, and by hot gas ingestion. The various types of stalls that may occur are discussed below.

### TAKEOFF OR LOW ALTITUDE AND HIGH AIR SPEED COMPRESSOR STALL.

Compressor stall may occur on takeoff with AB initiation or at low altitude and very high airspeed when in military or AB operation. The stall is recognized by a "pop" or "bang" followed by an audible "buzzing" sound and vibration, accompanied by a rapid RPM drop and high EGT. The stall should be cleared as soon as possible to prevent engine damage by overtemperature. This type of compressor stall can normally be recovered by rapidly retarding throttle to IDLE. Flameout usually does not occur.

### HIGH ALT-LOW AIRSPEED COMPRESSOR STALL.

Compressor stall/flameout at high altitude, low airspeed and very low ambient temperatures can occur during a throttle advance to military or afterburner power. The stall is recognized by an audible "chug" or "pop" accompanied by rapid unwinding of RPM and decreasing EGT. The stall may be cleared by rapidly retarding the throttle to idle. Immediately pressing the start button will provide a restart capability if the stall has progressed to a flameout.

## VARIABLE INLET GUIDE VANES.

Variable inlet guide vanes and air bleed valves have been incorporated in the J-85 engine to reduce the possibility of a compressor stall throughout the normal operating range of the engine. The vanes function automatically to direct the flow of air to the compressor blades at the proper angle. The bleed valves open and close automatically to provide proper control of compressor pressure. Prior to start, the bleed valves may be partly open or completely closed. This is due to residual servo hydro-pressure unbalance in the bleed valve actuator system and main fuel control after engine shutdown. The valves will open during normal engine start.

## ENGINE ENVELOPE (FIGURE 7-1).

Figure 7-1 depicts the operating envelope for the installed J85-5 engine. The chart is presented in terms of pressure altitude versus indicated mach number for standard day conditions with considerations for temperature deviation of ±10°C from standard. The chart illustrates the operating airspace at higher altitudes where colder temperatures and less dense air may cause the engine to stall or flameout. This operating restriction is further expanded as temperatures colder than standard are encountered. Conversely, the opposite is true as temperatures warmer than standard are encountered. These regions of flight require operator attention and have been portrayed on the chart as the black striped and shaded areas. Flight is not prohibited in these areas but merely requires the operator to acknowledge the engine limitations as indicated on the chart.

## THROTTLE MOVEMENT.

The engine stall margin and operating parameters decrease with increasing altitude where the air is less dense and colder. As a result, throttle movement must be more carefully controlled in the black striped and shaded areas shown in figure 7-1. Abrupt throttle movements, which are acceptable to the engine at low altitude, are not recommended in these areas and can result in a stall or flameout.

## AFTERBURNER INITIATION (HIGH ALTITUDE).

Afterburner initiation attempts in the black striped area as indicated in figure 7-1 are not recommended. Afterburner light-off is not guaranteed

*Note*

YAW AND ANGLE OF ATTACK HAVE AN UNPREDICTABLE EFFECT ON THE BOUNDARIES OF THE ENGINE ENVELOPE.

**ENGINE ENVELOPE** **J85-GE-5 SERIES ENGINES 1G LEVEL FLIGHT**

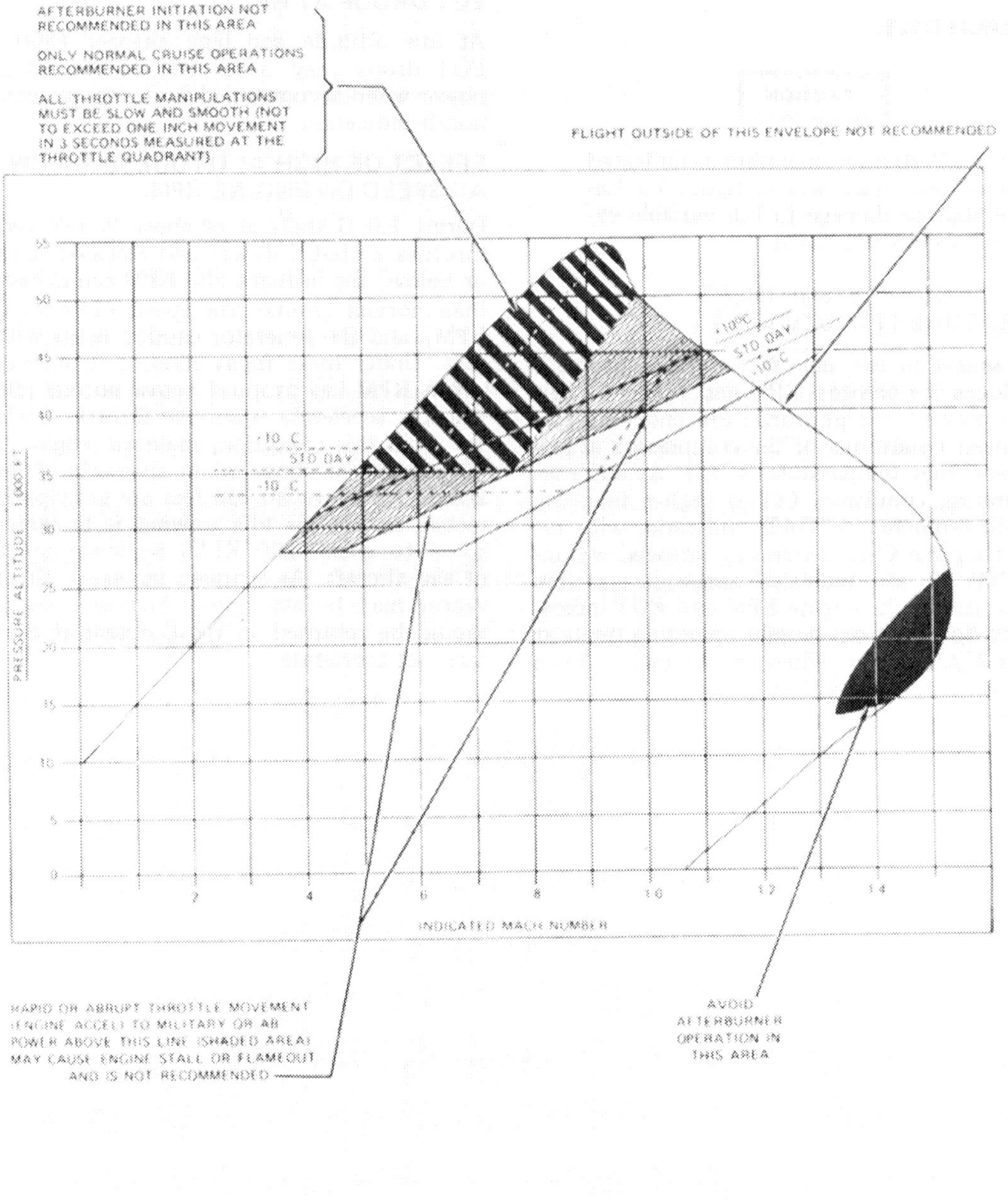


Figure 7-1.

and even if successful, may drive the engine RPM down (rollback) and possibly cause engine flame-out.

NOTE

To increase the probability of afterburner lightoff if required (in the black striped area), increase airspeed as much as practical before initiating afterburner.

## HIGH MACH DIVE.

CAUTION

Avoid afterburner operation as indicated in the solid black area of figure 7-1. Engine stall or damage to the variable exhaust nozzles may occur.

## EFFECT OF COMPRESSOR INLET TEMPERATURE ($T^2$ CUTBACK.)

The $T^2$ sensor in the main fuel control automatically reduces the physical RPM and EGT ($T^2$ cutback) to prevent overpressurization and high corrected speed conditions of the compressor at low compressor inlet temperatures (CIT). At any normal operating condition, CIT is higher than the outside air temperature (OAT) and varies with airspeed for a given OAT. Increasing airspeed will increase CIT. At low indicated airspeeds and low OAT conditions, the engine RPM and EGT indications may be below the normal operating limits at MIL and MAX power. When the aircraft is flown in the striped black area of the engine envelope, figure 7-1, $T^2$ cutback may be observed. In maneuvering flight, the CIT of each engine will vary depending on flight attitude. As a result, the engine sensing the lower CIT will have a decreased stall margin and increased probability of compressor stall if a throttle transient is made. If $T^2$ cutback is observed, the airspeed should be increased by exchanging altitude for airspeed to increase CIT prior to making a throttle movement.

## EGT DROOP AT HIGH-Q/MIL POWER.

At low altitude and high airspeed (500 KIAS), EGT droop may occur with engine at military power when accompanied by three percent or less nozzle indication.

## EFFECT OF HIGH ALTITUDE AND LOW AIRSPEED ON ENGINE RPM.

During 1.0 G stalls at or above 20,000 feet, with throttles at IDLE detent and airspeed 200 KIAS or below, the in-flight idle RPM can decay to less than normal ground idle speed (46.5% to 49.5% RPM), and the generator caution lights will illuminate. Under these flight conditions, an engine on which RPM has dropped below normal idle speed will not accelerate when the throttle is advanced. To avoid this condition, maintain engine RPM at 80 percent or above when airspeeds of less than 200 KIAS above 20,000 feet are anticipated. Corrective action for idle unwind is to increase airspeed to above 200 KIAS by lowering the nose of the aircraft. As airspeed increases, throttle advances may be attempted; however, the throttle should be returned to IDLE detent if the engine does not accelerate.

SECTION VIII

# CREW DUTIES

(Not Applicable)

# ALL-WEATHER OPERATION

## SECTION IX

T-38A 1-107

TABLE OF CONTENTS



## INSTRUMENT FLIGHT PROCEDURES

### INSTRUMENT TAKEOFF.

For an instrument takeoff, perform all normal pre-takeoff checks, and turn on pitot heat and engine anti-ice system, if necessary. Allow for increased takeoff roll if engine anti-ice is used. Check the horizontal situation indicator for proper heading, and align the index marker on the pitch trim knob with the reference index on the ADI case. On a level surface with proper strut inflation, this should give approximately a 3-degree nose low indication. This setting will give an approximate level flight indication for intermediate altitude level-offs during departures and at normal cruise conditions. Manual bank steering may be used to aid in maintaining directional control, but steering bar indications should be cross-checked with the compass card. Whenever visibility permits, runway features and lights should be used as an aid to maintain proper headings. Adjust back stick pressure to attain the takeoff attitude and allow the aircraft to fly off the runway. When vertical velocity indicator and altimeter indicate a definite climb, retract the landing gear. Raise the wing flaps immediately after the landing gear lever has been placed at LG UP.

### INSTRUMENT CLIMB.

Approaching 300 KIAS in a 5-degree climb indication, retard throttles to MIL thrust. Maintain a 2 to 5-degree climb indication and at least a 1000-fpm climb until reaching recommended climb schedule. A slow airspeed and/or low rate of climb may be required to comply with departure procedures. For this type climb, reduce power below MIL as required. Power settings between 90% and 95% RPM will provide comfortable climb rates at 300 KIAS for intermediate altitude level-offs. MAX thrust instrument climbs require extremely high pitch angles and are not normally used for instrument departures. If conditions require a MAX thrust climb, maintain a 2- to 5-degree climb indication until approaching recommended climb mach, then rotate to approximately a 20- to 25-degree initial climb indication.

### HOLDING PATTERNS.

Hold at 250 KIAS at all altitudes. To descend in holding patterns, reduce power and maintain holding airspeed in descent. The speed brake may be used for holding pattern descents, but higher descent rates must be anticipated.

### PENETRATION DESCENTS.

Prior to penetration descent, the canopy defog system should be operated at the highest flow possible (consistent with crewmembers' comfort) during high altitude flight to prevent the formation of frost or fog during descent. To enter a penetration descent, reduce power and lower the nose approximately 10 degrees on the attitude sphere. Open speed brake (if required) at 300 KIAS and maintain by adjusting pitch as required. Initiate the level-off from a penetration descent 1000 feet or more above the desired altitude by decreasing the the pitch attitude by approximately one half. Use normal lead point for level-off at the desired altitude. The speed brake may be left open or closed as required to obtain the desired airspeed at the final approach fix.

NOTE

For engine anti-ice operation, 80% RPM or above is recommended.

## INSTRUMENT APPROACHES.

NOTE

The aircraft is Category E for instrument approach purposes. However, the Category D minimums may be used when the final approach is flown 165 KIAS or below and the missed approach is flown 260 KTAS or less.

Figure 9-1 shows a typical TACAN penetration and approach. Normally, 240-280 KIAS will be maintained during approach maneuvering prior to extending the gear. Recommended final approach airspeed will depend upon the type of approach being made. AOA indexer will show a fast indication after final approach fix. For a straight-in approach, maintain 155 KIAS plus fuel minimum (AOA indexer on speed). When on final, 100% flaps may be used for the landing, if desired.

NOTE

Increase final approach and touchdown speeds by half the gust factor.

## CIRCLING APPROACHES.

A circling approach is a visual maneuver flown at a lower altitude than a normal VFR overhead traffic pattern. The pilots shallower look angle to the runway causes a tendency to fly a downwind and/or a base leg that is too close to the runway, thus increasing the possibility of an overshoot or steeper than normal final approach. Ensure sufficient downwind and/or base leg displacement prior to initiating the turn to final approach. As the circling maneuver may initially be a level turn, aircraft configuration will require higher power settings than those used in an overhead traffic pattern. Bank angles in excess of 45° may make a level turn impossible under some conditions of heavy gross weights, high temperatures and pressure altitudes. Maintain 175 KIAS plus fuel minimum until aligned with the landing runway. AOA indications will vary depending on airspeed, bank angle and back pressure applied during the circling maneuver. Refer to AFM 51-37 for illustrations of circling approach maneuvers.

## INSTRUMENT LANDING SYSTEM (ILS).

Refer to figures 9-5 and 9-6 for aircraft configuration. Refer to section IV for flight director procedures.

## MISSED APPROACH PROCEDURE.

To accomplish a missed approach, advance throttles to MIL, close speed brake (if open) as power is applied, and rotate the aircraft to normal instrument takeoff attitude. Retract landing gear and flaps as in an instrument takeoff and accelerate to 240-280 KIAS. Reduce power to 90% to 95% RPM, and climb at 240-280 KIAS to missed aproach altitude.

## SINGLE-ENGINE APPROACHES.

Refer to figures 9-2, 9-4, and 9-6 for recommended airspeed and configuration for single engine TACAN, radar, or ILS approach. Delay lowering landing gear until just prior to glide slope if heavy fuel loads, engine anti-ice operation, turbulence, or other conditions cause single-engine MIL thrust to be inadequate for gear down level flight at recommended airspeeds. MAX thrust should be used on single-engine approaches, if necessary.

### Single-Engine Missed Approach.

Refer to figures 9-2, 9-4, and 9-6 for single-engine instrument approach power settings and configurations. If a single-engine missed approach is necessary, use the procedure for single-engine go-around.

# TACAN HOLDING, PENETRATION, AND APPROACH (TYPICAL)

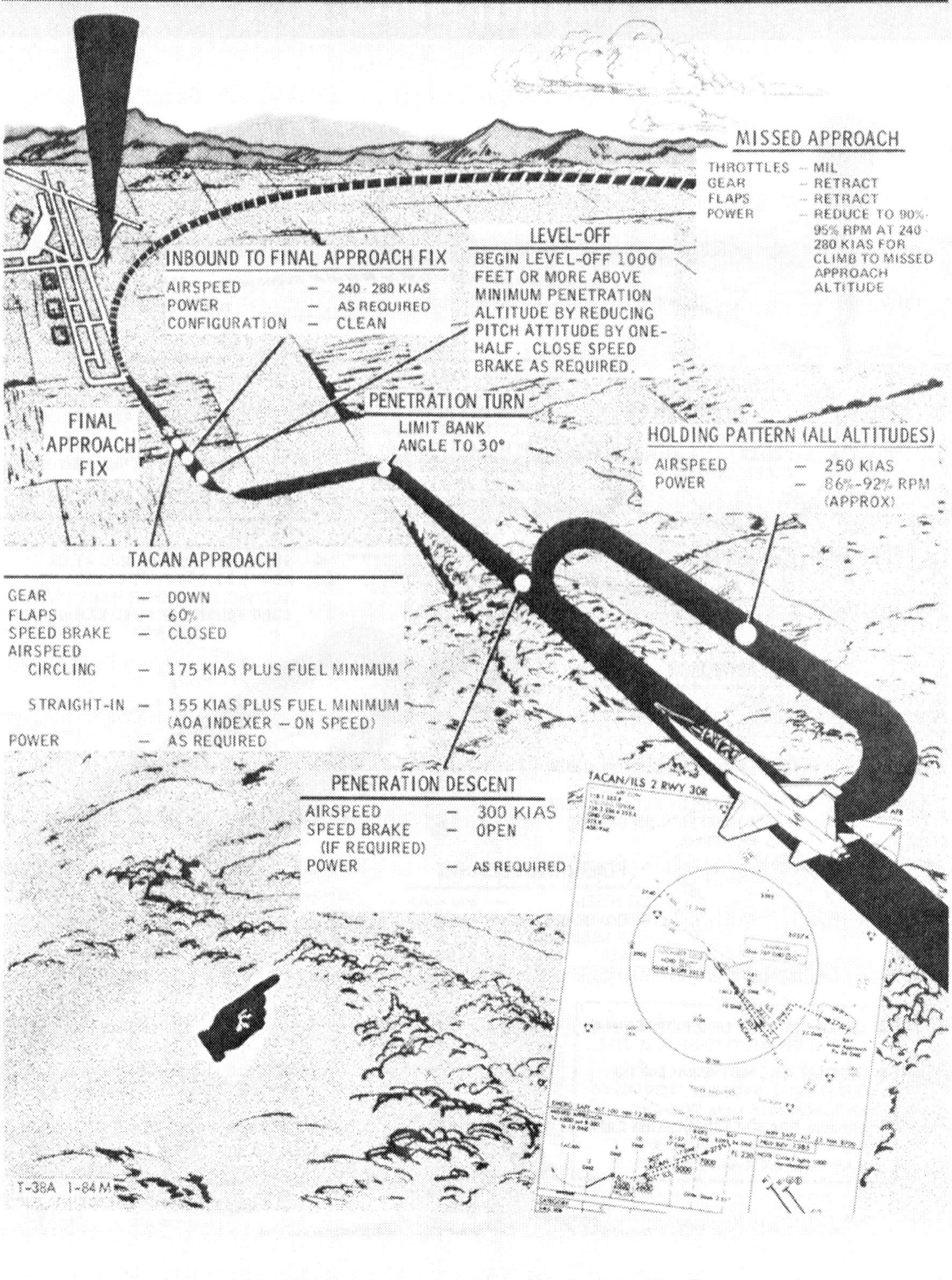


Figure 9-1.

# TACAN HOLDING, PENETRATION, AND APPROACH (TYPICAL)

## SINGLE-ENGINE

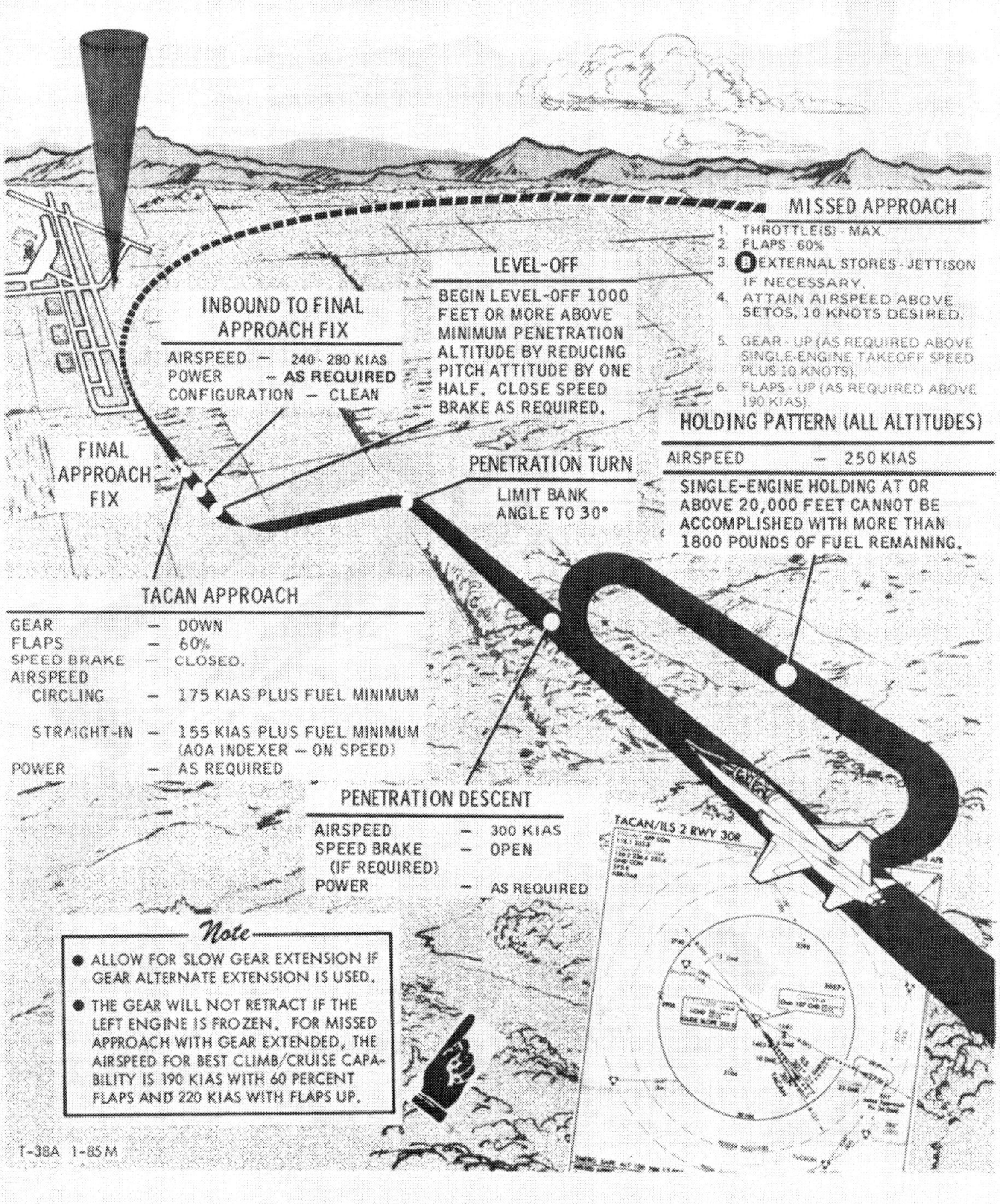


Figure 9-2.

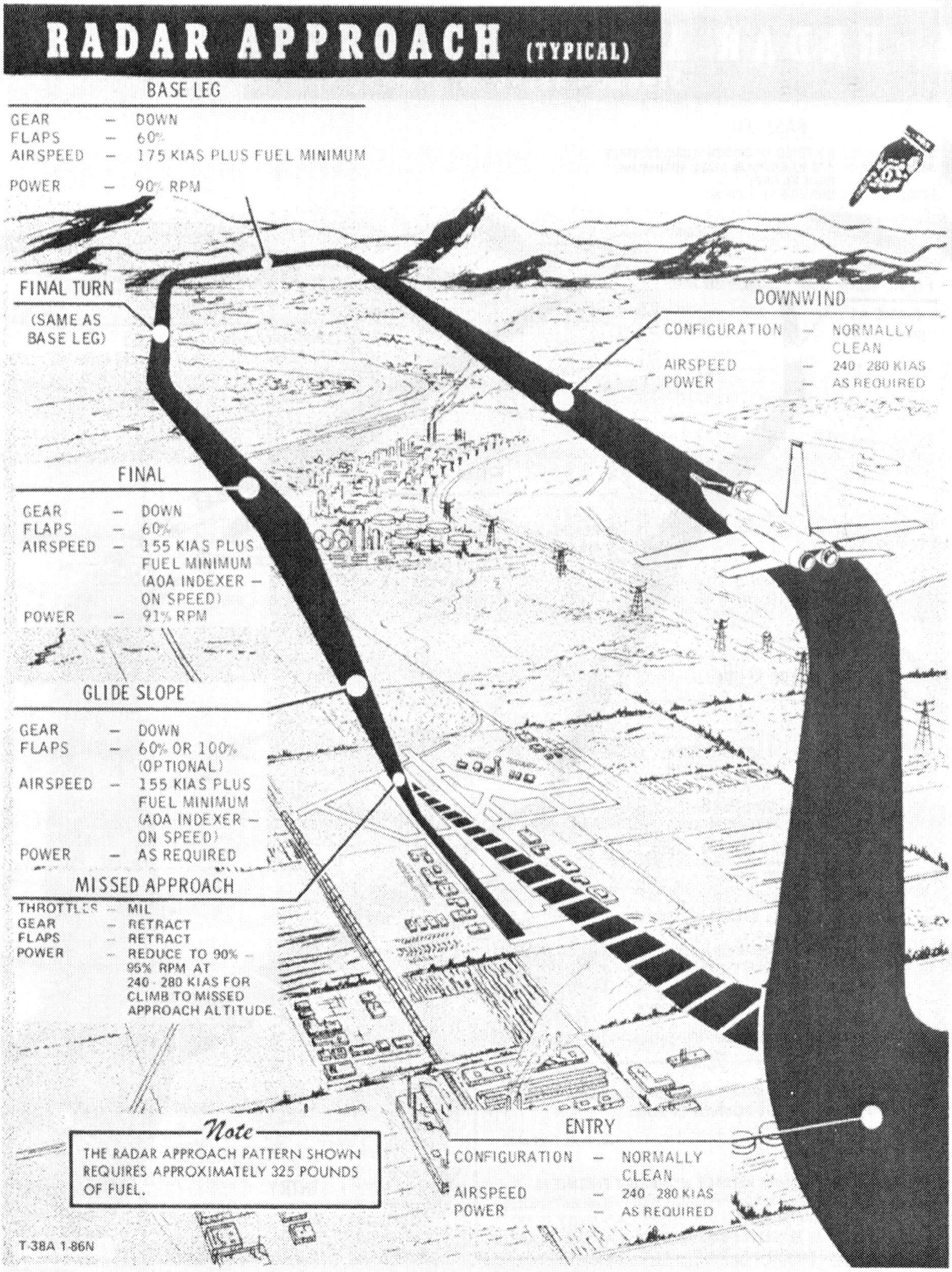
RADAR APPROACH (TYPICAL)
BASE LEG
GEAR — DOWN
FLAPS — 60%
AIRSPEED — 175 KIAS PLUS FUEL MINIMUM
POWER — 90% RPM
FINAL TURN
(SAME AS BASE LEG)
DOWNWIND
CONFIGURATION — NORMALLY CLEAN
AIRSPEED — 240 - 280 KIAS
POWER — AS REQUIRED
FINAL
GEAR — DOWN
FLAPS — 60%
AIRSPEED — 155 KIAS PLUS FUEL MINIMUM (AOA INDEXER — ON SPEED)
POWER — 91% RPM
GLIDE SLOPE
GEAR — DOWN
FLAPS — 60% OR 100% (OPTIONAL)
AIRSPEED — 155 KIAS PLUS FUEL MINIMUM (AOA INDEXER — ON SPEED)
POWER — AS REQUIRED
MISSED APPROACH
THROTTLES — MIL
GEAR — RETRACT
FLAPS — RETRACT
POWER — REDUCE TO 90% - 95% RPM AT 240 - 280 KIAS FOR CLIMB TO MISSED APPROACH ALTITUDE
ENTRY
CONFIGURATION — NORMALLY CLEAN
AIRSPEED — 240 - 280 KIAS
POWER — AS REQUIRED
Note
THE RADAR APPROACH PATTERN SHOWN REQUIRES APPROXIMATELY 325 POUNDS OF FUEL.
T-38A 1-86N

Figure 9-3.

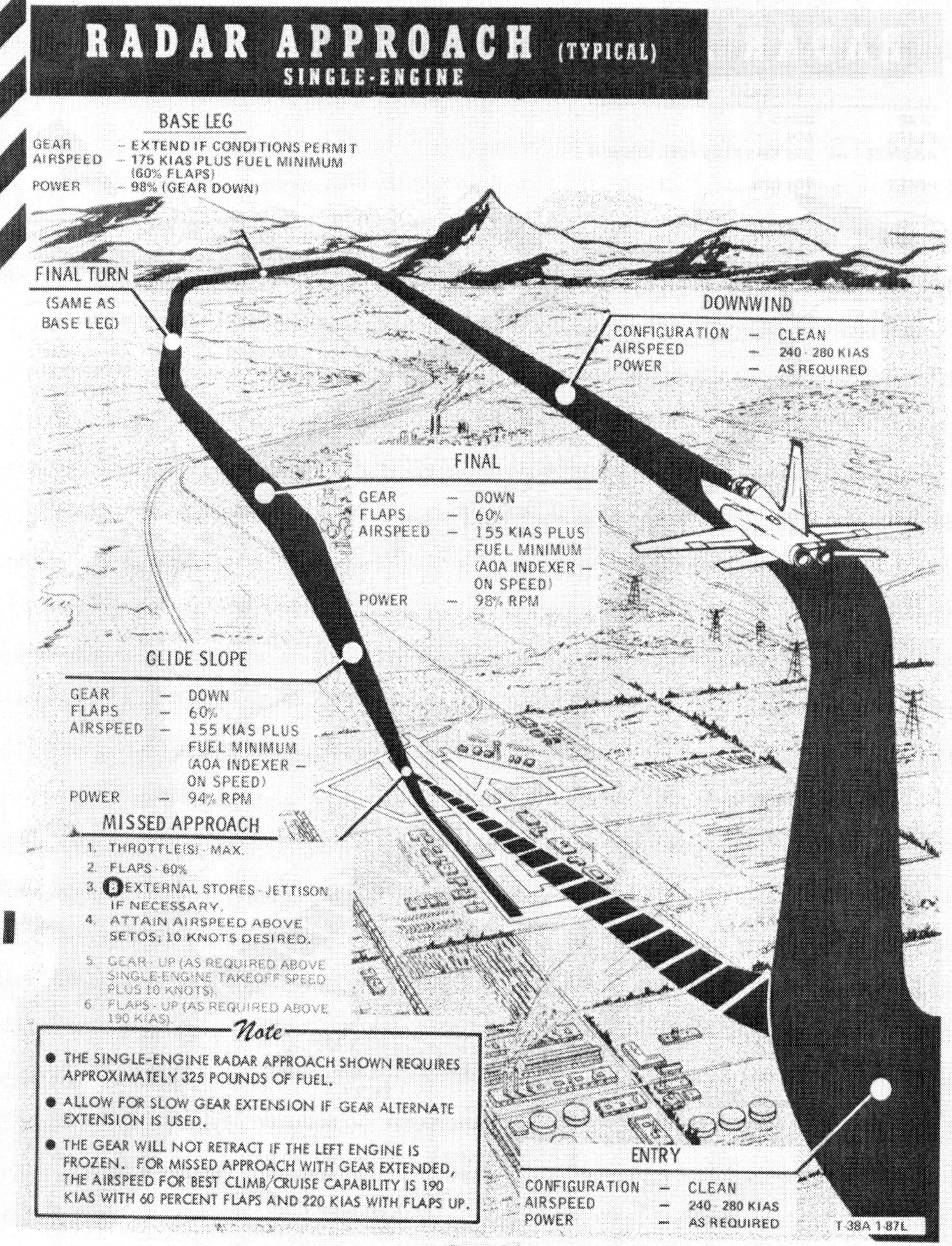
RADAR APPROACH (TYPICAL)
SINGLE-ENGINE
BASE LEG
GEAR – EXTEND IF CONDITIONS PERMIT
AIRSPEED – 175 KIAS PLUS FUEL MINIMUM (60% FLAPS)
POWER – 98% (GEAR DOWN)
FINAL TURN
(SAME AS BASE LEG)
DOWNWIND
CONFIGURATION – CLEAN
AIRSPEED – 240 - 280 KIAS
POWER – AS REQUIRED
FINAL
GEAR – DOWN
FLAPS – 60%
AIRSPEED – 155 KIAS PLUS FUEL MINIMUM (AOA INDEXER – ON SPEED)
POWER – 98% RPM
GLIDE SLOPE
GEAR – DOWN
FLAPS – 60%
AIRSPEED – 155 KIAS PLUS FUEL MINIMUM (AOA INDEXER – ON SPEED)
POWER – 94% RPM
MISSED APPROACH
1. THROTTLE(S) - MAX.
2. FLAPS - 60%
3. B EXTERNAL STORES - JETTISON IF NECESSARY.
4. ATTAIN AIRSPEED ABOVE SETOS; 10 KNOTS DESIRED.
5. GEAR - UP (AS REQUIRED ABOVE SINGLE-ENGINE TAKEOFF SPEED PLUS 10 KNOTS).
6. FLAPS - UP (AS REQUIRED ABOVE 190 KIAS).
Note
• THE SINGLE-ENGINE RADAR APPROACH SHOWN REQUIRES APPROXIMATELY 325 POUNDS OF FUEL.
• ALLOW FOR SLOW GEAR EXTENSION IF GEAR ALTERNATE EXTENSION IS USED.
• THE GEAR WILL NOT RETRACT IF THE LEFT ENGINE IS FROZEN. FOR MISSED APPROACH WITH GEAR EXTENDED, THE AIRSPEED FOR BEST CLIMB/CRUISE CAPABILITY IS 190 KIAS WITH 60 PERCENT FLAPS AND 220 KIAS WITH FLAPS UP.
ENTRY
CONFIGURATION – CLEAN
AIRSPEED – 240 - 280 KIAS
POWER – AS REQUIRED
T-38A 1-87L

Figure 9-4.

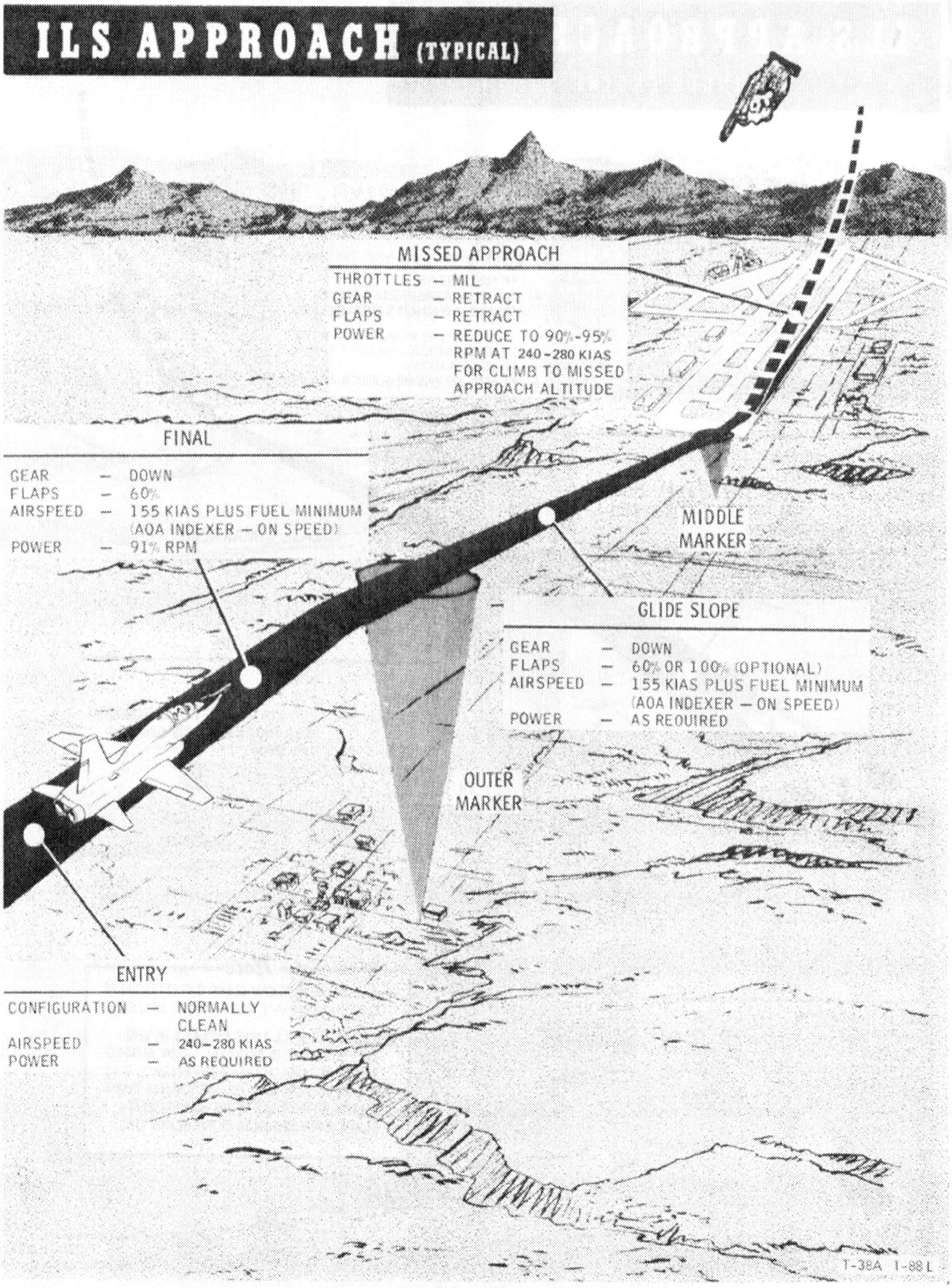
ILS APPROACH (TYPICAL)
MISSED APPROACH
THROTTLES – MIL
GEAR – RETRACT
FLAPS – RETRACT
POWER – REDUCE TO 90%-95% RPM AT 240-280 KIAS FOR CLIMB TO MISSED APPROACH ALTITUDE
FINAL
GEAR – DOWN
FLAPS – 60%
AIRSPEED – 155 KIAS PLUS FUEL MINIMUM (AOA INDEXER – ON SPEED)
POWER – 91% RPM
MIDDLE MARKER
GLIDE SLOPE
GEAR – DOWN
FLAPS – 60% OR 100% (OPTIONAL)
AIRSPEED – 155 KIAS PLUS FUEL MINIMUM (AOA INDEXER – ON SPEED)
POWER – AS REQUIRED
OUTER MARKER
ENTRY
CONFIGURATION – NORMALLY CLEAN
AIRSPEED – 240-280 KIAS
POWER – AS REQUIRED
T-38A 1-88 L

Figure 9-5.

# ILS APPROACH (TYPICAL)
## SINGLE-ENGINE

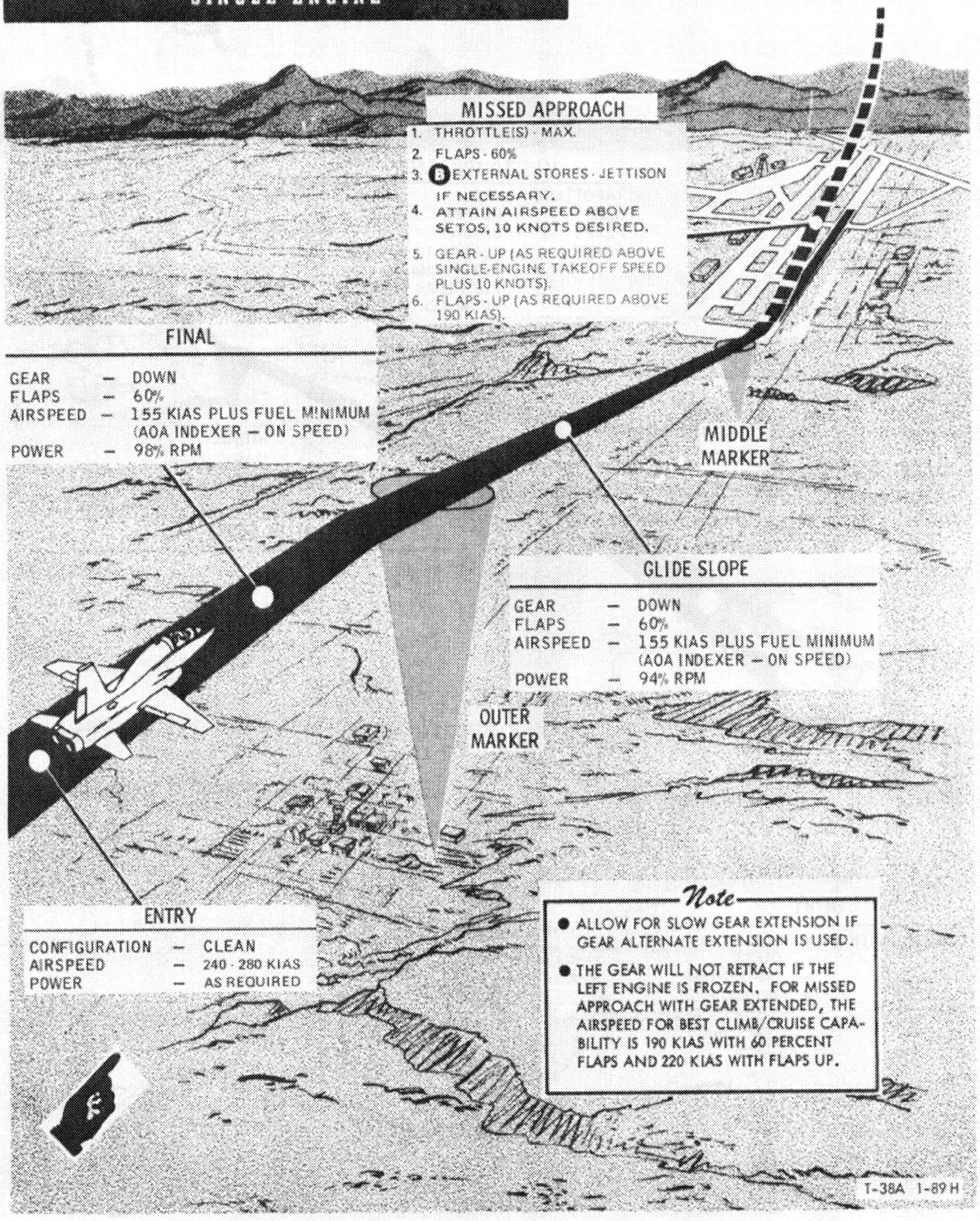


Figure 9-6.

# ICE AND RAIN

## TAKEOFF.

Monitor engine performance closely during takeoff on runways with large amounts of puddled water. Flameouts can be caused by water thrown up by the nosewheel.

## ICING.

Anti-icing equipment for the wings, empennage, and inlet ducts is not provided. The aircraft is provided with engine anti-ice, pitot heat, and canopy defog heat, which also provides windshield heat for adverse weather operation. Icing conditions which may be encountered are trace, light, moderate, and severe. Moderate and severe icing, particularly, can cause rapid buildup of ice on the aircraft surfaces and greatly affect performance.

**WARNING**

> The aircraft should not be flown in icing conditions. If icing is inadvertently encountered, leave the area of icing conditions as soon as possible.

When icing conditions are unavoidable, the pitot heat switch should be placed at PITOT HEAT and the canopy defog knob turned to full increase. The aircraft is not equipped with a windshield anti-icing or rain removal equipment. Instrument approaches in heavy rain are possible, but forward visibility thru the windshield may be marginal. Forward visibility in icing conditions is further reduced and may be completely obscured thru the windshield.

## ICE INGESTION.

Engine damage may occur if as little as 1/4 inch of ice accumulates on engine inlet duct lips. Ingestion of accumulated ice into an engine may be evidenced by a jar or noise in the engine and may result in damage to inlet guide vanes and first-stage compressor blades. Engine instrument indications may remain normal, even though engine damage from ice ingestion has been experienced.

**CAUTION**

- After ice ingestion, the affected engine should be operated at the lowest possible RPM necessary to make a safe landing, avoiding abrupt or rapid throttle movements.
- If flight in icing conditions results in ice accumulations on the aircraft, enter this information in Form 781, as the engines must be inspected for ice ingestion damage when this occurs.

## ENGINE ICING.

Engine inlet duct and/or guide vane icing may occur when the ambient temperature is at or slightly above freezing and either the humidity is high or when operating in visible moisture. Under these conditions, and when icing conditions are unavoidable, the engine anti-ice switch should immediately be placed at MAN ON, ensuring continuous anti-icing action.

**NOTE**

> To ensure effective anti-icing, maintain a minimum of 80% RPM when the engine anti-icing system is turned ON.

## RAIN.

**CAUTION**

> Flight in moderate precipitation may damage the nose cone or vertical stabilizer. Nose cone damage may result in in-flight engine FOD. If flight in moderate precipitation is unavoidable, slow to the minimum practical airspeed to negate or lessen damage.

# TURBULENCE AND THUNDERSTORMS

**WARNING**

> Intentional flight in thunderstorms should be avoided.

The recommended best penetration airspeed if turbulence and thunderstorms are experienced is **280** KIAS.

# NIGHT FLYING

When flying away from concentrations of ground lights, caution should be exercised to prevent spatial disorientation.

# COLD WEATHER OPERATION

Most cold weather operating difficulties are encountered on the ground. The following instructions are to be used in conjunction with the normal procedures given in section II when cold weather aircraft operation is necessary.

## BEFORE ENTERING AIRCRAFT.

Remove all protective covers and duct plugs; check to see that all surfaces, ducts, struts, drains, and vents are free of snow, ice, and frost. Brush off light snow and frost. Remove ice and encrusted snow either by a direct flow of air from a portable ground heater or by using de-icing fluid.

**WARNING**

- All ice, snow and frost must be removed from the aircraft before flight is attempted. Takeoff distance and climbout performance can be adversely affected by ice and snow accumulations. The roughness and distribution of these accumulations can vary stall speeds and alter flight characteristics to a degree extremely hazardous to safe flight.
- Ensure that water does not accumulate in control hinge areas or other critical areas where refreezing may cause damage or binding.

**CAUTION**

To avoid damage to aircraft surfaces, do not permit ice to be chipped or scraped away.

Check the fuel system vents on the vertical stabilizer for freedom from ice. Remove all dirt and ice from landing gear shock struts, actuating cylinder pistons, and limit switches. Wipe exposed parts of shock struts and pistons with a rag soaked in hydraulic fluid. Inspect aircraft carefully for fuel and hydraulic leaks caused by contraction of fittings or by shrinkage of packings. Inspect area behind aircraft to ensure that water or snow will not be blown onto personnel and equipment during engine start.

## ON ENTERING AIRCRAFT.

Use external power for starting to conserve the battery. No preheat or special starting procedures are required; however, at temperatures below −30°F (−34°C), allow the engines to idle 2 minutes before accelerating. Turn on cockpit heat and canopy defog system, as required, immediately after engine start. Check flight controls, speed brake, and aileron trim for proper operation. Cycle flight controls four to six times. Check hydraulic pressure and control reaction, and operation of all instruments.

## ENGINE OIL PRESSURE INDICATIONS.

Oil pressure indications above 55 psi will be observed after engine start. As the oil warms up, pressure should reduce to within operating limits. To reduce time for oil pressure to return to normal, the engine may be operated above idle up to military power until oil pressure is within limits. If oil pressure does not return to within operating limits, shut down engine and determine cause.

## ENGINE IDLE RPM.

Low engine idle RPM can be expected after engine start when the engines are cold and the ground ambient temperature is below −16°F (−26°C). Monitor EGT and increase engine RPM as necessary to cut in the ac generators. If engine RPM will not increase when the throttle is advanced, shut down engine and determine cause. Engine idle RPM should be within operating limits after the engine has warmed up and the oil pressure has decreased to the normal operating range.

## TAXIING.

Nosewheel steering effectiveness is reduced when taxiing on ice and hard packed snow. A combination of nosewheel steering and wheel braking should be used for directional control. The nosewheel will skid sideways easily, increasing the possibility of tire damage. Reduce taxi speeds and exercise caution at all times while operating on these surfaces. Increase the normal interval between aircraft both to ensure a safe stopping distance and to prevent icing of the aircraft from melted snow and ice caused by the jet blast of the preceding aircraft. Minimize taxi time to conserve fuel and reduce the amount of ice fog generated by the engines. If bare spots exist thru the snow, skidding onto them should be avoided. Check for sluggish instruments while taxiing.

## TAKEOFF.

Do not advance throttles into MAX range until the aircraft is rolling straight down the runway.

**WARNING**

> Do not take off on slush covered runway; the nosewheel may sling slush into the inlet ducts, causing engine flameout and/or damage.

## LANDING.

Use landing techniques given in section II. When landing on runways that have patches of dry surface, avoid locking the wheels. If the aircraft starts to skid, release brakes until recovery from skid is accomplished.

## ENGINE SHUTDOWN.

Use normal engine shutdown procedure.

# HOT WEATHER AND DESERT OPERATION

Operation of the aircraft in hot weather and in the desert requires that precautions be taken to protect the aircraft from damage caused by high temperatures, dust, and sand. Care must be taken to prevent the entrance of sand into aircraft parts and systems such as the engines, fuel system, pitot-static system, etc. All filters should be checked more frequently than under normal conditions. Plastic and rubber segments of the aircraft should be protected both from high temperatures and blowing sand. Canopy covers should be left off to prevent sand from accumulating between the cover and the canopy and acting as an abrasive on the plastic canopy. With a canopy closed, cockpit damage may result when ambient temperature is in excess of 110°F. Desert and hot weather operation require that in addition to normal procedures, the following precautions be observed.

## TAKEOFF.

1. Monitor pitch attitude closely to ensure a positive rate of climb during gear and flap retraction and to prevent an excessive angle of attack.
2. Be alert for gusts and wind shifts near the ground.

## APPROACH AND LANDING.

1. Monitor airspeed closely to ensure that recommended approach and touchdown airspeeds are maintained; high ambient temperatures cause speed relative to the ground to be higher than normal.
2. Anticipate a long landing roll due to higer ground speed at touchdown.
3. Utilize effective aerodynamic braking and all available runway for stopping the aircraft without overheating the wheel brakes.

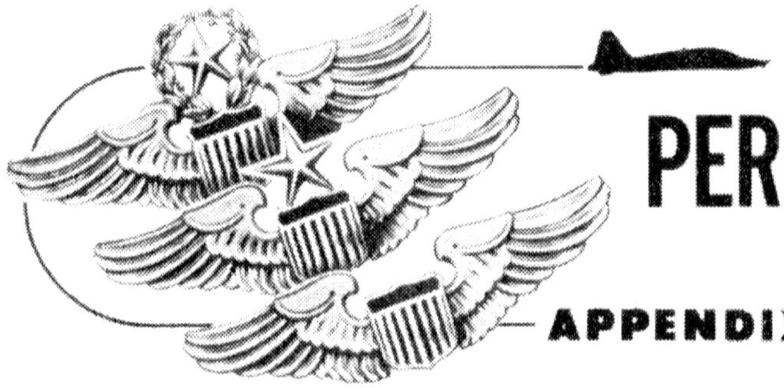

# APPENDIX I

# PERFORMANCE DATA

## TABLE OF CONTENTS



T-38A 1-108

# ABBREVIATIONS AND DEFINITIONS

KBAS — Basic airspeed (knots). Indicated airspeed corrected for instrument error.

KIAS — Indicated airspeed (knots). Airspeed indication uncorrected for instrument error.

KCAS — Calibrated airspeed (knots). Indicated airspeed corrected for instrument error and installation error (pitot-static system error and errors induced by aircraft attitude).

KEAS — Equivalent airspeed (knots). Calibrated airspeed corrected for compressibility.

KTAS — True airspeed (knots). Calibrated airspeed corrected for density and compressibility.

GS — Groundspeed (knots). Speed over the ground equal to true airspeed corrected for headwind (subtract) or tailwind (add).

Mach — A number expressing the ratio of the true airspeed of a moving body with the speed of sound in the air surrounding it.

IMN — Indicated mach number. Indicated mach reading uncorrected for instrument error.

TMN — True mach number. Indicated mach reading corrected for installation error.

ICAO — International Civil Aviation Organization.

IN. Hg — Inches of mercury.

$H_i$ — Indicated pressure altitude. Altimeter indicated pressure altitude with respect to the reference level set on barometric scale of the instrument. Standard pressure altitude is read by setting barometric scale at 29.92 inches of mercury.

$H_p$ — True pressure altitude. Altimeter reading corrected for installation error ($H_p = H_i - \Delta H_p$).

$\Delta H_p$ — Altimeter installation error correction.

$\Delta V_p$ — Airspeed installation error correction.

$\Delta V_c$ — Airspeed compressibility correction.

$\Delta V_M$ — Increase in takeoff speed above normal for military thrust.

$\Delta V_{SE}$ — Increase in takeoff speed above normal for single-engine maximum thrust.

% RPM — Engine speed expressed as a percentage of maximum engine speed (16,500 rpm = 100%).

$\Delta$ TEMP — Temperature correction.

a — Speed of sound at altitude.

$a_o$ — Speed of sound at sea level.

P — Static pressure at altitude.

$P_o$ — Static pressure at sea level.

$\delta$ — Pressure ratio, $P/P_o$

$\rho$ — Ambient air density.

$\rho_o$ — Air density at sea level.

$\sigma$ — Relative air density, $\rho/\rho_o$.

MAC — Mean aerodynamic chord.

CG — Center-of-gravity.

KN — Knot.

FPM — Feet-per-minute.

NMI — Nautical miles.

NMI/LB — Nautical miles per pound.

LB/MIN — Pounds per minute.

LB/HR/ENG — Pounds per hour per engine.

RCR — Runway Condition Reading.

SL — Sea level.

ALT — Altitude.

AMB TEMP — Ambient air temperature.

TEMP — Temperature.

STD — Standard.

G — Load factor or G-loading.

Q — Dynamic pressure.

PRESS — Pressure.

# PART 1 INTRODUCTION

TABLE OF CONTENTS



## INTRODUCTION.

The flight performance charts provide the pilot with flight test data for basic flight planning purposes. All charts are based on standard day conditions except when necessary, as in the takeoff and landing charts, to include temperature corrections for nonstandard days. These corrections are based on maintaining the recommended indicated mach number or indicated airspeed. Instrument error is assumed to be zero in all performance charts of this appendix.

## TAKEOFF FACTOR.

The takeoff factor is used to simplify the takeoff charts. The factor is based on atmospheric condition and the desired takeoff power setting. This factor reduces the time and effort required in takeoff planning.

## DESCRIPTION OF DRAG INDEX SYSTEM.

The Drag Index System permits the presentation of performance for a number of external store loadings on one chart and greatly reduces the number of charts required in flight planning work. In the drag index system, each item of the external store configuration, such as a bomb or pylon, is assigned a drag number whose value depends on the size and shape of the item and its location on the aircraft. These numbers are not drag coefficients. The summation of the store drag numbers for a particular loading defines a drag index for that configuration. This drag index, when used in the performance charts, determines the aircraft performance for that external store configuration. The T-38A, with no external stores capability, has a drag index of zero. The Weapon System Support Pod is assigned a drag number of 25.

## ALTIMETER AND AIRSPEED INSTALLATION ERROR CORRECTION.

Static pressure, which affects both airspeed and altimeter indications, is not always accurately measured because of the location of the static ports. This pressure error is a function of both airspeed and altitude. KCAS is obtained from KIAS by correcting for the installation error in static pressure (airspeed installation error). Knowing indicated airspeed and pressure altitude, both airspeed and altimeter installation corrections may be read from FA1-1.

## USE OF ALTIMETER CORRECTION CHART.

Consider the aircraft flying at 280 KIAS at 40,000 feet (FL 400). Read up the 280 KIAS line to intersect the 40,000-foot correction curve, and from this point, draw a horizontal line to the left margin of the chart. Read the correction, which is +60 feet. Since indicated altitude is pressure altitude plus correction, the proper indicated altitude is 40,060 feet.

## MACH NUMBER CORRECTION.

To convert true mach number to indicated mach number, use mach number correction chart FA1-2.

## COMPRESSIBILITY CORRECTION TO CALIBRATED AIRSPEED.

The compressibility correction chart (FA1-3) provides the necessary airspeed correction to convert KCAS to KEAS (KEAS = KCAS $- \Delta V_c$).

## AIRSPEED CONVERSION.

The chart in FA1-4 is used to convert between KCAS, true mach number, and KTAS. If KCAS is known, enter the chart at that value and move upward to the known pressure altitude. At that point, true mach number is read on the left-hand scale and KTAS for standard atmosphere conditions is interpolated between the sloping speed lines whose scale is located at the sea level pressure altitude line. To correct KTAS for nonstandard temperatures, move horizontally from the intersection of KCAS and the known altitude to the sea level pressure altitude line, then vertically downward to the known ambient air temperature, and read the corrected KTAS on the scale at the right.

## STANDARD ALTITUDE TABLE.

Significant properties of the ICAO standard atmosphere are tabulated at 1,000-foot increments between −2,000 and 65,000 feet altitude in FA1-5. Sea level values of the properties are listed in the top of the chart for use with the ratios shown in the table. As an example of the use of the table, find the equivalent airspeed in knots in standard atmosphere corresponding to 0.85 mach number at 30,000 feet pressure altitude. In FA1-5, at 30,000 feet read $a/a_o = 0.8909$, read $1/\sqrt{\sigma} = 1.6349$, and at the top of the table read $a_o = 661.7$ knots.

Then: $a = a_o \times a/a_o = 661.7 \times 0.8909 = 589.5$ knots.

KTAS = Mach $\times$ a = 0.85 $\times$ 589.5 - 501.1 knots.

KEAS = KTAS $\div 1/\sqrt{\sigma} = 501.1 \div 1.6349 =$ 306.5 knots.

## DENSITY ALTITUDE.

FA1-6 presents the variation of density altitude with ambient temperature for constant values of pressure altitude. Values of $1/\sqrt{\sigma}$ are tabulated at the right of the chart as a function of the density altitude scale on the left side. ICAO standard atmosphere conditions are defined by the line which slopes to the left and upward thru the chart. As an example of the use of the chart, find the value of $1/\sqrt{\sigma}$ at 8000 feet pressure altitude and 19° centigrade temperature. Move vertically upward to the 8000 feet pressure altitude line, then move horizontally right to the scale and read $1/\sqrt{\sigma} = 1.16$. The equivalent density altitude, if required, is 10,000 feet. Note that these conditions do not correspond to those of the standard atmosphere, since the true temperature at 8000 feet pressure altitude in standard atmosphere is approximately 0° and $1/\sqrt{\sigma} = 1.12$.

## STANDARD CONVERSION TABLE.

Linear scales for converting units of temperature, distance, and speed from one measurement system to another are provided in FA1-7. Additional conversion factors for volume, pressure, and weight are listed at the bottom of the table.

# ALTIMETER AND AIRSPEED INSTALLATION ERROR CORRECTIONS

CLEAN CONFIGURATION

MODEL: T-38A
DATE: 1 AUGUST 1965
DATA BASIS: **FLIGHT TEST**

ENGINE: (2) J85-GE-5
FUEL GRADE: JP-4
FUEL DENSITY: 6.5 LB/US GAL

ALTIMETER CORRECTION

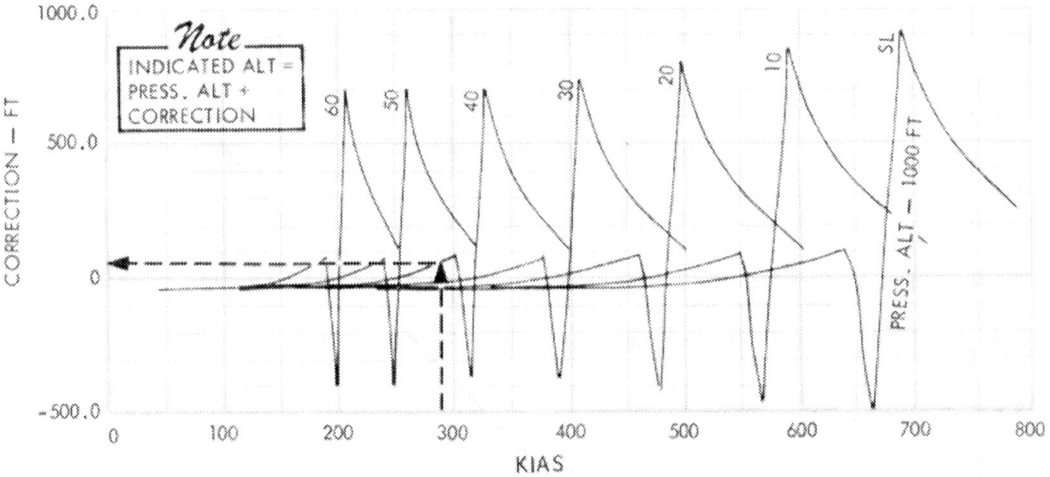


AIRSPEED CORRECTION

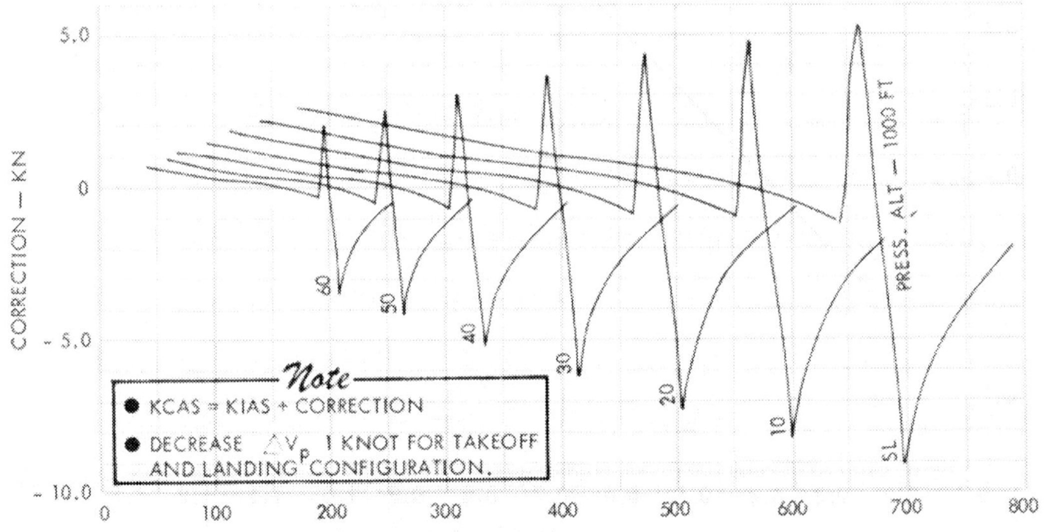


T-38A 1-300C

FA1-1

# MACH NUMBER CORRECTION

MODEL: T-38A
DATE: 1 AUGUST 1965
DATA BASIS: **FLIGHT TEST**

ENGINE: (2) J85-GE-5
FUEL GRADE: JP-4
FUEL DENSITY: 6.5 LB/US GAL

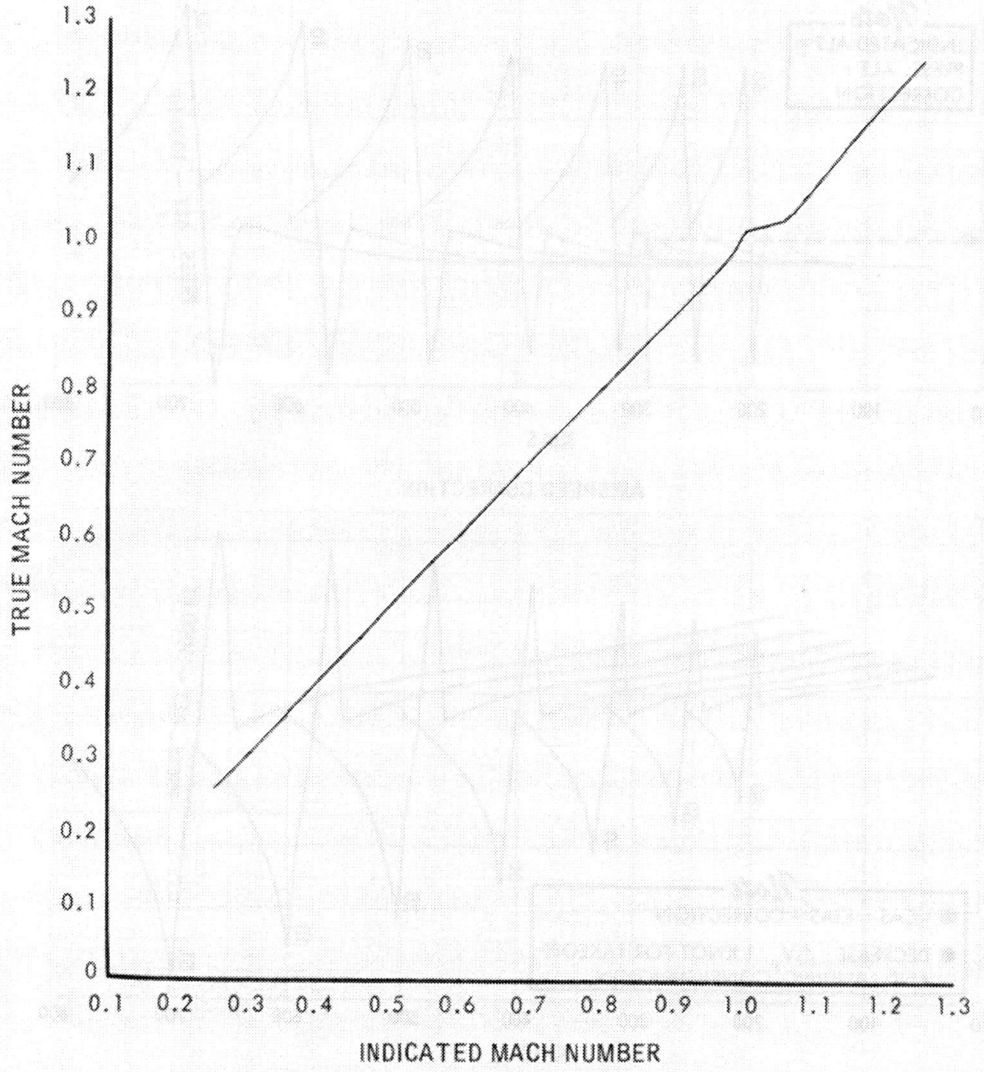


T-38A 1-301A

FA1-2

## COMPRESSIBILITY CORRECTION TO CALIBRATED AIRSPEED

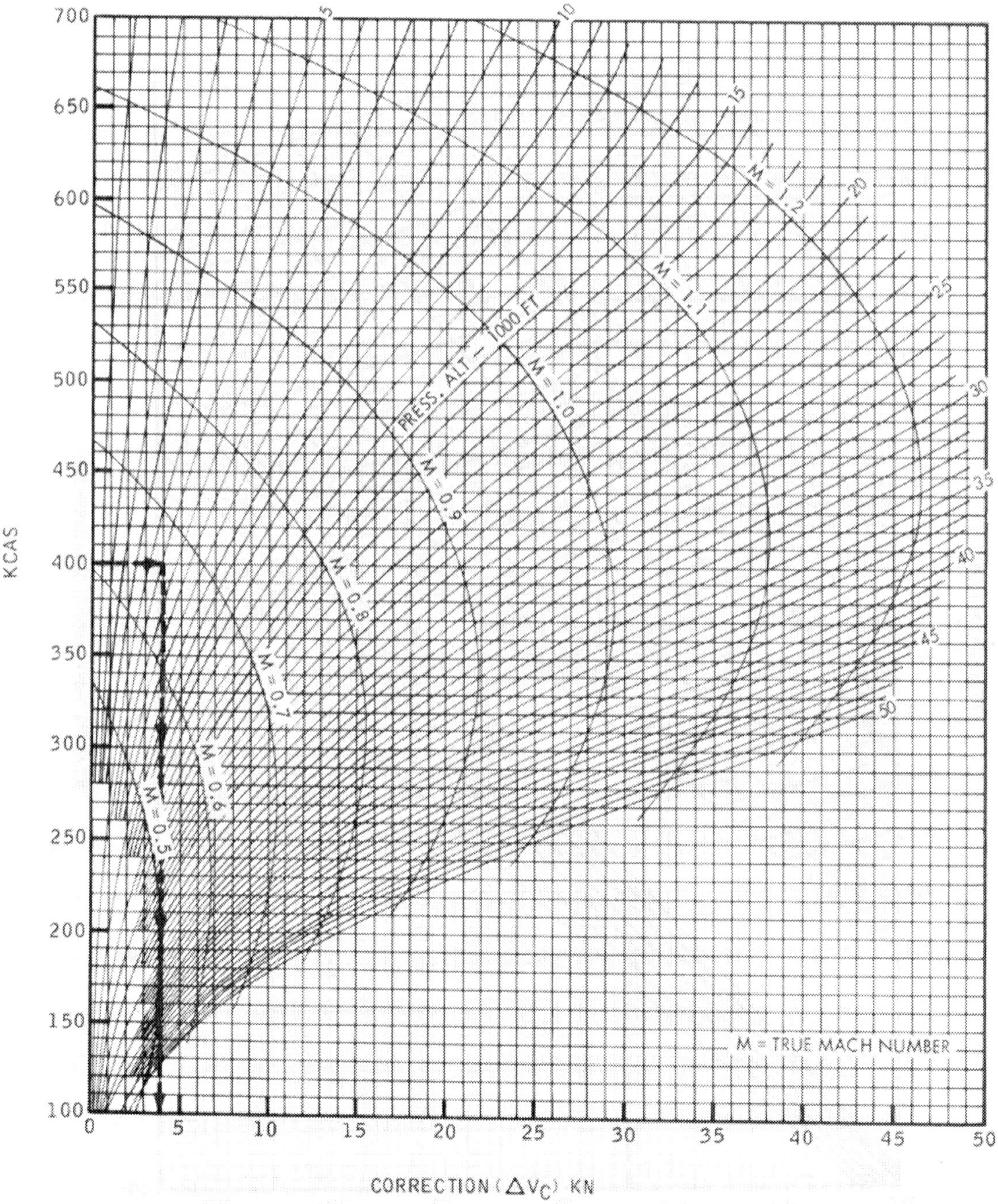


FA1-3

## AIRSPEED CONVERSION

EXAMPLE:
KCAS = 440
PRESS. ALT = 15,000 FT
TMN = 0.85
KTAS (STD DAY) = 530
KTAS (AT 20°C) = 565

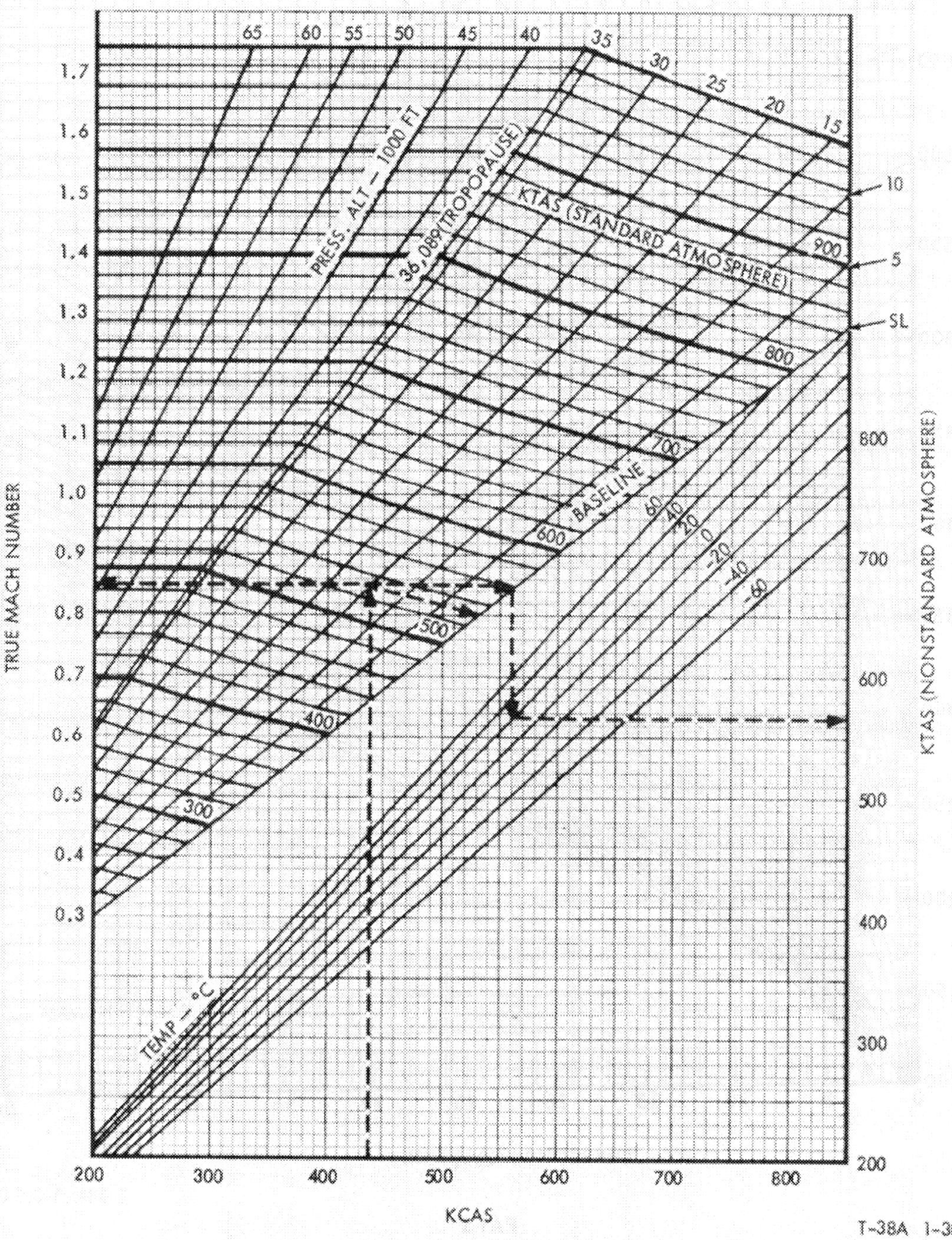


FA1-4

## STANDARD ALTITUDE TABLE

STANDARD SEA LEVEL AIR:

T = 59°F (15°C)

P = 29.921 IN. OF HG

W = 0.076475 LB/CU FT  $\rho_0$ = 0.0023769 SLUGS/CU FT

1" OF HG = 70.732 LB/SQ FT = 0.4912 LB/SQ IN.

$a_0$ = 1116.89 FT/SEC = 661.7 KN

### STANDARD ATMOSPHERE
### (NACA TECHNICAL REPORT NO. 1235)

| ALTITUDE FEET | DENSITY RATIO $\rho/\rho_0 = \sigma$ | $1/\sqrt{\sigma}$ | TEMPERATURE | | SPEED OF SOUND RATIO $a/a_0$ | PRESSURE | |
|---|---|---|---|---|---|---|---|
| | | | DEG. F | DEG. C | | IN. OF HG | RATIO $P/P_0 = \delta$ |
| −2,000 | 1.0598 | 0.9714 | 66.132 | 18.962 | 1.0064 | 32.15 | 1.0294 |
| −1,000 | 1.0296 | 0.9855 | 62.566 | 16.981 | 1.0030 | 31.02 | 1.0147 |
| 0 | 1.0000 | 1.0000 | 59.000 | 15.000 | 1.0000 | 29.92 | 1.0000 |
| 1,000 | 0.9711 | 1.0148 | 55.434 | 13.019 | 0.9966 | 28.86 | 0.9644 |
| 2,000 | 0.9428 | 1.0299 | 51.868 | 11.038 | 0.9931 | 27.82 | 0.9298 |
| 3,000 | 0.9151 | 1.0454 | 48.302 | 9.057 | 0.9896 | 26.82 | 0.8962 |
| 4,000 | 0.8881 | 1.0611 | 44.735 | 7.075 | 0.9862 | 25.84 | 0.8637 |
| 5,000 | 0.8617 | 1.0773 | 41.169 | 5.094 | 0.9827 | 24.90 | 0.8320 |
| 6,000 | 0.8359 | 1.0938 | 37.603 | 3.113 | 0.9792 | 23.98 | 0.8014 |
| 7,000 | 0.8106 | 1.1107 | 34.037 | 1.132 | 0.9756 | 23.09 | 0.7716 |
| 8,000 | 0.7860 | 1.1279 | 30.471 | −0.849 | 0.9721 | 22.22 | 0.7428 |
| 9,000 | 0.7620 | 1.1456 | 26.905 | −2.831 | 0.9686 | 21.39 | 0.7148 |
| 10,000 | 0.7385 | 1.1637 | 23.338 | −4.812 | 0.9650 | 20.58 | 0.6877 |
| 11,000 | 0.7156 | 1.1822 | 19.772 | −6.793 | 0.9614 | 19.79 | 0.6614 |
| 12,000 | 0.6932 | 1.2011 | 16.206 | −8.774 | 0.9579 | 19.03 | 0.6360 |
| 13,000 | 0.6713 | 1.2205 | 12.640 | −10.756 | 0.9543 | 18.29 | 0.6113 |
| 14,000 | 0.6500 | 1.2403 | 9.074 | −12.737 | 0.9507 | 17.58 | 0.5875 |
| 15,000 | 0.6292 | 1.2606 | 5.508 | −14.718 | 0.9470 | 16.89 | 0.5643 |
| 16,000 | 0.6090 | 1.2815 | 1.941 | −16.699 | 0.9434 | 16.22 | 0.5420 |
| 17,000 | 0.5892 | 1.3028 | −1.625 | −18.681 | 0.9397 | 15.57 | 0.5203 |
| 18,000 | 0.5699 | 1.3246 | −5.191 | −20.662 | 0.9361 | 14.94 | 0.4994 |
| 19,000 | 0.5511 | 1.3470 | −8.757 | −22.643 | 0.9324 | 14.34 | 0.4791 |
| 20,000 | 0.5328 | 1.3700 | −12.323 | −24.624 | 0.9287 | 13.75 | 0.4595 |
| 21,000 | 0.5150 | 1.3935 | −15.889 | −26.605 | 0.9250 | 13.18 | 0.4406 |
| 22,000 | 0.4976 | 1.4176 | −19.456 | −28.587 | 0.9213 | 12.64 | 0.4223 |
| 23,000 | 0.4807 | 1.4424 | −23.022 | −30.568 | 0.9175 | 12.11 | 0.4046 |
| 24,000 | 0.4642 | 1.4678 | −26.588 | −32.549 | 0.9138 | 11.60 | 0.3876 |
| 25,000 | 0.4481 | 1.4938 | −30.154 | −34.530 | 0.9100 | 11.10 | 0.3711 |
| 26,000 | 0.4325 | 1.5206 | −33.720 | −36.511 | 0.9062 | 10.63 | 0.3552 |
| 27,000 | 0.4173 | 1.5480 | −37.286 | −38.492 | 0.9024 | 10.17 | 0.3398 |
| 28,000 | 0.4025 | 1.5762 | −40.852 | −40.473 | 0.8986 | 9.725 | 0.3250 |
| 29,000 | 0.3881 | 1.6052 | −44.419 | −42.455 | 0.8948 | 9.297 | 0.3107 |
| 30,000 | 0.3741 | 1.6349 | −47.985 | −44.436 | 0.8909 | 8.885 | 0.2970 |
| 31,000 | 0.3605 | 1.6654 | −51.551 | −46.417 | 0.8871 | 8.488 | 0.2837 |
| 32,000 | 0.3473 | 1.6968 | −55.117 | −48.398 | 0.8832 | 8.106 | 0.2709 |
| 33,000 | 0.3345 | 1.7291 | −58.683 | −50.379 | 0.8793 | 7.737 | 0.2586 |
| 34,000 | 0.3220 | 1.7623 | −62.249 | −52.361 | 0.8754 | 7.382 | 0.2467 |
| 35,000 | 0.3099 | 1.7964 | −65.816 | −54.342 | 0.8714 | 7.041 | 0.2353 |
| 36,000 | 0.2981 | 1.8315 | −69.382 | −56.323 | 0.8675 | 6.712 | 0.2243 |
| 37,000 | 0.2844 | 1.8753 | −69.700 | −56.500 | 0.8671 | 6.397 | 0.2138 |
| 38,000 | 0.2710 | 1.9209 | −69.700 | −56.500 | 0.8671 | 6.097 | 0.2038 |
| 39,000 | 0.2583 | 1.9677 | −69.700 | −56.500 | 0.8671 | 5.811 | 0.1942 |
| 40,000 | 0.2462 | 2.0155 | −69.700 | −56.500 | 0.8671 | 5.538 | 0.1851 |
| 41,000 | 0.2346 | 2.0645 | −69.700 | −56.500 | 0.8671 | 5.278 | 0.1764 |
| 42,000 | 0.2236 | 2.1148 | −69.700 | −56.500 | 0.8671 | 5.030 | 0.1681 |
| 43,000 | 0.2131 | 2.1662 | −69.700 | −56.500 | 0.8671 | 4.794 | 0.1602 |
| 44,000 | 0.2031 | 2.2189 | −69.700 | −56.500 | 0.8671 | 4.569 | 0.1527 |
| 45,000 | 0.1936 | 2.2728 | −69.700 | −56.500 | 0.8671 | 4.355 | 0.1455 |
| 46,000 | 0.1845 | 2.3281 | −69.700 | −56.500 | 0.8671 | 4.151 | 0.1387 |
| 47,000 | 0.1758 | 2.3848 | −69.700 | −56.500 | 0.8671 | 3.956 | 0.1322 |
| 48,000 | 0.1676 | 2.4428 | −69.700 | −56.500 | 0.8671 | 3.770 | 0.1260 |
| 49,000 | 0.1597 | 2.5022 | −69.700 | −56.500 | 0.8671 | 3.593 | 0.1201 |
| 50,000 | 0.1522 | 2.5630 | −69.700 | −56.500 | 0.8671 | 3.425 | 0.1145 |
| 51,000 | 0.1451 | 2.6254 | −69.700 | −56.500 | 0.8671 | 3.264 | 0.1091 |
| 52,000 | 0.1383 | 2.6892 | −69.700 | −56.500 | 0.8671 | 3.111 | 0.1040 |
| 53,000 | 0.1318 | 2.7546 | −69.700 | −56.500 | 0.8671 | 2.965 | 0.09909 |
| 54,000 | 0.1256 | 2.8216 | −69.700 | −56.500 | 0.8671 | 2.826 | 0.09444 |
| 55,000 | 0.1197 | 2.8903 | −69.700 | −56.500 | 0.8671 | 2.693 | 0.09001 |
| 56,000 | 0.1141 | 2.9606 | −69.700 | −56.500 | 0.8671 | 2.567 | 0.08578 |
| 57,000 | 0.1087 | 3.0326 | −69.700 | −56.500 | 0.8671 | 2.446 | 0.08176 |
| 58,000 | 0.1036 | 3.1063 | −69.700 | −56.500 | 0.8671 | 2.331 | 0.07792 |
| 59,000 | 0.09877 | 3.1819 | −69.700 | −56.500 | 0.8671 | 2.222 | 0.07426 |
| 60,000 | 0.09414 | 3.2593 | −69.700 | −56.500 | 0.8671 | 2.118 | 0.07078 |
| 61,000 | 0.08972 | 3.3386 | −69.700 | −56.500 | 0.8671 | 2.018 | 0.06746 |
| 62,000 | 0.08551 | 3.4198 | −69.700 | −56.500 | 0.8671 | 1.924 | 0.06429 |
| 63,000 | 0.08150 | 3.5029 | −69.700 | −56.500 | 0.8671 | 1.833 | 0.06127 |
| 64,000 | 0.07767 | 3.5881 | −69.700 | −56.500 | 0.8671 | 1.747 | 0.05840 |
| 65,000 | 0.07403 | 3.6754 | −69.700 | −56.500 | 0.8671 | 1.665 | 0.05566 |

T-38A 1-304A

FA1-5

# DENSITY ALTITUDE

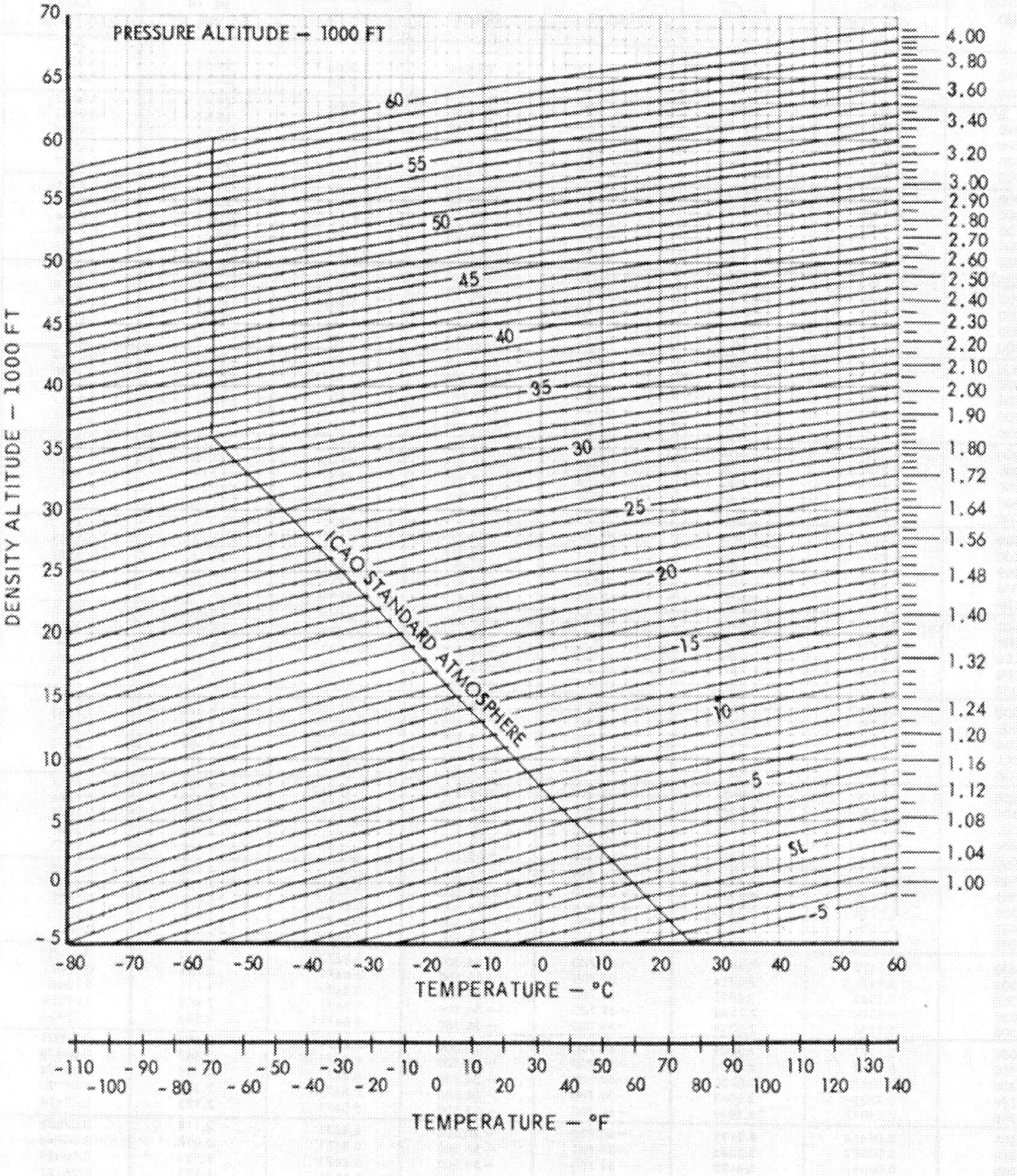


T-38A 1-305A

FA1-6

# STANDARD CONVERSION TABLE

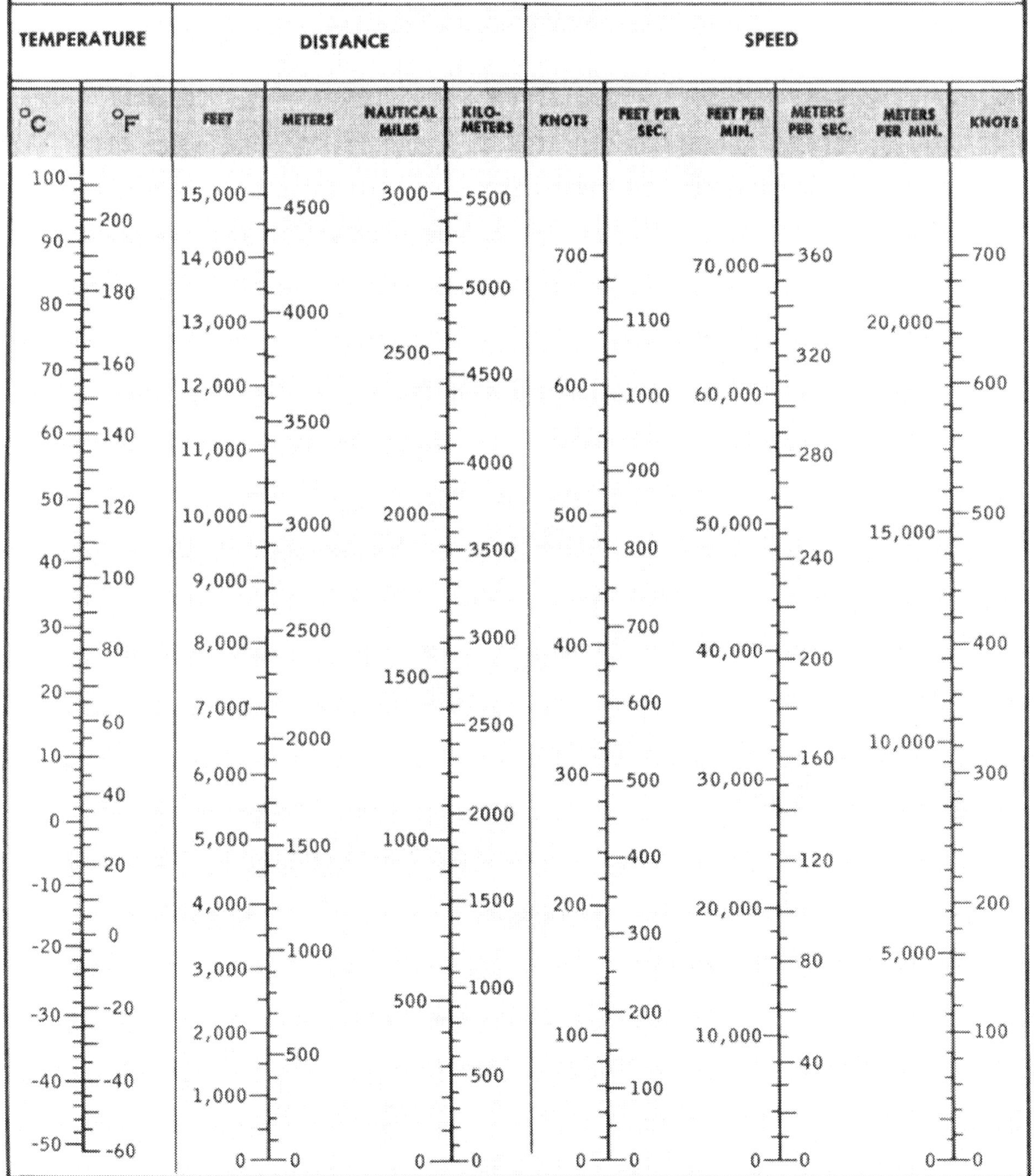


## *Note*

- TO OBTAIN US GALLONS MULTIPLY LITERS BY 0.264
- TO OBTAIN IMPERIAL GALLONS MULTIPLY LITERS BY 0.220
- TO OBTAIN INCHES OF MERCURY MULTIPLY MILLIBARS BY 0.0295
- TO OBTAIN POUNDS MULTIPLY KILOGRAMS BY 2.20

FA1-7

T-38A 1-306A

TABLE OF CONTENTS



## WIND COMPONENTS.

A takeoff and landing wind components chart (FA2-1) is provided to enable the pilot to convert surface winds to headwind and crosswind components. The headwind component is used to compute takeoff and landing data. The crosswind component is used to determine the feasibility of operations.

### USE OF WIND COMPONENTS CHART.

The chase-thru lines (FA2-1) show a 30-degree right crosswind of 39 knots. The headwind component is 34 knots, and the crosswind component is 20 knots.

## TAKEOFF FACTOR.

The takeoff factor is a number which is common to all takeoff charts for a given thrust rating and atmospheric condition. The takeoff factor chart (FA2-2) shows the takeoff factor as a function of pressure altitude, runway air temperature, and thrust rating, including the effect of the anti-ice system operation.

### USE OF TAKEOFF FACTOR CHART.

The chase-thru lines on FA2-2 show a runway air temperature of 15°C and a pressure altitude of 4,000 feet which give takeoff factors of 3.45 and 5.25 for MAX and MIL thrust respectively.

## TAKEOFF SPEED.

Takeoff speed is the speed at which the main gear lifts off the runway. The takeoff speed chart (FA2-3) enables the pilot to determine normal takeoff speed and the climb speed to be attained to clear a 50-foot obstacle.

### USE OF TAKEOFF SPEED CHART.

The chase-thru lines on FA2-3 show the normal takeoff speed for an aircraft with a gross weight of 11,800 pounds is 154 KIAS.

## TAKEOFF DISTANCE.

Takeoff distance is ground run distance in feet to liftoff. Takeoff distance to clear a 50-foot obstacle is ground run distance in feet to liftoff plus the air distance to clear a 50-foot obstacle. The takeoff distance charts (FA2-4, FA2-5), show ground run distance and total distance to clear a 50-foot obstacle as a function of takeoff factor, gross weight, wind velocity and runway slope, for takeoff on a dry, hard surface runway. The charts show data for normal takeoffs at MAX or MIL thrust, using the normal takeoff procedures given in section II. For large takeoff factors and heavy gross weights which occur with MIL thrust, the normal takeoff speed is increased by $\Delta V_M$ to assure 100 ft/min rate of climb with two engines operating.

#### USE OF TAKEOFF DISTANCE CHARTS.

The chase-thru lines on FA2-4 show a maximum thrust takeoff for a gross weight of 11,800 pounds, headwind of 10 knots, and a takeoff factor of 3.45. The resulting normal ground run distance is 3050 feet. The corresponding total distance to clear a 50-foot obstacle is 4600 feet (FA2-5).

### EFFECT OF RUNWAY CONDITION READING (RCR).

Runway Condition Reading (RCR) is a number that indicates the degree of braking friction on the runway surface. RCR 5 is icy, RCR 12 is wet, and RCR 23 and above is dry. On slippery runways, the critical field length is increased, which may cause an increase in the minimum acceleration speed check. The refusal speed and critical engine failure speeds are decreased (when compared to dry, hard-surfaced runways).

### CRITICAL FIELD LENGTH.

Critical field length is the total runway length required to accelerate with both engines operating to the critical engine failure speed, experience an engine failure, then either continue to takeoff or stop. The critical field length is shown for MAX thrust on FA2-6. For single-engine takeoff at large takeoff factors and heavy gross weights, the normal takeoff speed is increased by $\Delta V_{SE}$ to assure 100 feet per minute rate of climb.

#### USE OF CRITICAL FIELD LENGTH CHART.

The chase-thru lines on FA2-6 show that at a takeoff factor of 3.45 and a gross weight of 11,800 pounds, $\Delta V_{SE}$ is 9 knots. The chase-thru lines further show that with a 10-knot headwind, the Critical Field Length for an RCR of 23 is 5800 feet, and is increased to 6500 feet for an RCR of 12. The Single Engine Takeoff Speed is normal takeoff speed plus $\Delta V_{SE}$ ; 154 + 9 = 163 KIAS.

### REFUSAL SPEED.

Refusal speed is the maximum speed to which the aircraft can accelerate and then stop in the remaining runway length. Stopping distance data in the refusal speed charts (FA2-7, FA2-8) are for two engines at MAX and MIL thrust, using max braking in 3 point attitude and include a pilot reaction time of 3 seconds.

### CRITICAL ENGINE FAILURE SPEED.

Critical engine failure speed is the speed to which the aircraft will accelerate with both engines, experience an engine failure, and permit either acceleration to takeoff or deceleration to a stop in the same distance. Data for critical engine failure speed is presented in FA2-7. If a critical engine failure speed computes to less than 110 KIAS, use 110 KIAS as the critical engine failure speed. When an RCR factor is present, use the full computed critical engine failure speed corrected for RCR.

#### USE OF REFUSAL SPEED CHARTS OR CRITICAL ENGINE FAILURE SPEED CHARTS.

The chase-thru lines on FA2-7 show a refusal speed of 133 KIAS for a takeoff factor of 3.45 at gross weight of 11,800 pounds on a 7,000-foot runway for an RCR of 23. Correcting for an RCR of 12 reduces the refusal speed to 106 KIAS. A 10-knot headwind increases the refusal speed to 143 KIAS for an RCR of 23 and 116 KIAS for an RCR of 12. For a takeoff factor of 3.45 at gross weight 11,800 and a critical field length of 5,800 feet, the critical engine failure speed is 122 KIAS for an RCR of 23. An RCR of 12 results in a critical engine failure speed of 102 KIAS. Adding the 10-knot headwind increases these speeds to 132 KIAS for an RCR of 23 and 112 KIAS for an RCR of 12. If critical engine failure speed computes to less than 110 KIAS, use 110 KIAS as critical engine failure speed. (Exception: If an RCR factor is present, use the actual computed speed as critical engine failure speed.)

### TAKEOFF ABORT CHARTS (GENERAL).

The takeoff abort charts contained in FA2-6 thru FA2-10 provide the means of planning for a GO—NO-GO decision should an engine fail during takeoff. A discussion is provided to illustrate the factors which influence the decision to stop or go. A detailed description of each abort chart is provided in the preceding paragraphs. The principal factor affecting aborted takeoff is the relationship of actual runway length to critical field length. This relationship falls into three categories as follows:

CATEGORY I. Runway Length Greater than Critical Field Length. (Refusal speed exceeds critical engine failure speed.)

a. If engine failure occurs below critical engine failure speed:

Aircraft should be stopped, as runway length will be sufficient for stopping. Takeoff distance increases as engine failure speed decreases and may exceed the runway length under certain conditions.

b. If engine failure occurs between critical engine failure speed and refusal speed:

Takeoff should normally be continued; however aircraft can take off or stop within remaining dis-distance.

c. If engine failure occurs above refusal speed:

Aircraft must continue takeoff as it would overrun runway in stopping. Sufficient runway for takeoff will be available.

CATEGORY II. Runway Length Same as Critical Field Length. Refusal speed and critical engine failure speed coincide; therefore, aircraft must be stopped if below critical engine failure speed and should continue takeoff if above the coincidence speed. Runway will be adequate for either condition.

CATEGORY III. Runway Length Less than Critical Field Length. (Refusal speed less than critical engine failure speed.) This is the most critical category. If flight operations are to be conducted under these conditions, decision speed (FA2-9) must be used as the Go-No-Go factor. If engine failure occurs between refusal speed and decision speed, the takeoff must be aborted, even though barrier engagement can be expected. If engine failure occurs after decision speed, sufficient runway for takeoff should be available, and takeoff should be continued.

## DECISION SPEED.

Decision speed is the minimum speed at which the aircraft can experience an engine failure and still accelerate to single-engine takeoff speed in the remaining runway. The decision speed is found on FA2-9.

### USE OF DECISION SPEED CHART.

The chase-thru line on FA2-9 shows a decision speed of 102 KIAS for a takeoff factor of 3.45 at a gross weight of 11,800 pounds on a 7000-foot runway with a 10-knot headwind, and $\Delta V_{SE}$ = 9 knots (FA2-6).

## VELOCITY DURING TAKEOFF GROUND RUN.

The velocity during takeoff ground run chart shows the relationship between KIAS and distance traveled during ground run on a dry, hard surface runway. The two-engine velocity during takeoff ground run chart (FA2-10), is used to check acceleration performance. Compute the minimum acceleration check speed for a point 2000 feet from brake release. If the takeoff run is less than 3000 feet, compute minimum acceleration check speed for a point 1000 feet from brake release. Under certain slippery runway conditions, the minimum acceleration check speed may be above the corrected critical engine failure speed. When this occurs, adjust the acceleration check distance to any usable value up to 2000 feet from brake release that will result in a minimum acceleration check speed equal to or less than the critical engine failure speed corrected for RCR. The forecast speed at this point is the normal acceleration check speed. Minimum accleration check speed is the minimum acceptable speed at the check distance with which takeoff should be continued. Minimum acceleration check speed is computed by subtracting 3 knots for each 1000 feet of runway in excess of the critical field length or 10 knots from normal acceleration check speed, whichever is less. The single-engine velocity during ground run (FA2-11) is used to evaluate single-engine takeoff acceleration performance.

### USE OF TWO-ENGINE VELOCITY DURING TAKEOFF GROUND RUN CHART.

Assume a takeoff weight of 11,800 pounds, a runway temperature of 15°C, RCR of 12, a pressure altitude of 4000 feet, and a 10-knot headwind. Enter the chart at the takeoff speed 155 KIAS (FA2-3) and ground run distance of 3000 feet (FA2-4). From the point of intersection of these lines, draw a line parallel to the guideline. Enter FA2-10 at ground run distance of 2000 feet. Proceed vertically to intersection with constructed airspeed guideline and read airspeed of 129 knots from the left side of the chart. This is the velocity at a point 2000 feet from brake release. Since this acceleration check speed is above the corrected critical engine failure speed of 112 KIAS (FA2-7), reenter chart at ground run distance of 1500 feet. Proceed vertically to intersection with constructed airspeed guideline and read airspeed of 110 knots from the left side of the chart. This is the velocity at 1500 feet from brake release.

**USE OF SINGLE-ENGINE VELOCITY DURING TAKEOFF GROUND RUN CHART.**

Assume a takeoff weight of 11,800 pounds, a runway air temperature of 15°C, a pressure altitude of 4000 feet, and a 10-knot headwind. Enter FA2-11 at the runway temperature 15°C, right horizontally to the pressure altitude of 4000 feet, down vertically to the aircraft gross weight 11,800 pounds, left horizontally to the baseline. Draw a line that parallels the guideline. Assume an engine failure at 120 KIAS and it is desired to find the distance necessary to accelerate to 160 KIAS. Enter the chart at the no-wind groundspeed of 110 knots (120 minus 10 knots headwind) and 150 knots (160 minus 10 knots headwind). Read the distances for 110 knots no wind (3600) and 150 knots no wind (7400). The difference between the noted distances (7400 minus 3600) is 3800 feet and is the distance necessary to accelerate to 160 KIAS.

TAKEOFF AND LANDING
WIND COMPONENTS

MODEL: T-38A
DATE: 1 AUGUST 1965
DATA BASIS: **FLIGHT TEST**

ENGINE: (2) J85-GE-5
FUEL GRADE: JP-4
FUEL DENSITY: 6.5 LB/US GAL

***Note***

ENTER CHART WITH STEADY WIND TO DETERMINE HEADWIND COMPONENT AND WITH MAXIMUM GUST VELOCITY TO DETERMINE CROSSWIND COMPONENT.

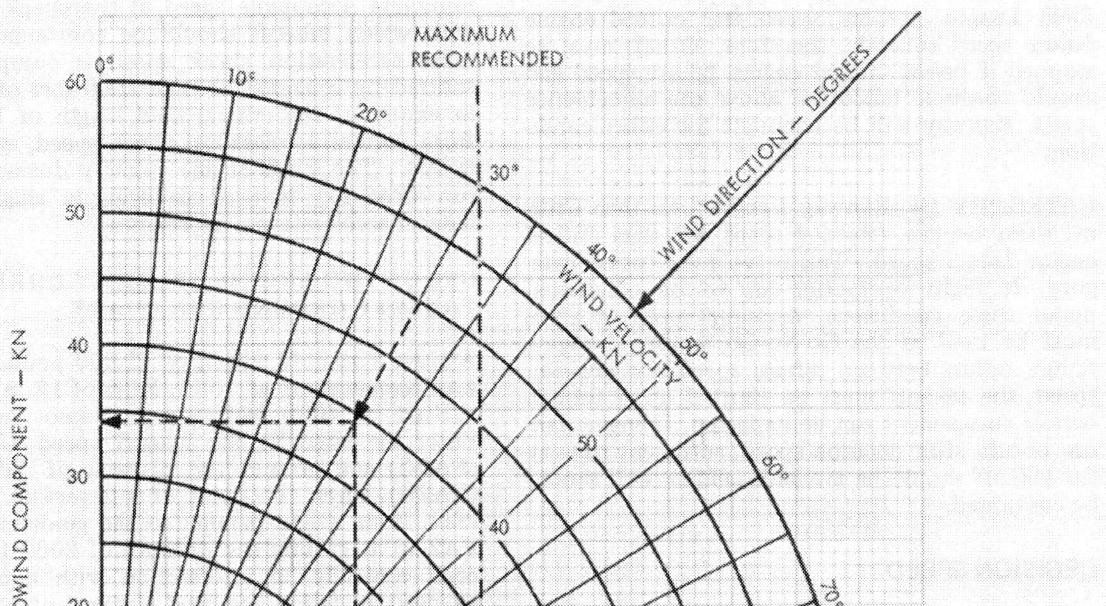


FA2-1

# TAKEOFF FACTOR

MODEL: T-38A
DATE: 1 AUGUST 1965
DATA BASIS: **FLIGHT TEST**

ENGINE: (2) J85-GE-5
FUEL GRADE: JP-4
FUEL DENSITY: 6.5 LB/US GAL

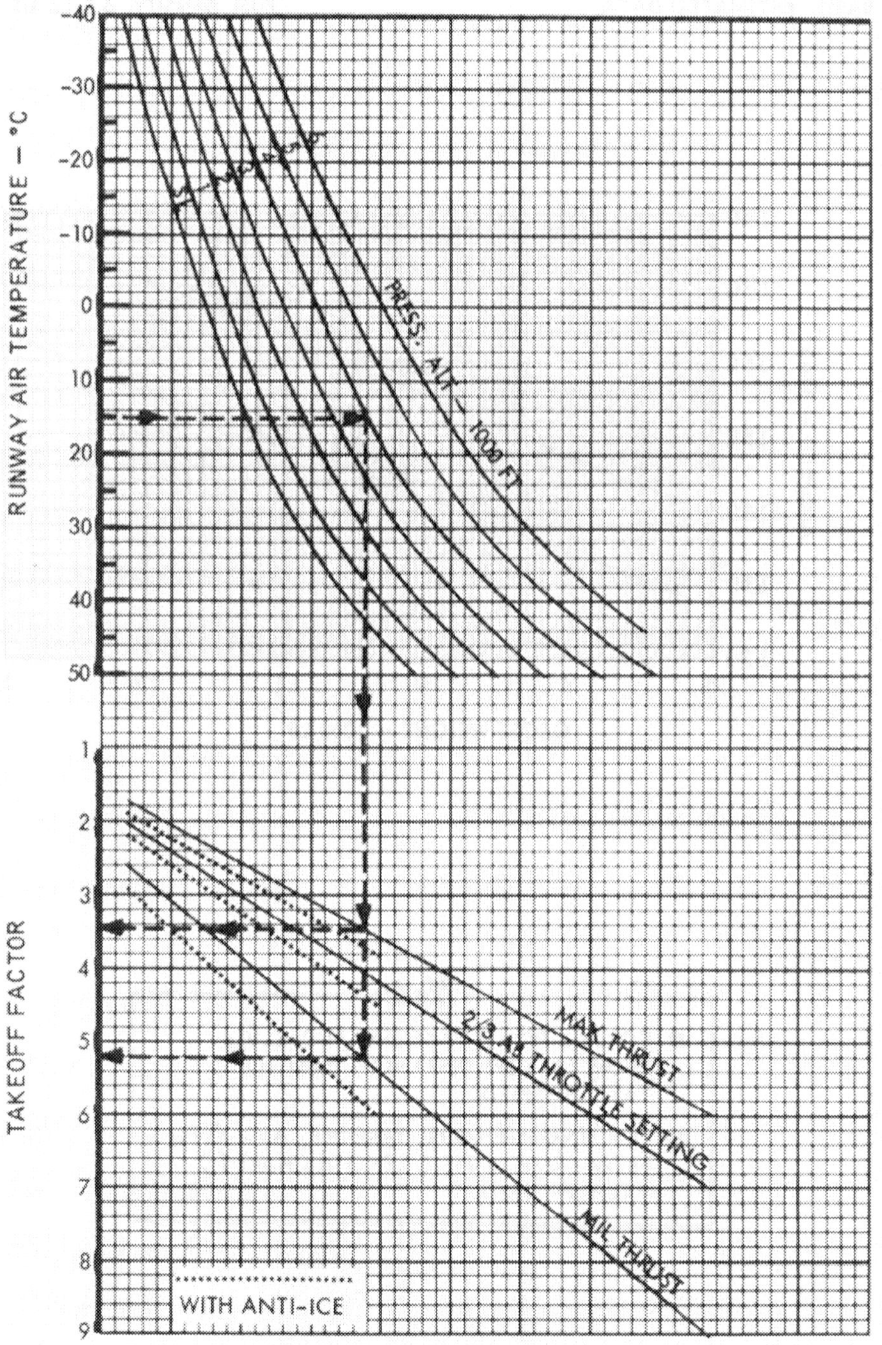


T-38A 1-401B

FA2-2

# TAKEOFF SPEED

FLAPS — 60%

MODEL: T-38A
DATE: 1 OCTOBER 1976
DATA BASIS: ESTIMATED DATA

ENGINE: (2) J85-GE-5
FUEL GRADE: JP-4
FUEL DENSITY: 6.5 LB/US GAL

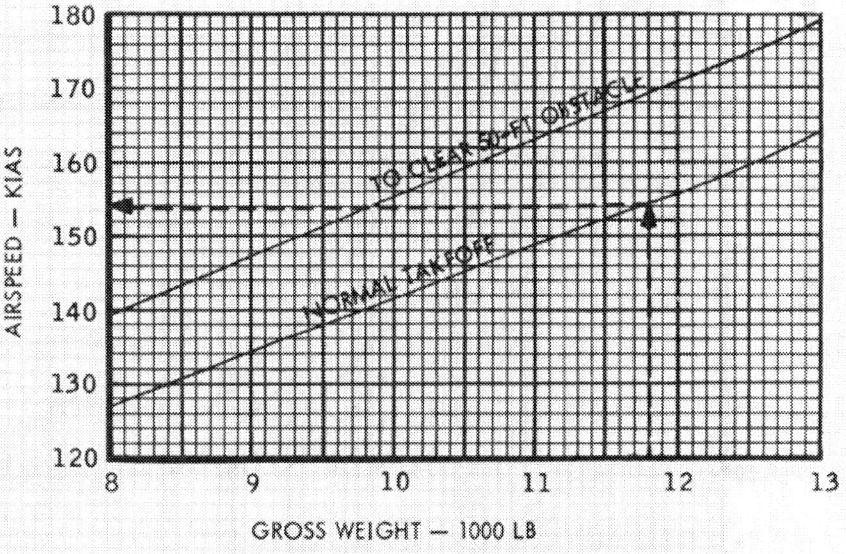


***Note***

- FOR MIL THRUST TAKEOFF, ADD $\Delta V_M$ FROM TAKEOFF DISTANCE CHART TO TAKEOFF SPEED.
- FOR SINGLE-ENGINE TAKEOFF, ADD $\Delta V_{SE}$ FROM CRITICAL FIELD LENGTH CHART TO TAKEOFF SPEED.

T-38A 1-402E

FA2-3

# TAKEOFF DISTANCE

DRY, HARD-SURFACED RUNWAY
FLAPS — 60%

MODEL: T-38A
DATE: 1 OCTOBER 1976
DATA BASIS: ESTIMATED DATA

ENGINE: (2) J85-GE-5
FUEL GRADE: JP-4
FUEL DENSITY: 6.5 LB/US GAL

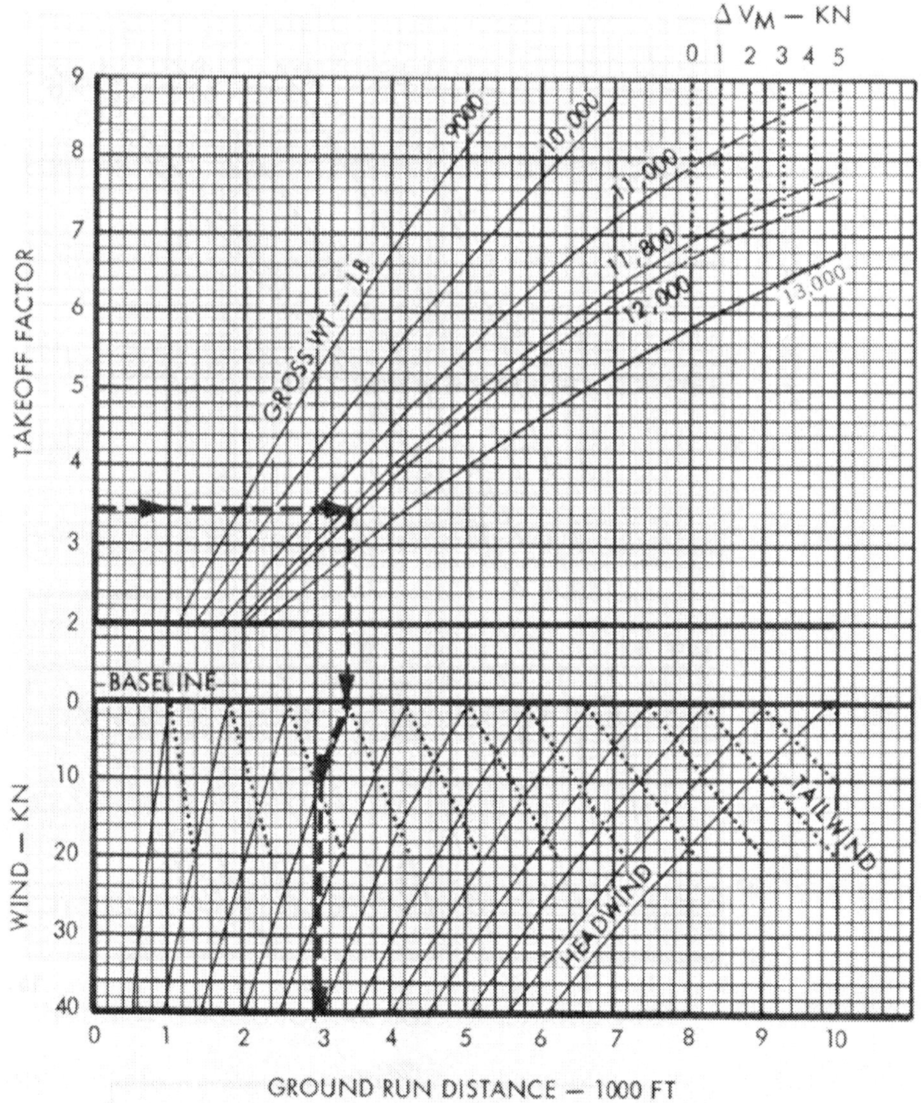


**Note**

- ADD $\Delta V_M$ TO TAKEOFF SPEED (FA2-3)
- INCREASE GROUND RUN DISTANCE 5 PERCENT FOR EACH PERCENT OF UPHILL RUNWAY SLOPE.
- WITH THE WSSP INSTALLED, ADD 6 PERCENT TO THE COMPUTED GROUND RUN DISTANCE

T-38A 1-403 D

FA2-4

# TAKEOFF DISTANCE TO CLEAR 50-FOOT OBSTACLE

DRY, HARD-SURFACED RUNWAY
FLAPS — 60%

MODEL: T-38A
DATE: 1 AUGUST 1965
DATA BASIS: **FLIGHT TEST**

ENGINE: (2) J85-GE-5
FUEL GRADE: JP-4
FUEL DENSITY: 6.5 LB/US GAL

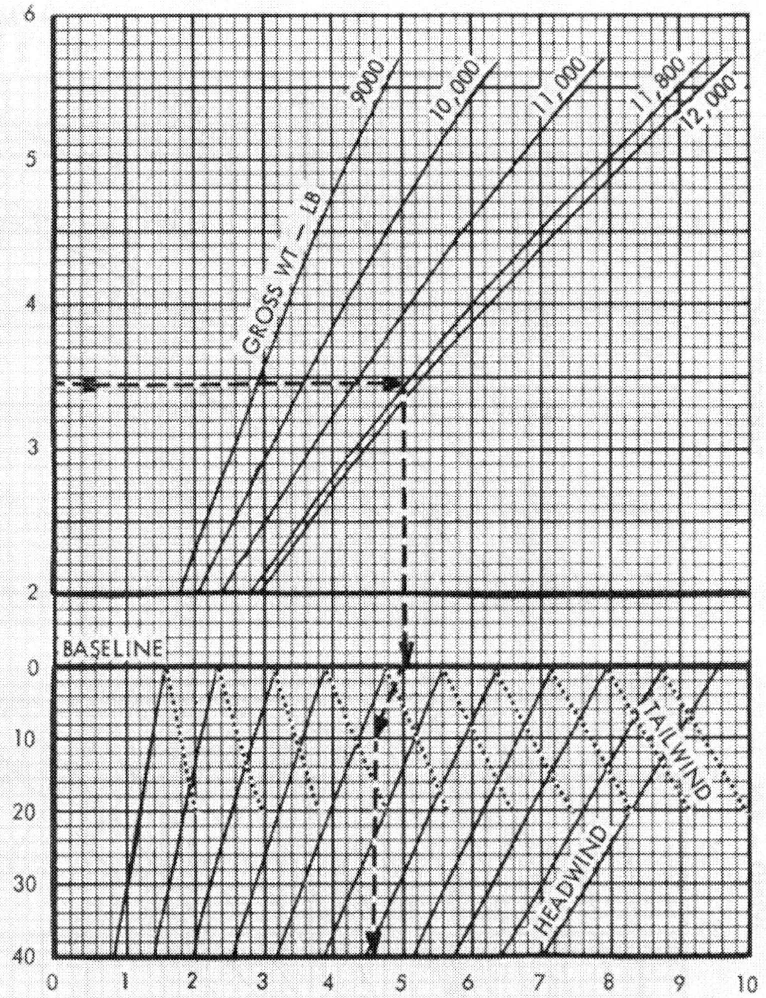


*Note*

INCREASE TOTAL DISTANCE 5 PERCENT FOR EACH PERCENT OF UPHILL RUNWAY SLOPE.

T-38A 1-404 D

FA2-5

# CRITICAL FIELD LENGTH

MAX THRUST
FLAPS — 60%

MODEL: T-38A
DATE: 1 JULY 1978
DATA BASIS: ESTIMATED DATA

ENGINES: (2) J85-GE-5
FUEL GRADE: JP-4
FUEL DENSITY: 6.5 LB/US GAL

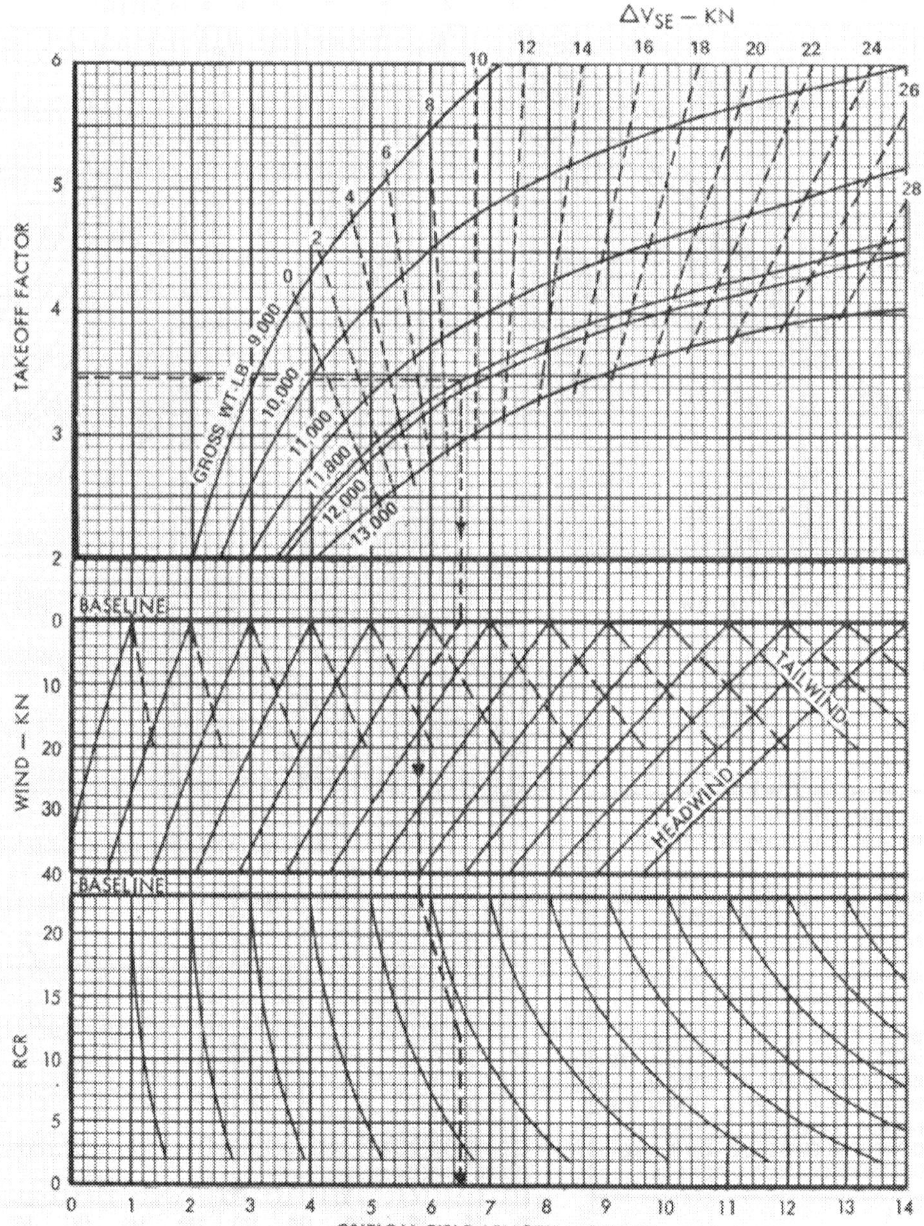


*Note*

- FOR SINGLE-ENGINE TAKEOFF SPEED, ADD $\Delta V_{SE}$ TO TAKEOFF SPEED.
- INCREASE CRITICAL FIELD LENGTH 5 PERCENT FOR EACH PERCENT OF UPHILL RUNWAY SLOPE.

T-38A 1-405D

FA2-6.

# REFUSAL SPEED OR CRITICAL ENGINE FAILURE SPEED

MAX THRUST
FLAPS — 60%

MODEL: T-38A
DATE: 1 JULY 1978
DATA BASIS: ESTIMATED DATA

ENGINES: (2) J85-GE-5
FUEL GRADE: JP-4
FUEL DENSITY: 6.5 LB/US GAL

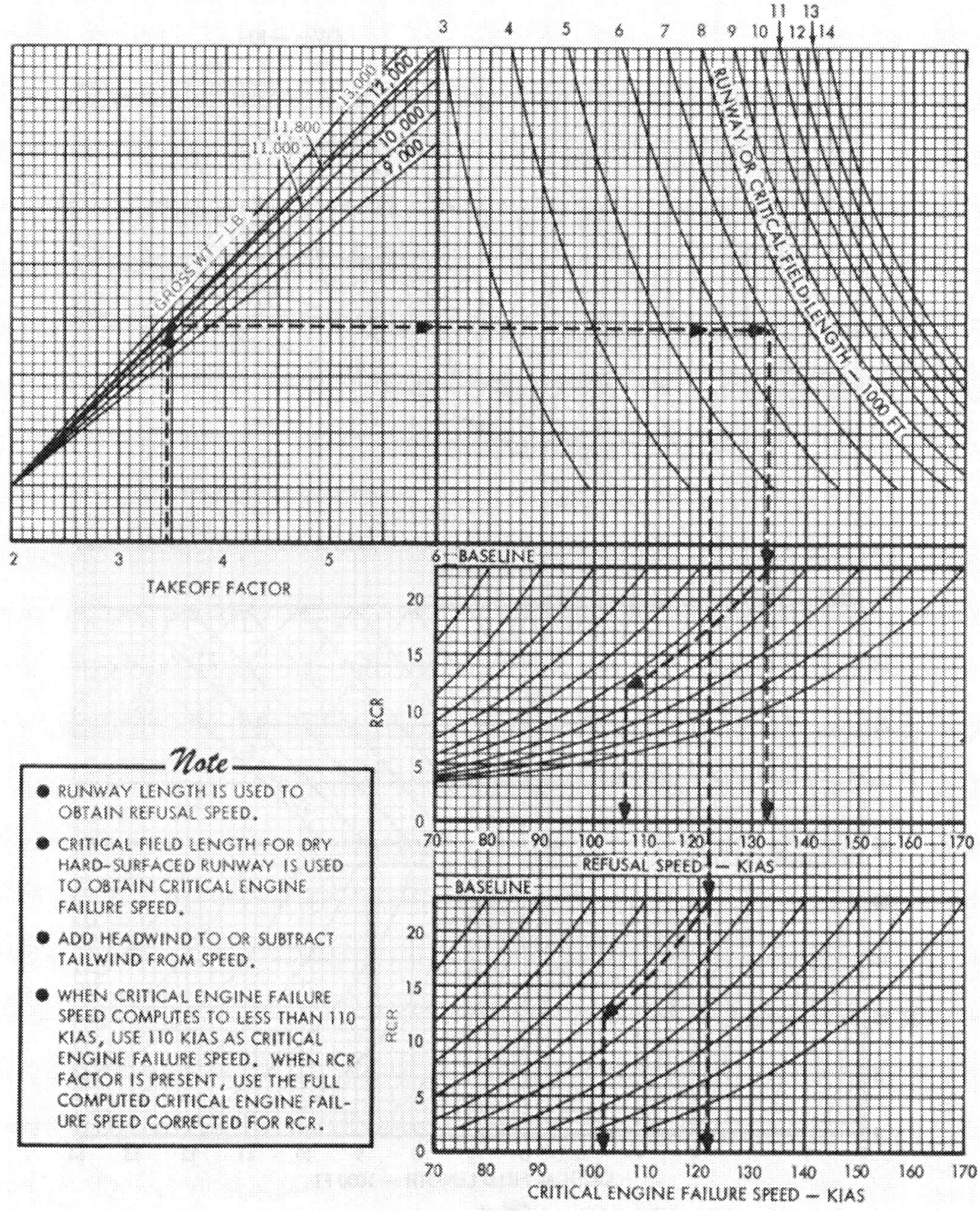


*Note*

- RUNWAY LENGTH IS USED TO OBTAIN REFUSAL SPEED.
- CRITICAL FIELD LENGTH FOR DRY HARD-SURFACED RUNWAY IS USED TO OBTAIN CRITICAL ENGINE FAILURE SPEED.
- ADD HEADWIND TO OR SUBTRACT TAILWIND FROM SPEED.
- WHEN CRITICAL ENGINE FAILURE SPEED COMPUTES TO LESS THAN 110 KIAS, USE 110 KIAS AS CRITICAL ENGINE FAILURE SPEED. WHEN RCR FACTOR IS PRESENT, USE THE FULL COMPUTED CRITICAL ENGINE FAILURE SPEED CORRECTED FOR RCR.

T-38A 1-406E

FA2-7

## REFUSAL SPEED

MIL THRUST
FLAPS — 60%

MODEL: T-38A
DATE: 1 JULY 1978
DATA BASIS: **ESTIMATED DATA**

ENGINES: (2) J85-GE-5
FUEL GRADE: JP-4
FUEL DENSITY: 6.5 LB/US GAL

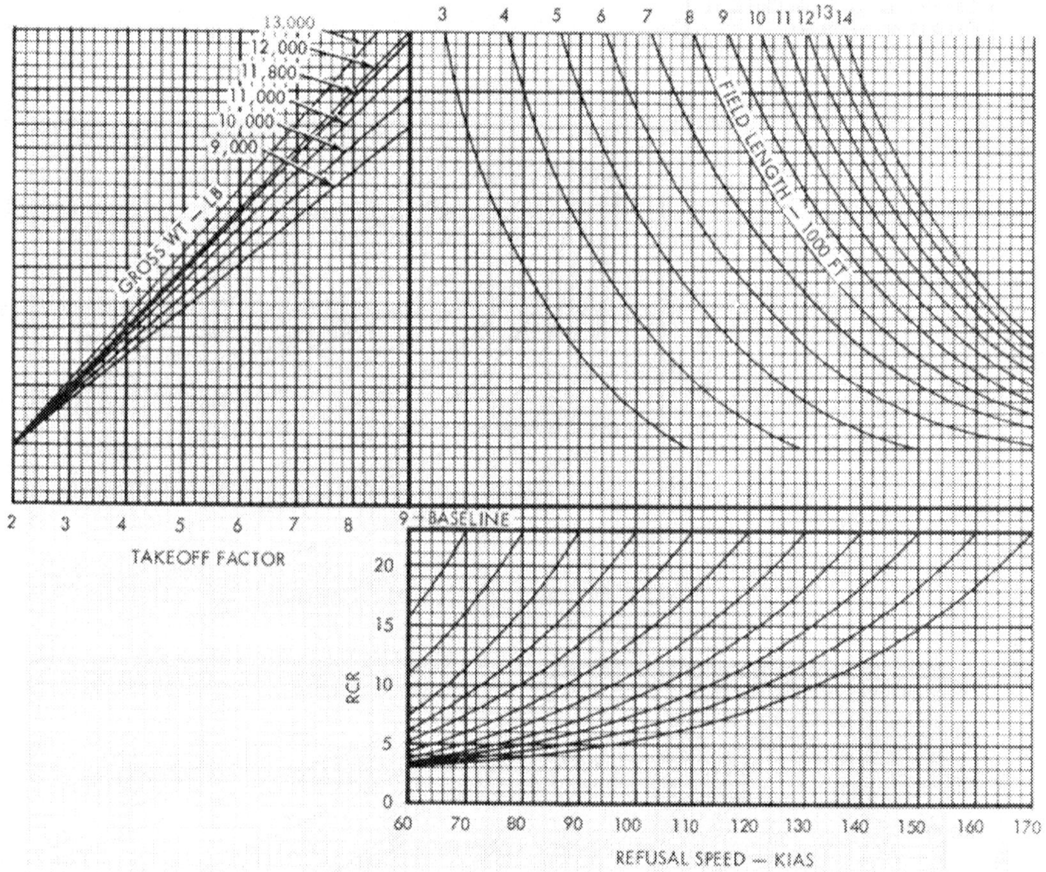


**Note**

ADD HEADWIND TO OR SUBTRACT TAILWIND FROM SPEED.

T-38A 1-407C

FA2-8

# DECISION SPEED

MAX THRUST
DRY, HARD-SURFACED RUNWAY
FLAPS – 60%

MODEL: T-38A
DATE: 1 JULY 1978
DATA BASIS: ESTIMATED DATA

ENGINE: (2) J85-GE-5
FUEL GRADE: JP-4
FUEL DENSITY: 6.5 LB/US GAL

**Note**

OBTAIN $\Delta V_{SE}$ FROM CRITICAL FIELD LENGTH CHART.

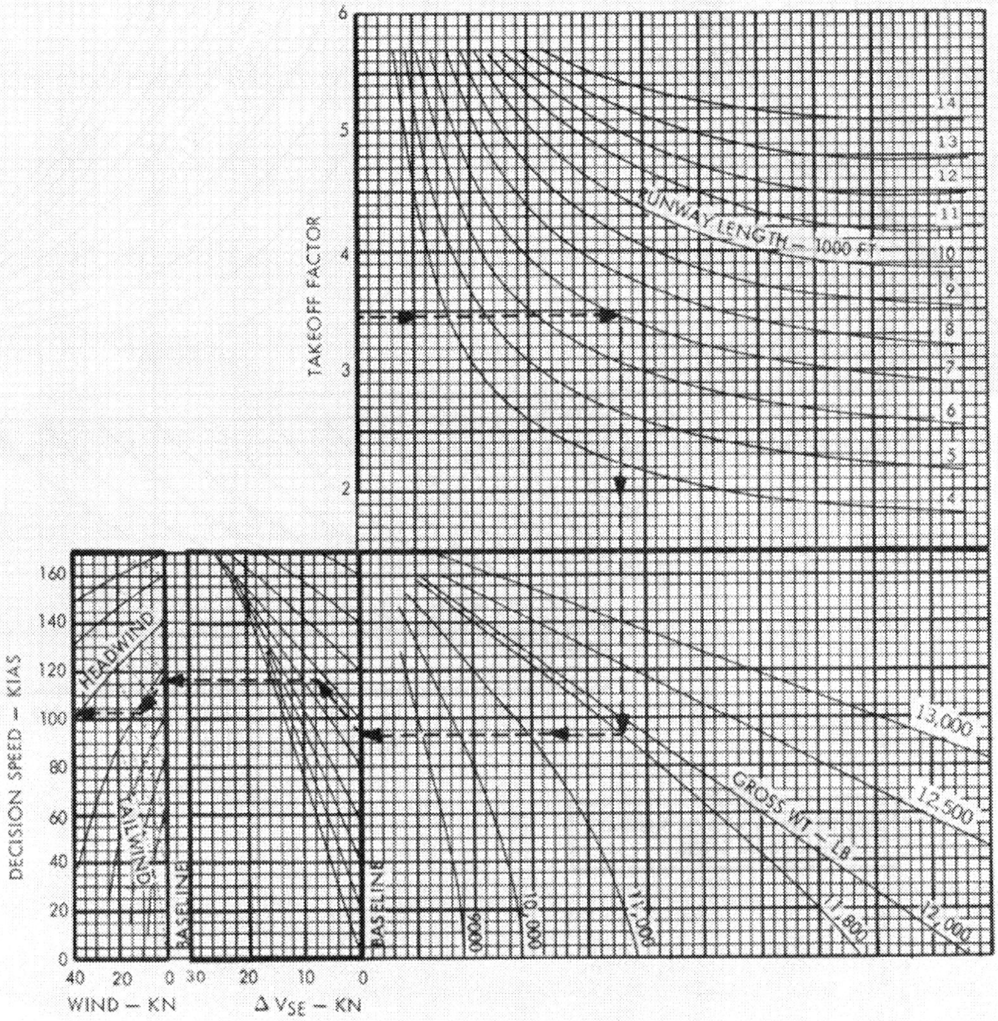


T-38A 1-409 A

FA2-9

# VELOCITY DURING TAKEOFF GROUND RUN

DRY, HARD-SURFACED RUNWAY
FLAPS — 60%

MODEL: T-38A
DATE: 1 JULY 1978
DATA BASIS: ESTIMATED DATA

ENGINE: (2) J85-GE-5
FUEL GRADE: JP-4
FUEL DENSITY: 6.5 LB/US GAL

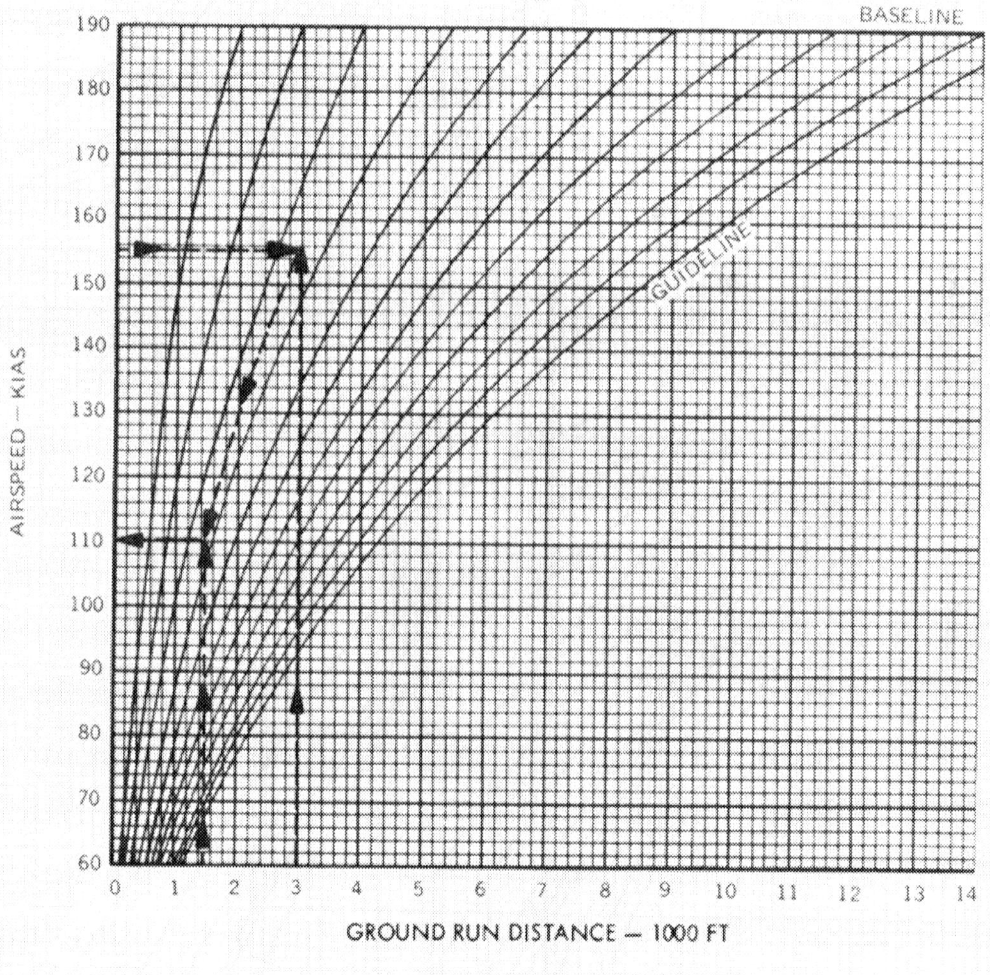


**Note**

SUBTRACT 3 KNOTS FOR EACH 1000 FEET OF RUNWAY IN EXCESS OF THE CRITICAL FIELD LENGTH, NOT TO EXCEED 10 KNOTS.

T-38A 1-410 F

FA2-10

# VELOCITY DURING TAKEOFF GROUND RUN

MAXIMUM THRUST
FLAPS – 60%

SINGLE ENGINE

MODEL: T-38A
DATE: 1 JULY 1978
DATA BASIS: ESTIMATED DATA

ENGINE: (2) J85-GE-5
FUEL GRADE: JP-4
FUEL DENSITY: 6.5 LB/US GAL

**Note**

TO OBTAIN AIRSPEED WITH WIND, ADD HEADWIND VELOCITY OR SUBTRACT TAILWIND VELOCITY.

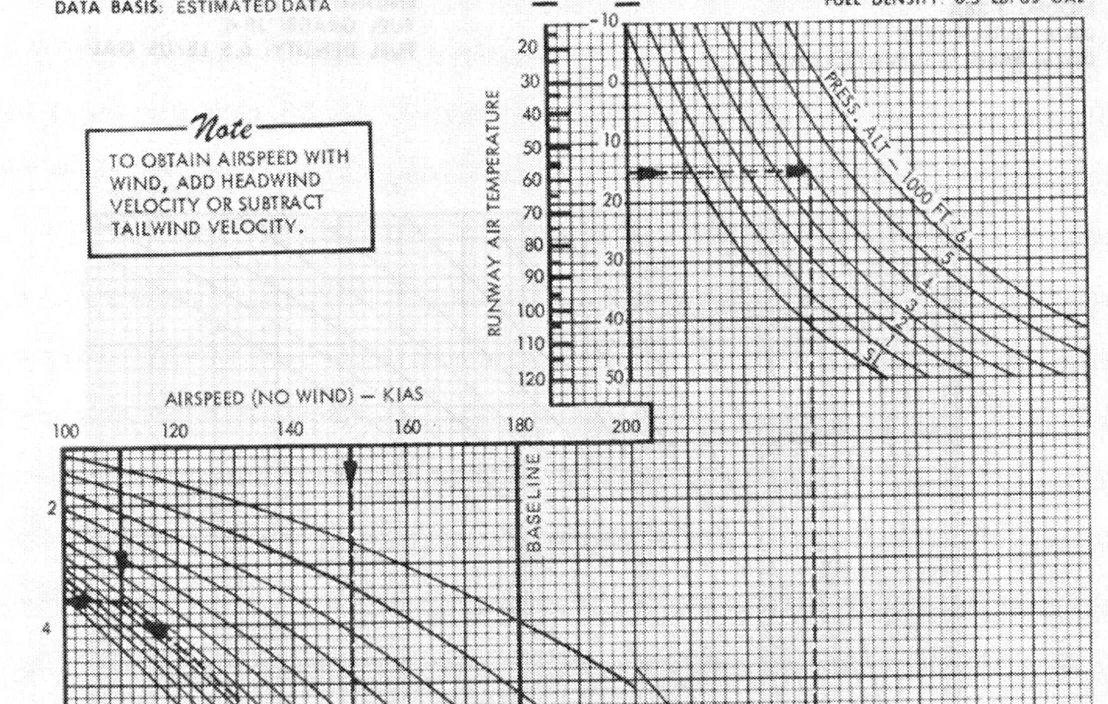


T-38A 1-411B

FA2-11

TABLE OF CONTENTS



## PURPOSE OF CHARTS.

The charts provide a means of determining the aircraft climb performance. Included are ceilings to which the aircraft may climb in the performance of missions.

## CLIMB CHARTS.

The climb charts (FA3-1 thru FA3-5) show the climb performance for MIL thrust for both two engines and single engine and MAX thrust for two engines. Two-engine MIL and MAX thrust climb charts are included for both restricted and unrestricted climb schedules. The restricted climb charts (FA3-1 and FA3-3) show performance data which reflects a MIL thrust climb at 300 KCAS to 10,000 feet followed by a level acceleration to unrestricted climb speed and continuation of climb. The restricted climb charts should be used for all climbs not performed in a military climb corridor. The unrestricted climb charts MIL and MAX THRUST CLIMB (FA3-2 and FA3-4) are used when a military climb corridor is available. All of the charts show climb performance in terms of gross weight versus fuel used, time, and distance. Climb speed schedules and allowances prior to climb are provided on each chart. The charts require successive approximations when climbing from an altitude other than sea level. The fuel, air distance, and time shown include the effects of kinetic energy change and weight reduction during climb. The fuel allowance for taxi, takeoff, and acceleration to climb speed is noted and should be subtracted from gross weight before entering the chart when climb follows a takeoff.

## USE OF CLIMB CHARTS.

The chase-thru lines on the MIL thrust restricted climb chart (FA3-1) show 565 pounds of fuel used in climb from sea level to 35,000 feet pressure altitude at an initial gross weight of 11,500 pounds and a temperature 10°C hotter than standard day. The corresponding time and air distance are 8.3 minutes and 67 nautical miles, respectively. Had the initial altitude been 15,000 feet and the gross weight 11,270 pounds, by using successive approximations, the sea level gross weight would be 11,500 pounds (same as above). From sea level to 15,000 feet, the fuel used, time, and distance are 290 pounds, 3.0 minutes, and 23 miles, respectively. Then from 15,000 feet to 35,000 feet, the fuel used is 275 pounds (565 − 290), 5.3 minutes (8.3 − 3.0), and the distance is 44 nautical miles (67 − 23). The MIL thrust climb charts (FA3-2) show that 480 pounds of fuel are required in climb from sea level to 35,000 feet, and correspondingly, it takes 7.3 minutes and 62 nautical miles. This climb is started at 11,400 pounds; however, since

95 more pounds of fuel are required for acceleration to climb speed, the MIL thrust restricted climb and the MIL thrust climb from 15,000 feet to 35,000 feet are identical.

## OPTIMUM CRUISE-CLIMB ALTITUDE.

The optimum cruise-climb altitude chart (FA3-6) shows this altitude versus gross weight for two-engine and single-engine operation. Normal thrust cruise ceilings are included and show the limitations of the optimum cruise-climb altitude.

### USE OF OPTIMUM CRUISE-CLIMB ALTITUDE CHART.

Assume two-engine operation and a gross weight of 10,500 pounds at end of climb. The chase-thru lines show optimum cruise-climb altitude from FA3-6 is 41,200. The optimum cruise-climb altitude will increase as the fuel is used in cruise. This altitude is not limited by the normal thrust cruise ceiling.

## SINGLE-ENGINE SERVICE CEILING.

The single-engine service ceiling chart (FA3-7) shows the service ceiling that can be attained by flying with MAX or MIL thrust at the climb schedules shown.

### USE OF SINGLE-ENGINE SERVICE CEILING CHART.

The chase-thru lines in FA3-7 show a single-engine service ceiling of 24,500 feet for MIL thrust and a gross weight of 10,500 pounds.

# MIL THRUST CLIMB

MODEL: T-38A
DATE: **1 OCTOBER 1976**
DATA BASIS: **ESTIMATED DATA**

RESTRICTED CLIMB SCHEDULE

FUEL USED

ENGINES: (2) J85-GE-5
FUEL GRADE: JP-4
FUEL DENSITY: 6.5 LB/US GAL

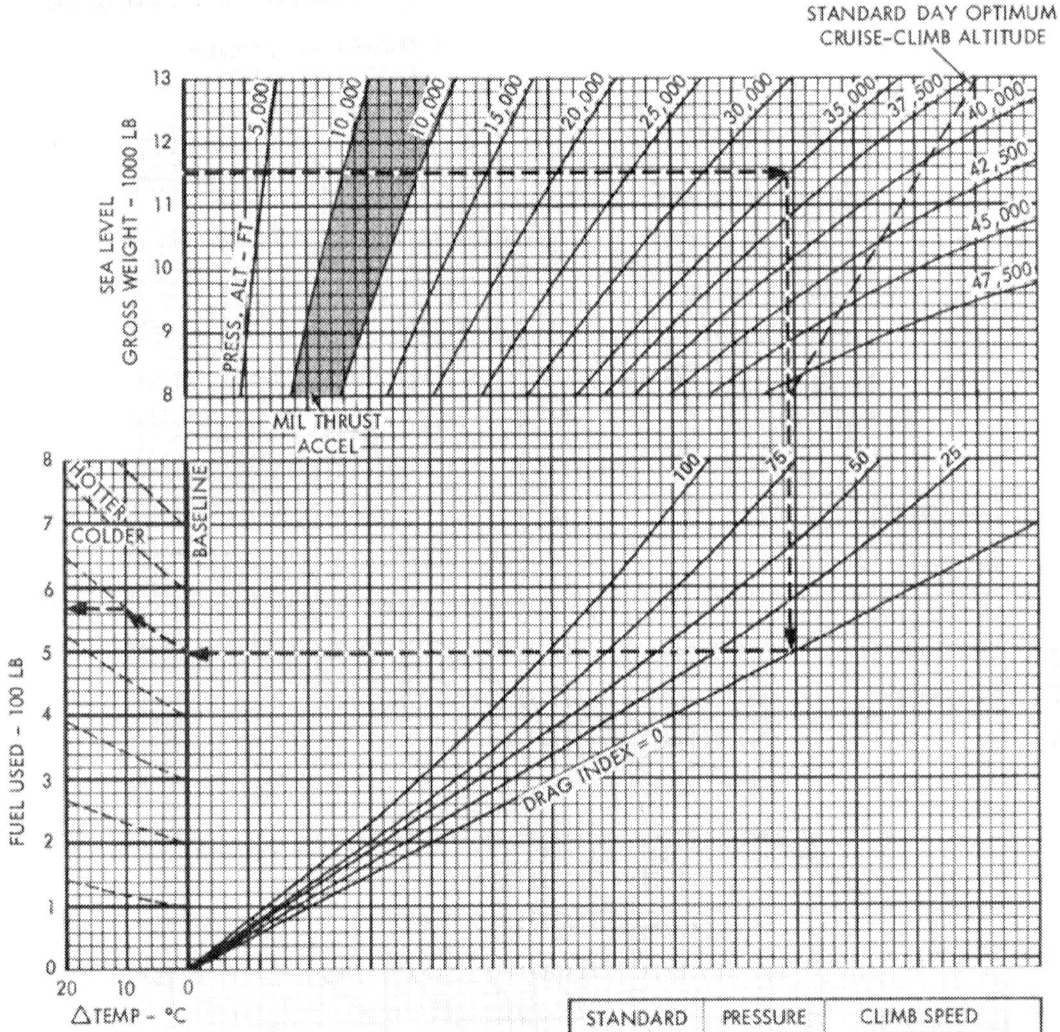


| FUEL, TIME, AND DISTANCE ALLOWANCE PRIOR TO CLIMB | |
|---|---|
| GROUND TAXI: 18 LB/MIN | |
| FROM BRAKE RELEASE TO BEGIN CLIMB USING MAXIMUM THRUST | |
| FUEL (LB) | 190 |
| TIME (MIN) | 0.6 |
| DISTANCE (NMI) | 2.0 |

| STANDARD DAY TEMP - °C | PRESSURE ALTITUDE (FEET) | CLIMB SPEED SCHEDULE | |
|---|---|---|---|
| | | KCAS | TMN |
| 15.0 | SL | 300 | 0.45 |
| 5.1 | 5,000 | 300 | 0.50 |
| -4.8 | 10,000 | 300 | 0.55 |
| -4.8 | 10,000 | 435 | 0.78 |
| -14.7 | 15,000 | 406 | 0.79 |
| -24.6 | 20,000 | 377 | 0.81 |
| -34.5 | 25,000 | 349 | 0.83 |
| -44.4 | 30,000 | 322 | 0.84 |
| -54.3 | 35,000 | 295 | 0.86 |
| -56.5 AT | 40,000 | 264 | 0.87 |
| 37,000 FT | 45,000 | 236 | 0.87 |
| & ABOVE | 50,000 | 210 | 0.87 |

T-38A 1-506(1)

*FA3-1. (Sheet 1)*

## MIL THRUST CLIMB

RESTRICTED CLIMB SCHEDULE

MODEL: T-38A
DATE: 1 OCTOBER 1976
DATA BASIS: ESTIMATED DATA

TIME TO CLIMB AND DISTANCE TRAVELED

ENGINES: (2) J85-GE-5
FUEL GRADE: JP-4
FUEL DENSITY: 6.5 LB/US GAL

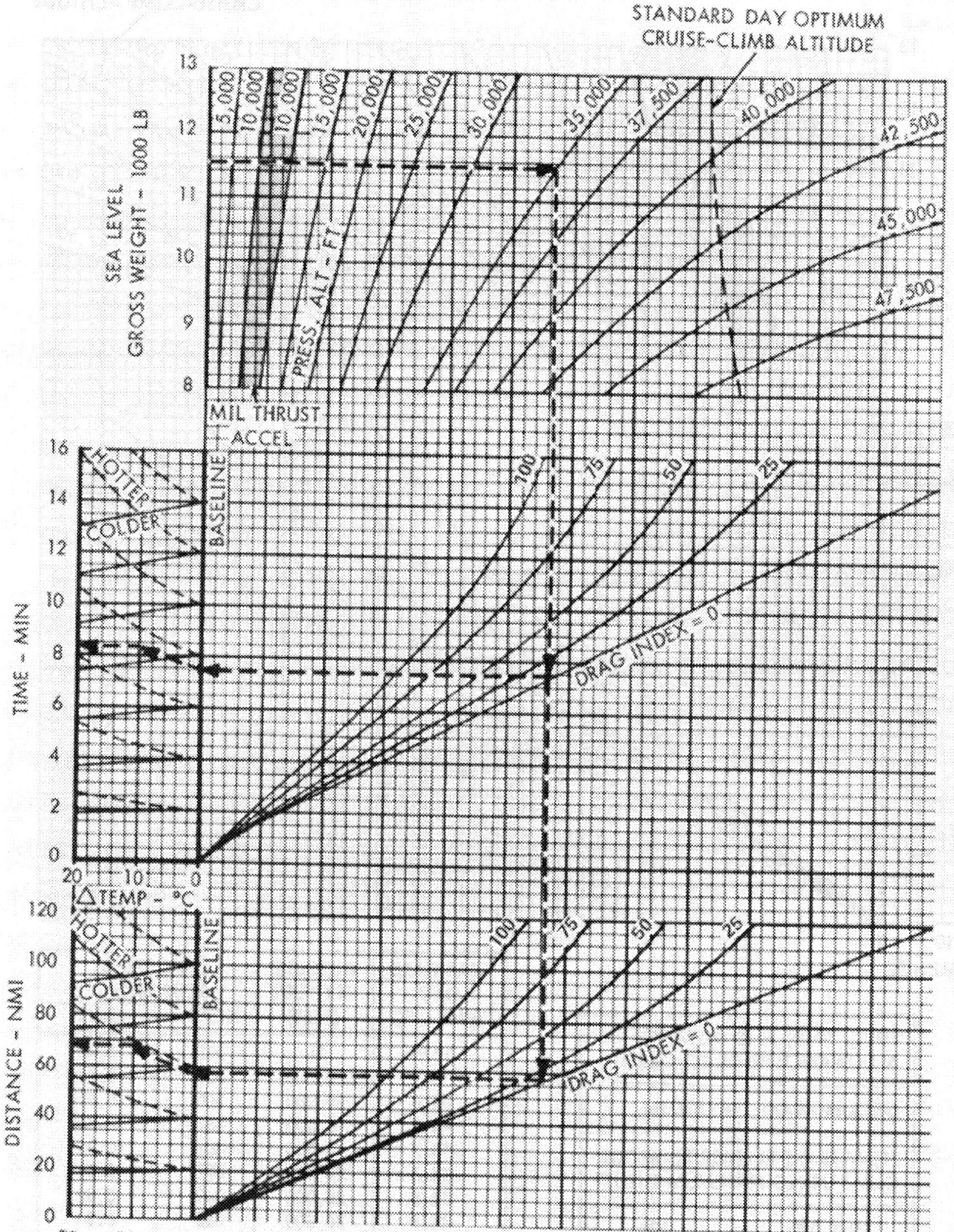


FA3-1. (Sheet 2)

T-38A 1-506(2)

# MIL THRUST CLIMB
## FUEL USED

**MODEL:** T-38A
**DATE:** 1 OCTOBER 1976
**DATA BASIS:** ESTIMATED DATA

**ENGINE:** (2) J85-GE-5
**FUEL GRADE:** JP-4
**FUEL DENSITY:** 6.5 LB/US GAL

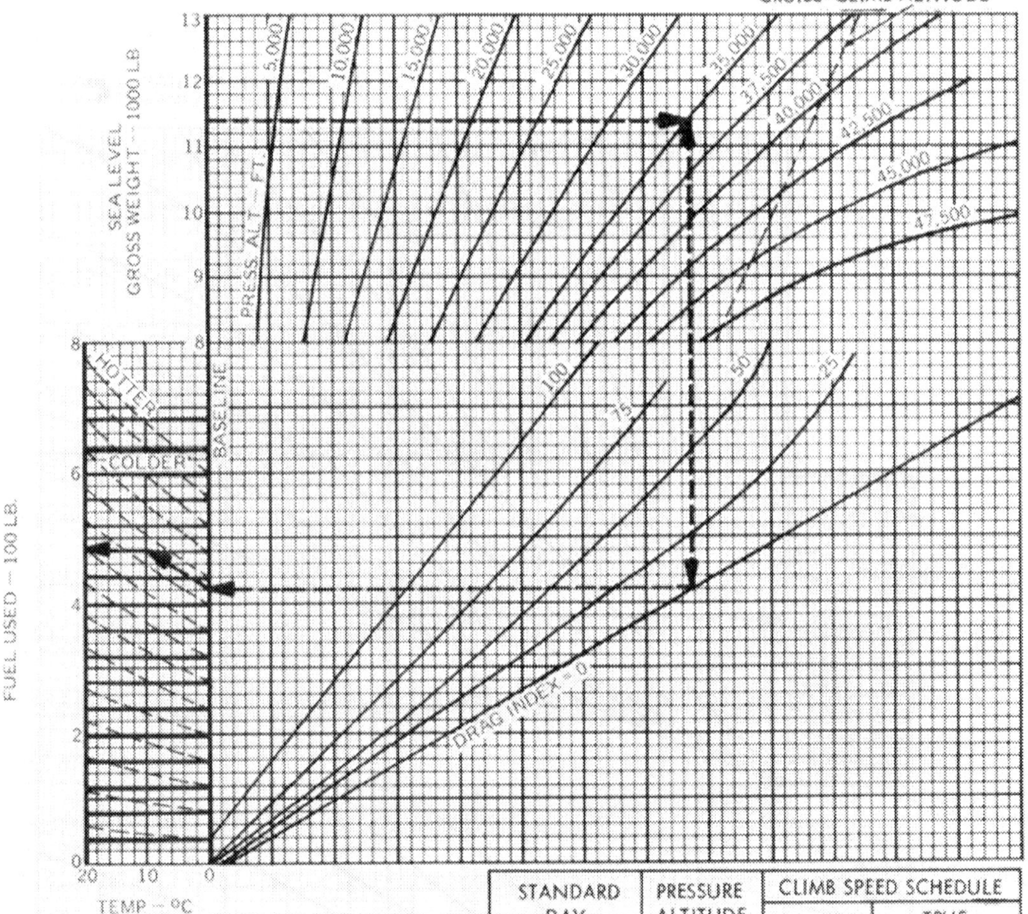


| STANDARD DAY TEMP – °C | PRESSURE ALTITUDE – FEET | CLIMB SPEED SCHEDULE KCAS | CLIMB SPEED SCHEDULE TRUE MACH NO. |
|---|---|---|---|
| 15.0 | SEA LEVEL | 496 | 0.75 |
| 5.1 | 5,000 | 466 | 0.76 |
| -4.8 | 10,000 | 435 | 0.78 |
| -14.7 | 15,000 | 406 | 0.79 |
| -24.6 | 20,000 | 377 | 0.81 |
| -34.5 | 25,000 | 349 | 0.83 |
| -44.4 | 30,000 | 322 | 0.84 |
| -54.3 | 35,000 | 295 | 0.86 |
| -56.5 AT | 40,000 | 264 | 0.87 |
| 37,000 FT | 45,000 | 236 | 0.87 |
| AND ABOVE | 50,000 | 210 | 0.87 |

| FUEL, TIME, AND DISTANCE ALLOWANCE PRIOR TO CLIMB | |
|---|---|
| GROUND TAXI: 18 LB/MIN | |
| FROM BRAKE RELEASE TO BEGIN CLIMB USING MAXIMUM THRUST | |
| FUEL (LB) | 285 |
| TIME (MIN) | 1.0 |
| DISTANCE (NMI) | 4.0 |

T-38A 1-501(1)C

**FA3-2. (Sheet 1)**

## MIL THRUST CLIMB
TIME TO CLIMB AND DISTANCE TRAVELED

MODEL: T-38A
DATE: 1 OCTOBER 1976
DATA BASIS: ESTIMATED DATA

ENGINE: (2) J85-GE-5
FUEL GRADE: JP-4
FUEL DENSITY: 6.5 LB/US GAL

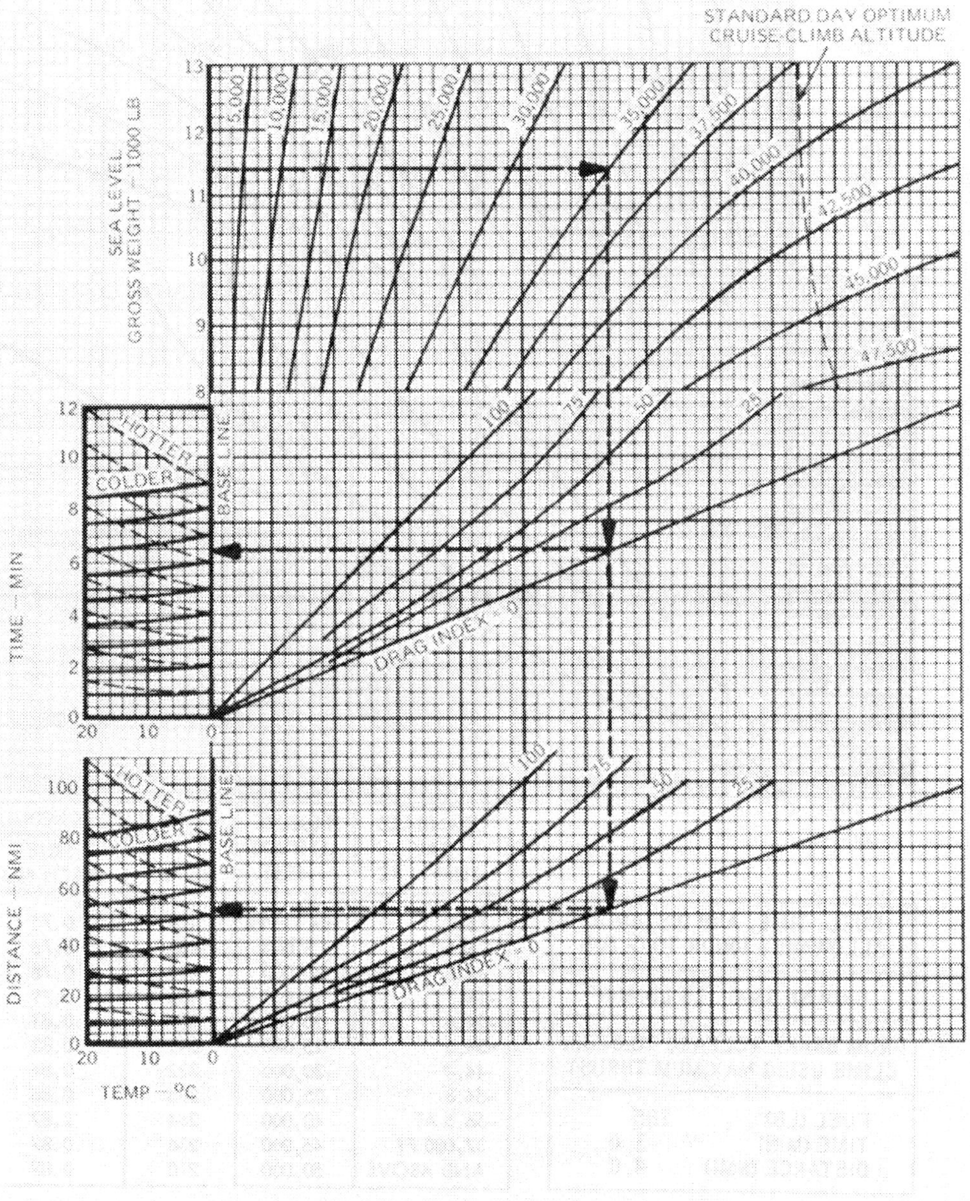


FA3-2. (Sheet 2)

# MAX THRUST CLIMB

RESTRICTED CLIMB SCHEDULE

MODEL: T-38A
DATE: 1 OCTOBER 1976
DATA BASIS: ESTIMATED DATA

FUEL USED

ENGINES: (2) J85-GE-5
FUEL GRADE: JP-4
FUEL DENSITY: 6.5 LB/US GAL

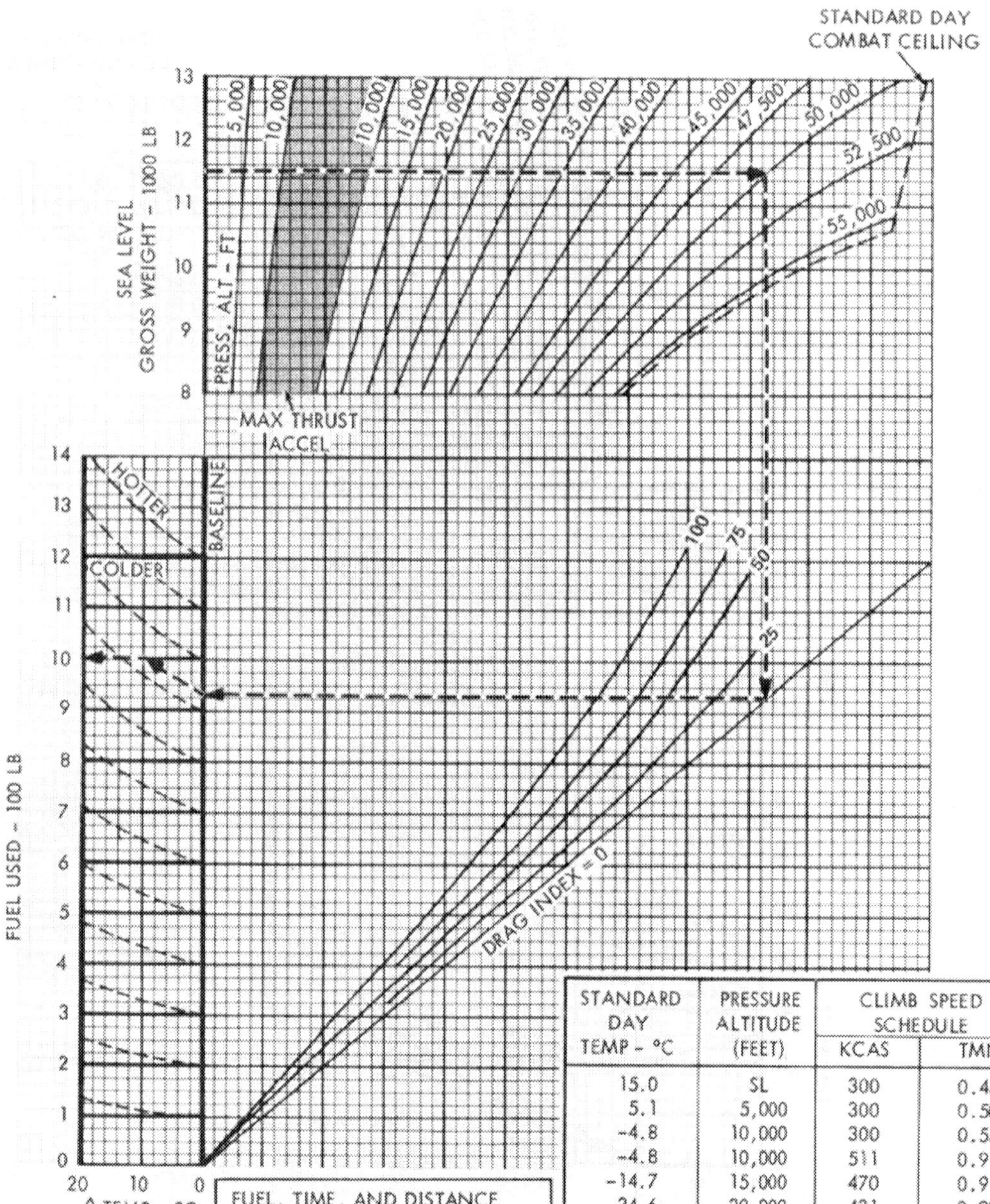


| FUEL, TIME, AND DISTANCE ALLOWANCE PRIOR TO CLIMB | |
|---|---|
| GROUND TAXI: 18 LB/MIN | |
| FROM BRAKE RELEASE TO BEGIN CLIMB USING MAXIMUM THRUST | |
| FUEL (LB) | 190 |
| TIME (MIN) | 0.6 |
| DISTANCE (NMI) | 2.0 |

| STANDARD DAY TEMP – °C | PRESSURE ALTITUDE (FEET) | CLIMB SPEED SCHEDULE | |
|---|---|---|---|
| | | KCAS | TMN |
| 15.0 | SL | 300 | 0.45 |
| 5.1 | 5,000 | 300 | 0.50 |
| -4.8 | 10,000 | 300 | 0.55 |
| -4.8 | 10,000 | 511 | 0.91 |
| -14.7 | 15,000 | 470 | 0.91 |
| -24.6 | 20,000 | 431 | 0.92 |
| -34.5 | 25,000 | 393 | 0.92 |
| -44.4 | 30,000 | 357 | 0.93 |
| -54.3 | 35,000 | 322 | 0.93 |
| -56.5 AT 37,000 FT & ABOVE | 40,000 | 287 | 0.93 |
| | 45,000 | 256 | 0.93 |
| | 50,000 | 228 | 0.93 |
| | 55,000 | 203 | 0.93 |

T-38A 1-507(1)

*FA3-3. (Sheet 1)*

# MAX THRUST CLIMB

RESTRICTED CLIMB SCHEDULE

TIME TO CLIMB AND DISTANCE TRAVELED

MODEL: T-38A
DATE: 1 OCTOBER 1976
DATA BASIS: ESTIMATED DATA

ENGINES: (2) J85-GE-5
FUEL GRADE: JP-4
FUEL DENSITY: 6.5 LB/US GAL

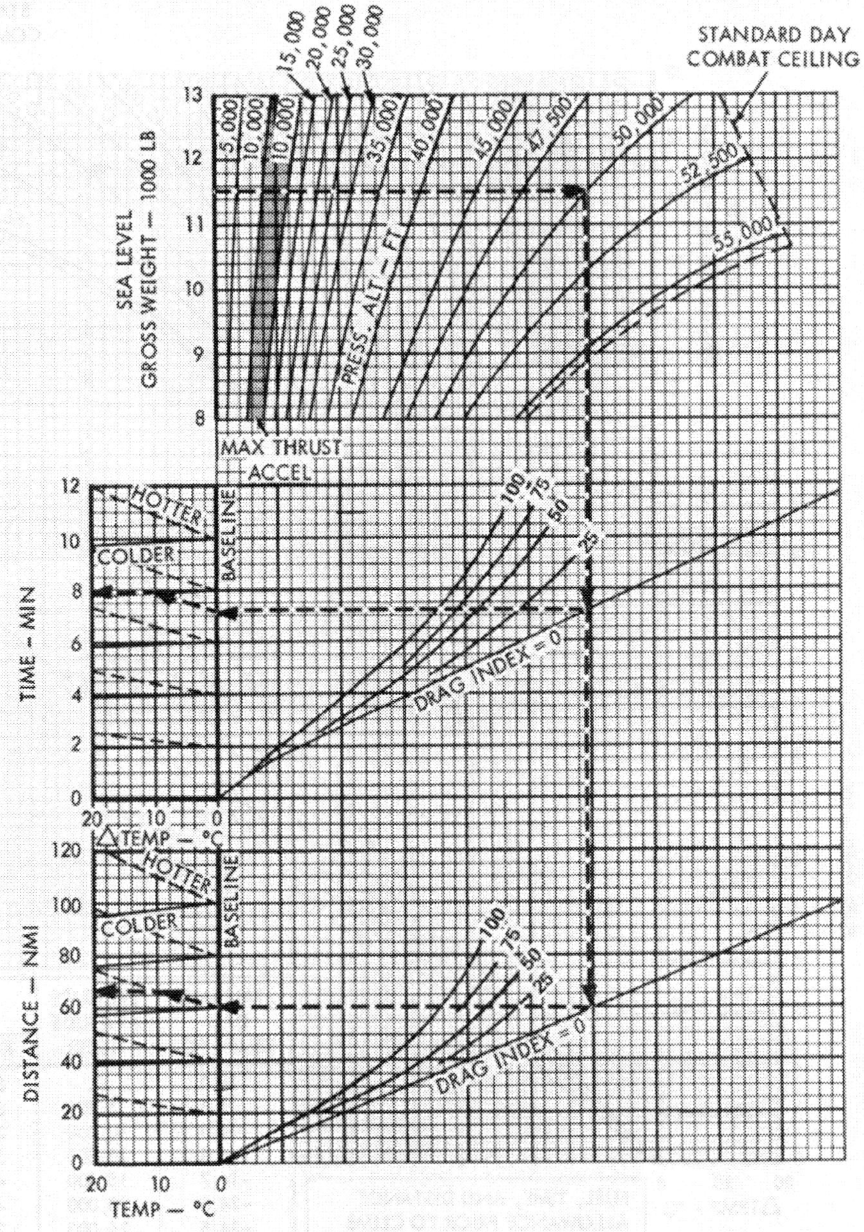


*FA3-3. (Sheet 2)*

T-38A 1-507(2)

# MAX THRUST CLIMB
## FUEL USED

MODEL: T-38A
DATE: 1 OCTOBER 1976
DATA BASIS: ESTIMATED DATA

ENGINE: (2) J85-GE-5
FUEL GRADE: JP-4
FUEL DENSITY: 6.5 LB/US GAL

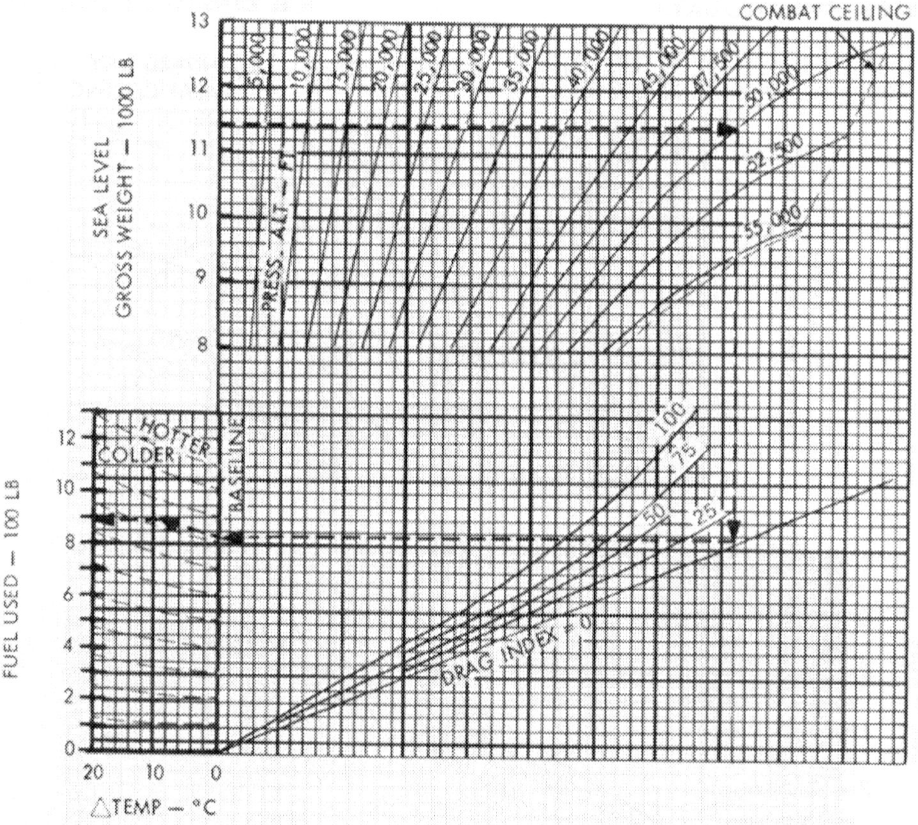


| FUEL, TIME, AND DISTANCE ALLOWANCE PRIOR TO CLIMB | |
|---|---|
| GROUND TAXI: 18 LB/MIN | |
| FROM BRAKE RELEASE TO BEGIN CLIMB USING MAXIMUM THRUST | |
| FUEL (LB) | 315 |
| TIME (MIN) | 1.2 |
| DISTANCE (NMI) | 6.0 |

| STANDARD DAY TEMP — °C | PRESSURE ALTITUDE — FEET | CLIMB SPEED SCHEDULE | |
|---|---|---|---|
| | | KCAS | TRUE MACH NO. |
| 15.0 | SEA LEVEL | 596 | 0.90 |
| 5.1 | 5,000 | 553 | 0.90 |
| -4.8 | 10,000 | 511 | 0.91 |
| -14.7 | 15,000 | 470 | 0.91 |
| -24.6 | 20,000 | 431 | 0.92 |
| -34.5 | 25,000 | 393 | 0.92 |
| -44.4 | 30,000 | 357 | 0.93 |
| -54.3 | 35,000 | 322 | 0.93 |
| -56.5 AT 37,000 FT AND ABOVE | 40,000 | 287 | 0.93 |
| | 45,000 | 256 | 0.93 |
| | 50,000 | 228 | 0.93 |
| | 55,000 | 203 | 0.93 |

T-38A 1-500(1)C

FA3-4. (Sheet 1)

# MAX THRUST CLIMB
## TIME TO CLIMB AND DISTANCE TRAVELED

MODEL: T-38A
DATE: 1 OCTOBER 1976
DATA BASIS: ESTIMATED DATA

ENGINE: (2) J85-GE-5
FUEL GRADE: JP-4
FUEL DENSITY: 6.5 LB/US GAL

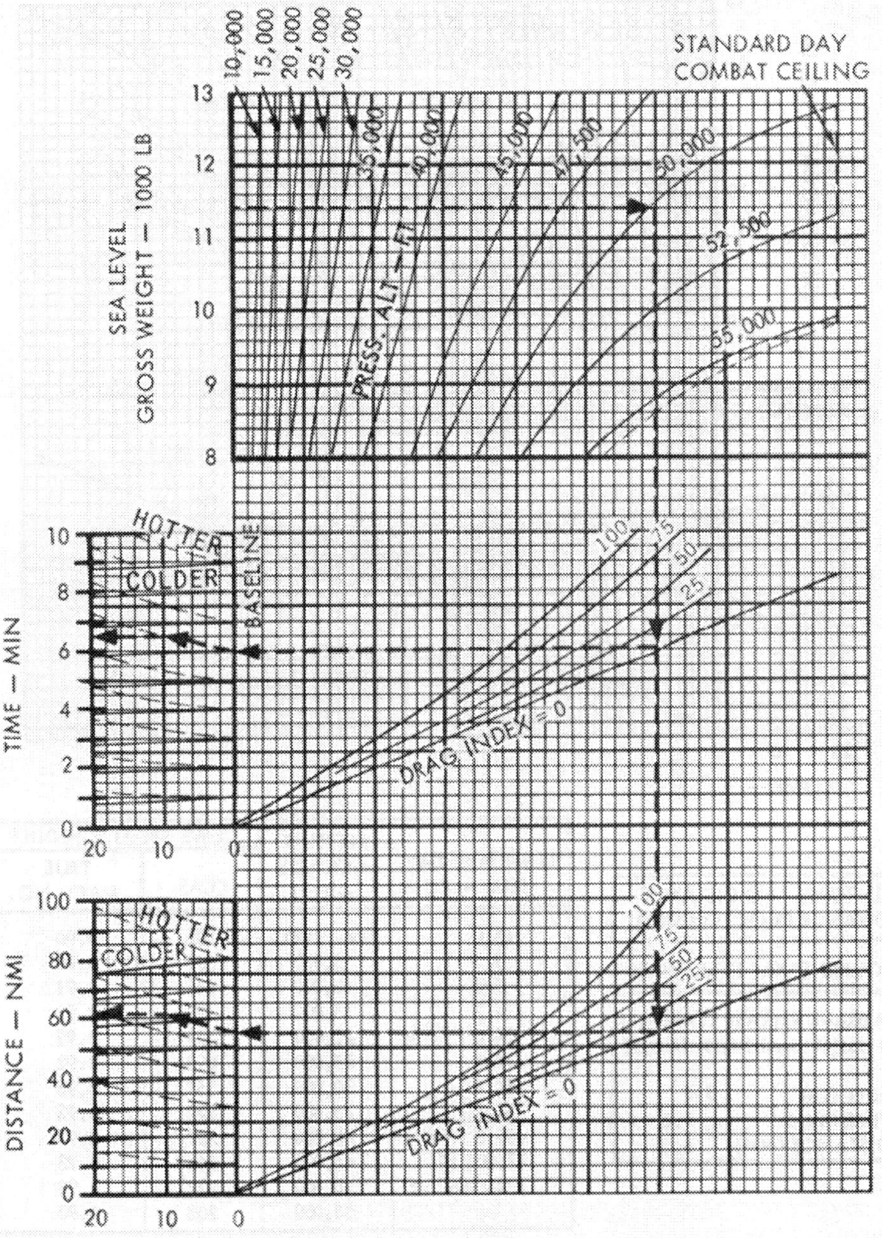


T-38A 1-500(2) C

FA3-4. (Sheet 2)

# MIL THRUST CLIMB
## FUEL USED

SINGLE ENGINE

MODEL: T-38A
DATE: 1 OCTOBER 1976
DATA BASIS: ESTIMATED DATA

ENGINE: (2) J85-GE-5
FUEL GRADE: JP-4
FUEL DENSITY: 6.5 LB/US GAL

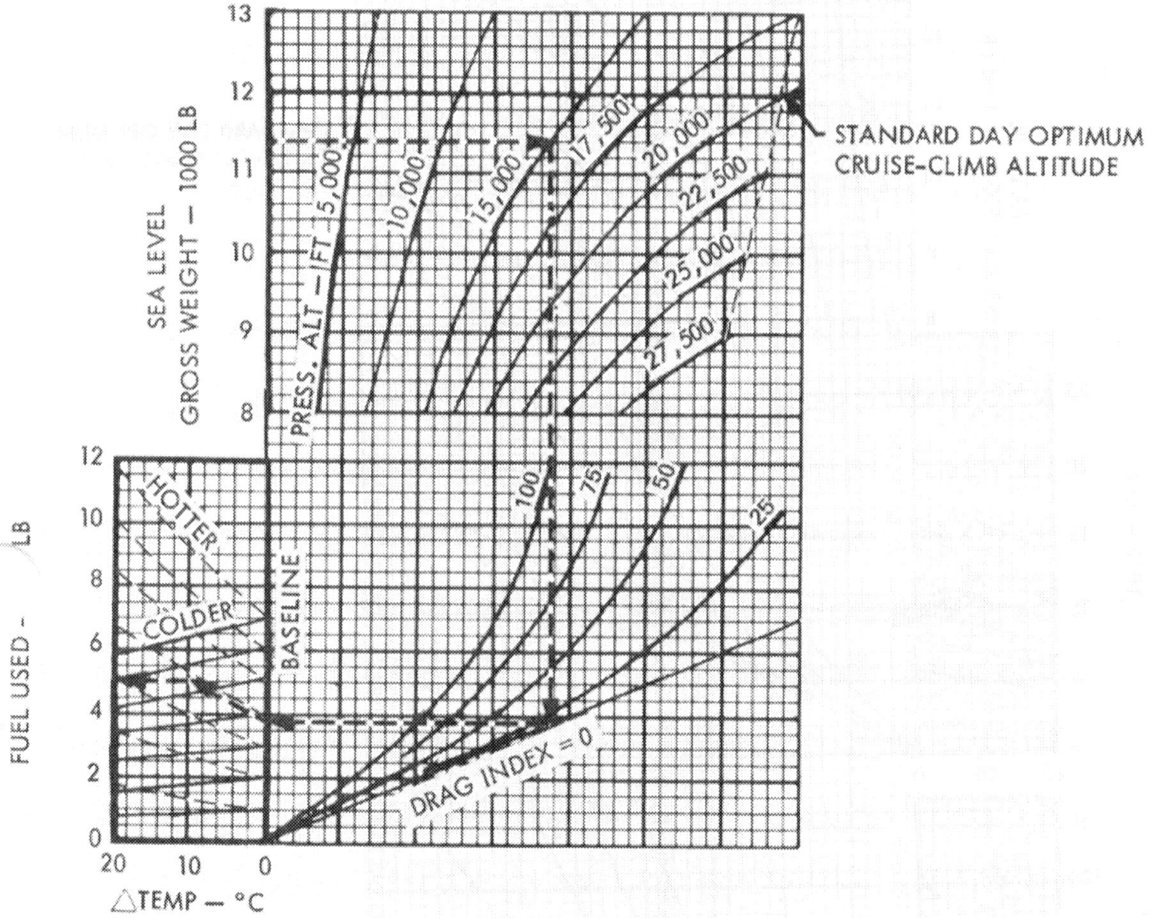


| STANDARD DAY TEMP — °C | PRESSURE ALTITUDE — FEET | CLIMB SPEED SCHEDULE | |
|---|---|---|---|
| | | KCAS | TRUE MACH NO. |
| 15.0 | SEA LEVEL | 281 | 0.43 |
| 5.1 | 5,000 | 278 | 0.46 |
| - 4.8 | 10,000 | 271 | 0.49 |
| -14.7 | 15,000 | 264 | 0.52 |
| -24.6 | 20,000 | 256 | 0.56 |
| -34.5 | 25,000 | 246 | 0.59 |
| -44.4 | 30,000 | 227 | 0.61 |

*FA3-5. (Sheet 1)*

T-38A 1-502(1) C

## MIL THRUST CLIMB

TIME TO CLIMB AND DISTANCE TRAVELED

SINGLE ENGINE

MODEL: T-38A
DATE: 1 OCTOBER 1978
DATA BASIS: ESTIMATED DATA

ENGINE: (2) J85-GE-5
FUEL GRADE: JP-4
FUEL DENSITY: 6.5 LB/US GAL

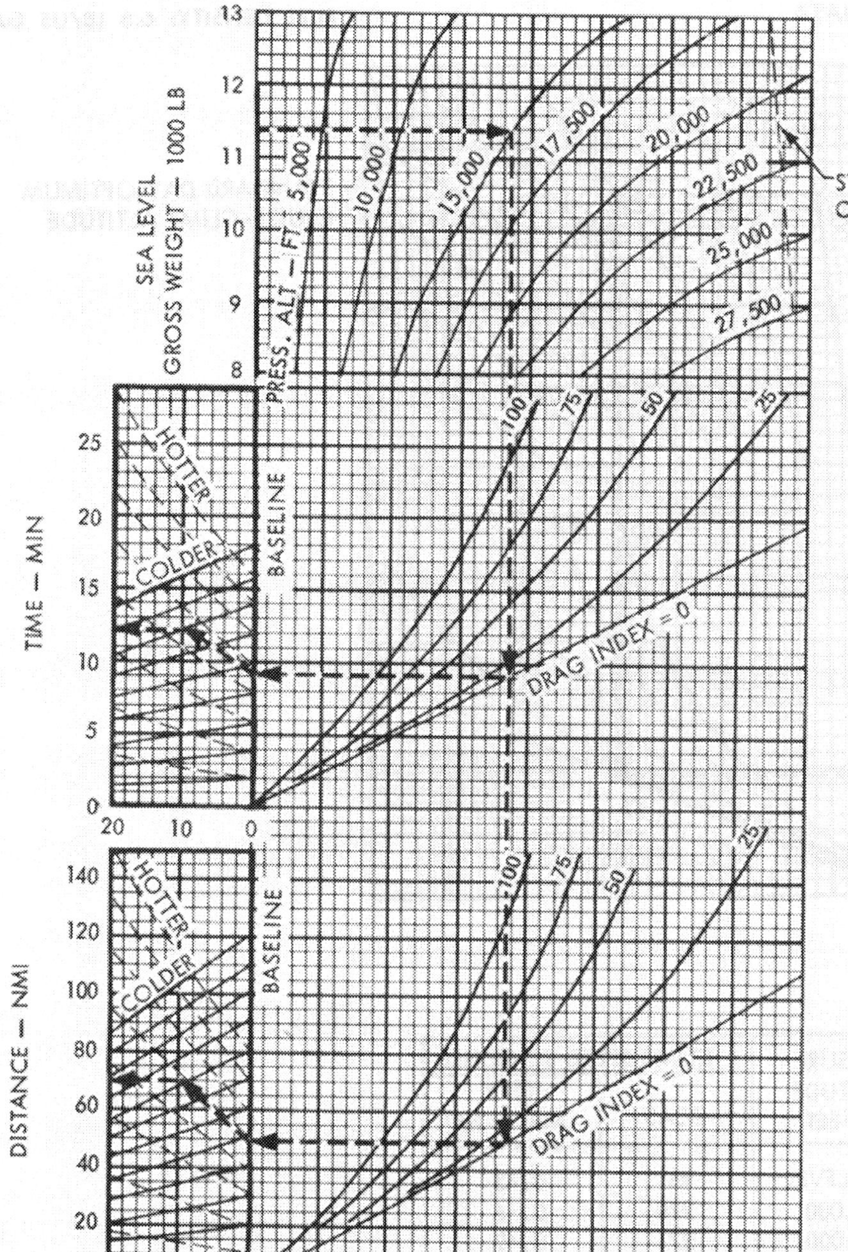


FA3-5. (Sheet 2)

# OPTIMUM CRUISE-CLIMB ALTITUDE

DRAG INDEX = 0

MODEL: T-38A
DATE: 1 AUGUST 1965
DATA BASIS: **FLIGHT TEST**

ENGINE: (2) J85-GE-5
FUEL GRADE: JP-4
FUEL DENSITY: 6.5 LB/US GAL

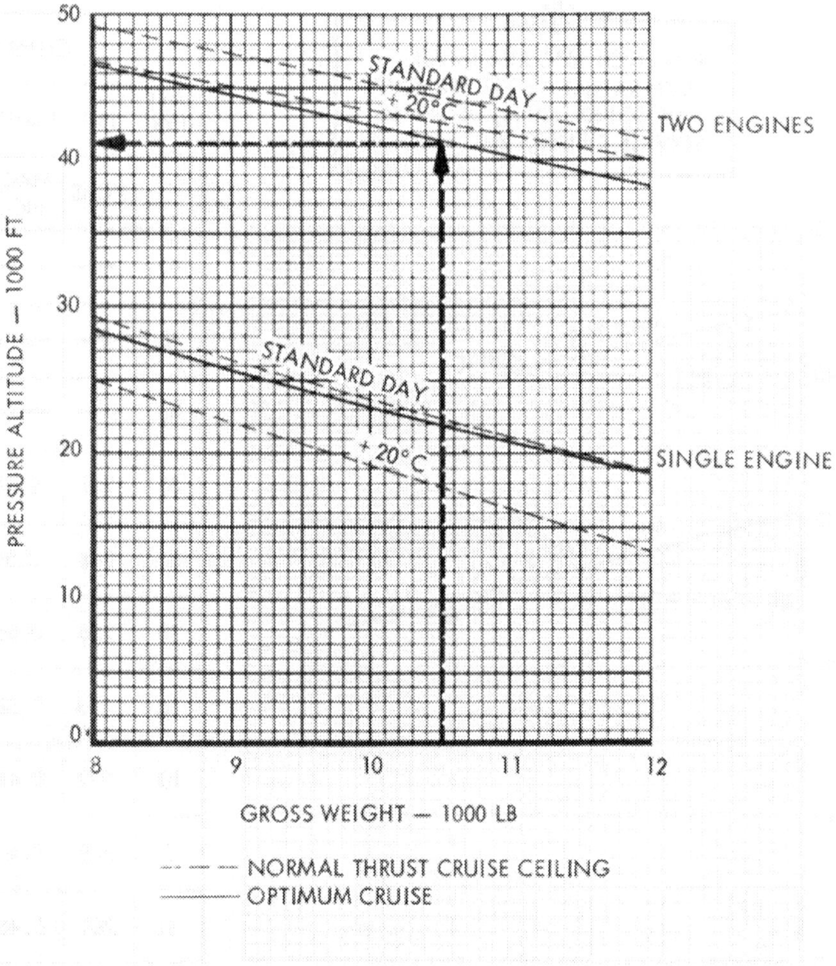


T-38A 1-503B

FA3-6.

# SINGLE ENGINE SERVICE CEILING

STANDARD DAY
DRAG INDEX = 0

MODEL: T-38A
DATE: 1 AUGUST 1965
DATA BASIS: **FLIGHT TEST**

ENGINE: (2) J85-GE-5
FUEL GRADE: JP-4
FUEL DENSITY: 6.5 LB/US GAL

***Note***
FOR EACH 10°C RISE IN AIR TEMPERATURE ABOVE STANDARD DAY CONDITIONS, DECREASE SERVICE CEILING 2600 FEET.

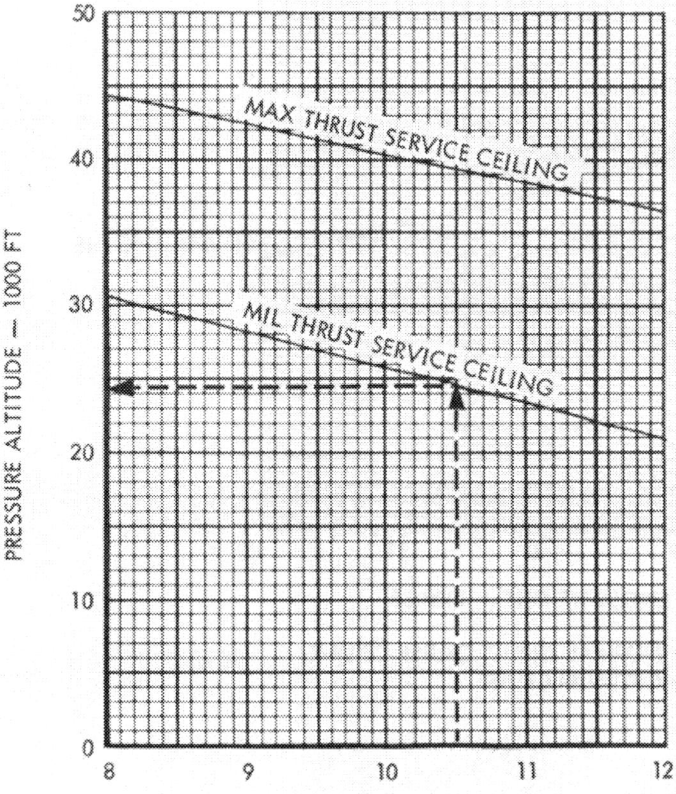


| ALT 1000 FT | CLIMB SCHEDULE | | | |
|---|---|---|---|---|
| | MIL THRUST | | MAX THRUST | |
| | KCAS | MACH NO. | KCAS | MACH NO. |
| 45 | — | — | 216 | 0.80 |
| 40 | — | — | 242 | 0.80 |
| 35 | — | — | 270 | 0.80 |
| 30 | 227 | 0.61 | 295 | 0.78 |
| 25 | 246 | 0.59 | 313 | 0.75 |
| 20 | 256 | 0.56 | 331 | 0.72 |
| 15 | 264 | 0.52 | 349 | 0.69 |
| 10 | 272 | 0.49 | 365 | 0.65 |
| 5 | 278 | 0.46 | 377 | 0.62 |
| SL | 281 | 0.43 | 394 | 0.60 |

T-38A 1-505B

FA3-7.

# PART 4 CRUISE

TABLE OF CONTENTS



## PURPOSE OF CHARTS.

The cruise charts provide cruise and loiter data which can be used to determine the subsonic cruise and loiter portions of any type of flight plan. Charts for constant altitude cruise and optimum cruise altitude for short range missions are included. Diversion range summary tables are provided in tabular form for two-engine and single-engine operation.

## CRUISE CHARTS.

The cruise charts (FA4-1 and FA4-2) are for two-engine and single-engine operation. They provide cruise and loiter data throughout the speed range from maximum endurance to military thrust. Each chart is composed of three pages whose parameters are weight, altitude, mach number, ambient temperature, true airspeed, fuel flow, drag number, and nautical miles per pound of fuel. The average gross weight used in the charts is the average of the gross weights at the beginning and the end of the cruise or cruise interval. This average gross weight is equal to the gross weight at the beginning of cruise less one half of the fuel necessary for cruise. An ICAO standard day temperature table is included on sheet 3 of each chart.

### USE OF CRUISE CHART.

Assume a constant altitude cruise at 0.8 mach number and a pressure altitude of 20,000 feet when the temperature is −20°C and the average gross weight is 10,400 pounds. The chase-thru lines on sheet 1 of FA4-1 show the maximum range mach number of 0.702. Then, by following the guidelines from the intersection with the baseline (maximum range) to the assumed mach number (0.8), the basic reference number is 2.75. The chase-thru lines on sheet 3 show 0.225 nautical mile per pound of fuel for the assumed mach number and the basic reference number determined on sheet 1 and 2. Entering sheet 4 with the assumed mach number and the nautical miles per pound from sheet 3 the chase-thru lines show a true airspeed of 495 knots and fuel flow of 1100 pounds per hour per engine. If the fuel available is 1000 pounds, the cruise distance is 225 nautical miles (0.225 X 1000) and the time is 27 minutes (1000 X 60 ÷ 1100 X 2). When the distance is known instead of the fuel available, the fuel required is computed by the reverse process (225 ÷ 0.225 = 1000) and the average gross weight is obtained by successive approximations, knowing the gross weight at the start of the cruise.

## CONSTANT ALTITUDE CRUISE.

The constant altitude cruise charts (FA4-3 and FA4-4) are for two-engine and single-engine operation. The charts are used to determine cruise performance at a particular pressure altitude, temperature, wind velocity, and average gross weight. The charts provide data for air and ground speeds, time, nautical miles per pound of fuel, fuel flow, and fuel required. When the fuel required is unknown, the average gross weight is obtained by successive approximations.

### USE OF CONSTANT ALTITUDE CRUISE CHARTS.

The chase-thru lines on FA4-3 are for an average gross weight of 10,020 pounds, a constant altitude cruise of 35,000 feet, a temperature of −46°C, a headwind of 50 knots, and a distance of 400 nautical miles. On sheet 1, the chase-thru lines show a mach number of 0.83, a true airspeed (airspeed reflector) and groundspeed of 485 knots and 435 knots, respectively, and a time of 55 minutes. Using the airspeed of 485 knots and the time of 55 minutes, the chase-thru lines on sheet 2 show 0.338 nautical mile per pound of fuel, a fuel flow of 720 pounds per hour per engine, and 1320 pounds of fuel. Since 1320 pounds of fuel is required, the gross weight at the start of the cruise is 10,680 pounds (10,020 + 1/2 X 1320).

## OPTIMUM CRUISE ALTITUDE FOR SHORT RANGE MISSIONS.

For short-range flights, it is not economical to climb to the same optimum cruise altitude as used for long range missions. FA4-5 presents the optimum constant altitude cruise for short-range missions and also indicates when the mission is in the short-range category; that is, below the optimum cruise-climb altitude.

### USE OF OPTIMUM CRUISE ALTITUDE FOR SHORT RANGE MISSIONS CHART.

For a short-range mission 100 nautical miles from base and a start climb gross weight of 11,400 pounds, the chase-thru lines show the optimum cruise at constant altitude (FA4-5) is 28,000 feet. Had the distance been 150 nautical miles, the optimum cruise-climb altitude would be the most economical.

## DIVERSION RANGE SUMMARY TABLES.

Diversion range summary tables are presented in FA4-6 thru FA4-13 for two-engine and single-engine operation. These tables show, in quick reference form, the range available and the time required to return to base with 600, 800, 1000 or 1400 pounds of fuel available. The range is based on having 300 pounds of fuel remaining for the approach and landing after the descent is completed. The 300 pounds of fuel is ample for one missed approach. Range and time data are shown in the tables for three optional return profiles, together with the optimum altitudes for cruise. The optimum altitude is the constant cruise altitude which provides the maximum range for the particular type of flight profile. Climb is made to the cruise altitude, using military thrust.

**NOTE**

The Mil Thrust Climb Speed Schedule at the bottom of each table must be used to obtain the maximum ranges in the table.

Cruise speeds and descent data are provided at the bottom of the tables.

The three types of flight profiles are:

1. a. Cruise at initial altitude to base.
   b. Descend to sea level with idle thrust and speed brake closed after arrival over base.

2. a. Climb on course to optimum cruise altitude.
   b. Cruise at optimum altitude to base.
   c. Descend to sea level with idle thrust and speed brake open after arrival over base.

3. a. Climb on course to optimum cruise altitude.
   b. Cruise at optimum altitude.
   c. Descend on course to sea level with idle thrust and speed brake closed.

### USE OF DIVERSION RANGE SUMMARY TABLES.

Assume the following conditions prevail: Single-engine operation, fuel remaining is 1240 pounds, and the aircraft is 200 nmi from the base at 15,000 feet altitude. Drag Index = 0.

Determine which flight profiles in FA4-10 provide necessary range in return to base.

1. In FA4-10, enter the chart at the top of the column marked 15,000 feet initial altitude.

2. Proceed downward to the section of the chart for 1000 pounds of fuel shown at the left side of the page.

3. The ranges available with the three profile options are as follows:

| | |
|---|---|
| First option | 168 nmi. |
| Second option | 179 nmi. |
| Third option | 211 nmi. |

4. As the required range is 200 nmi, the flight profile for the third option must be used.

5. Climb with MIL thrust from 15,000 at mach number 0.52 (footnote number 5) to 25,000 at mach number 0.59. At 25,000 feet, cruise at 0.62 mach; engine fuel flow will be approximately 1275 lb/hr. At 40 nmi from the base, descend on course at 240 KCAS, idle thrust, with the speed brake closed.

6. The time required with no wind is 36 minutes for 211 nmi, and the fuel used is 1000 pounds by the time the landing is completed. As the fuel available was 1240 pounds, 240 pounds of this amount would be available for headwind conditions.

## CRUISE

### MACH NUMBER AND REFERENCE NUMBER

MODEL: T-38A
DATE: 1 OCTOBER 1976
DATA BASIS: ESTIMATED DATA

ENGINE: (2) J85-GE-5
FUEL GRADE: JP-4
FUEL DENSITY: 6.5 LB/US GAL

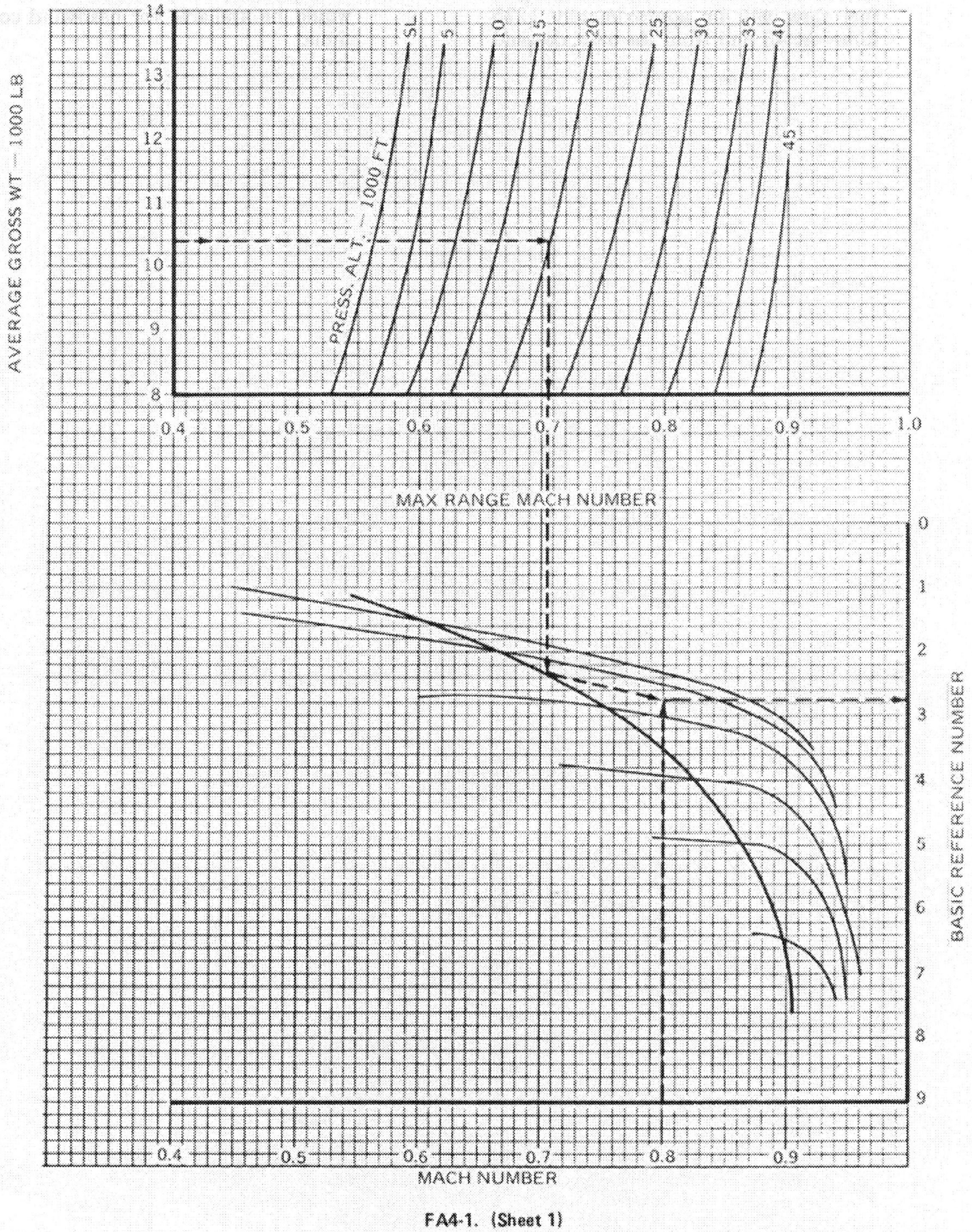


FA4-1. (Sheet 1)

# CRUISE

CORRECTED REFERENCE NUMBER
FOR EXTERNAL STORES

MODEL: T-38A
DATE: 1 OCTOBER 1976
DATA BASIS: ESTIMATED DATA

ENGINE: (2) J85-GE-5
FUEL GRADE: JP-4
FUEL DENSITY: 6.5 LB/US GAL

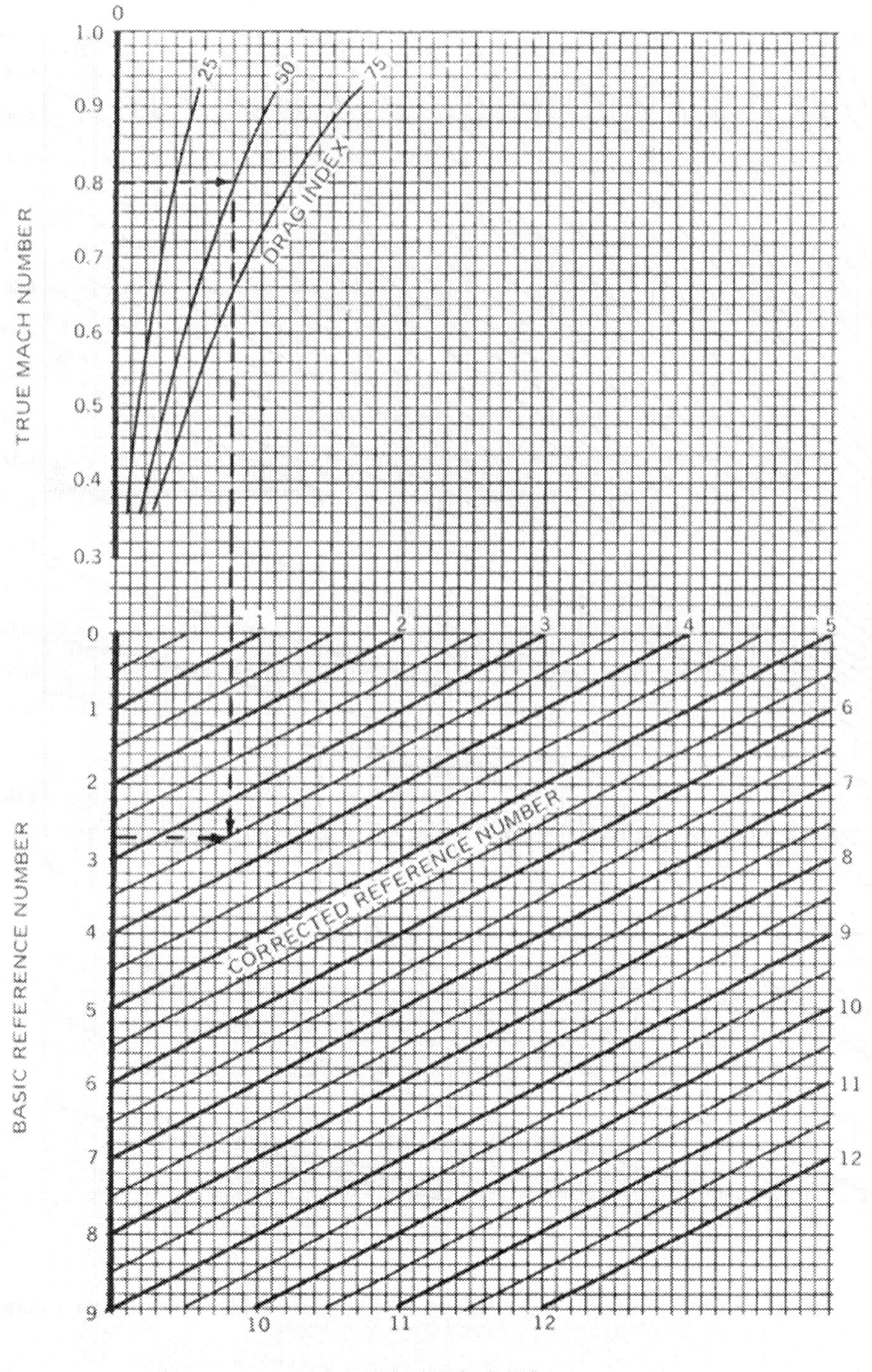


FA4-1. (Sheet 2)

# CRUISE

NAUTICAL MILES PER POUND

MODEL: T-38A
DATE: 1 OCTOBER 1976
DATA BASIS: ESTIMATED DATA

ENGINES: (2) J85-GE-5
FUEL GRADE: JP-4
FUEL DENSITY: 6.5 LB/US GAL

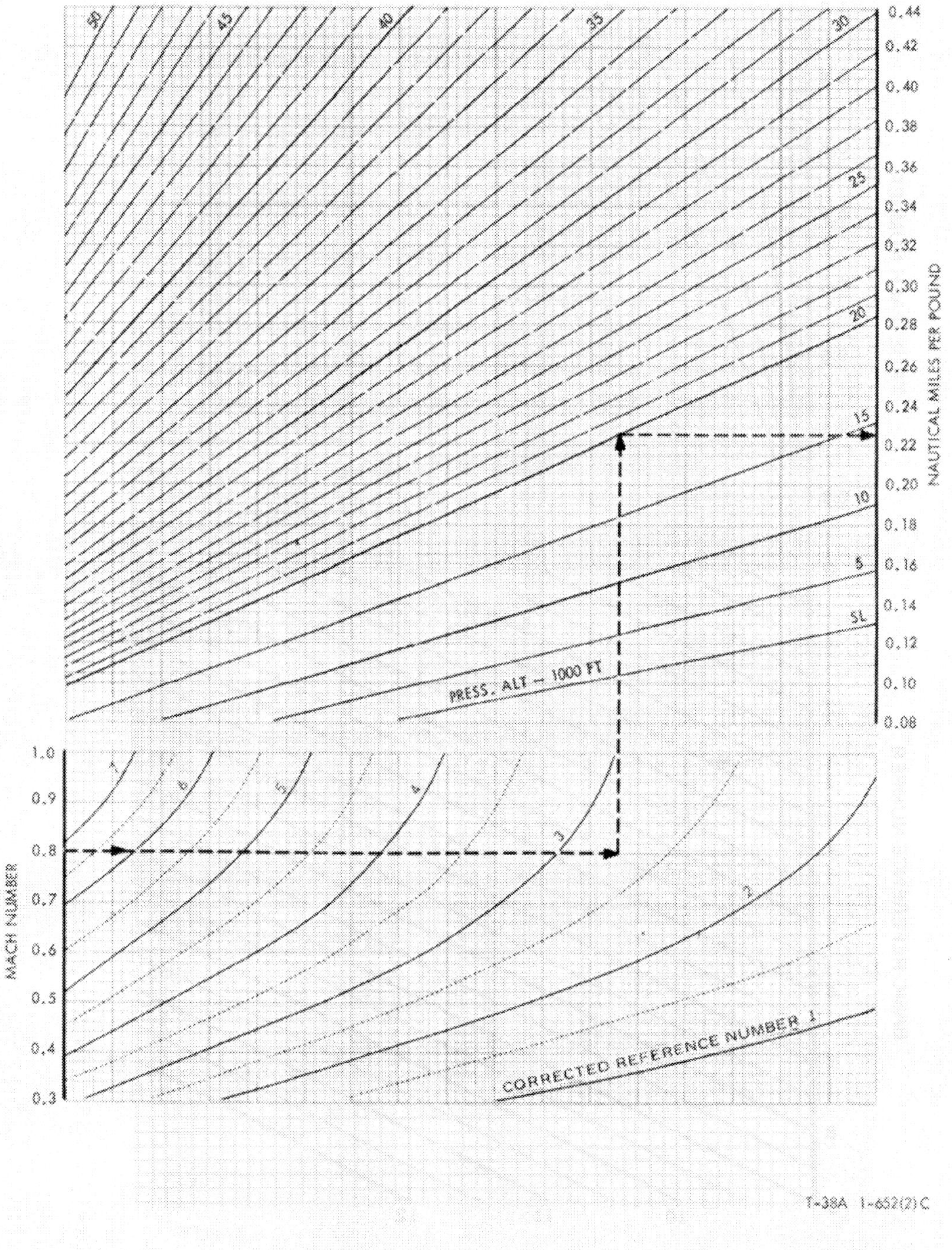


FA4-1. (Sheet 3)

# CRUISE
## FUEL FLOW AND TRUE AIRSPEED

MODEL: T-38A
DATE: 1 OCTOBER 1976
DATA BASIS: ESTIMATED DATA

ENGINES: (2) J85-GE-5
FUEL GRADE: JP-4
FUEL DENSITY: 6.5 LB/US GAL

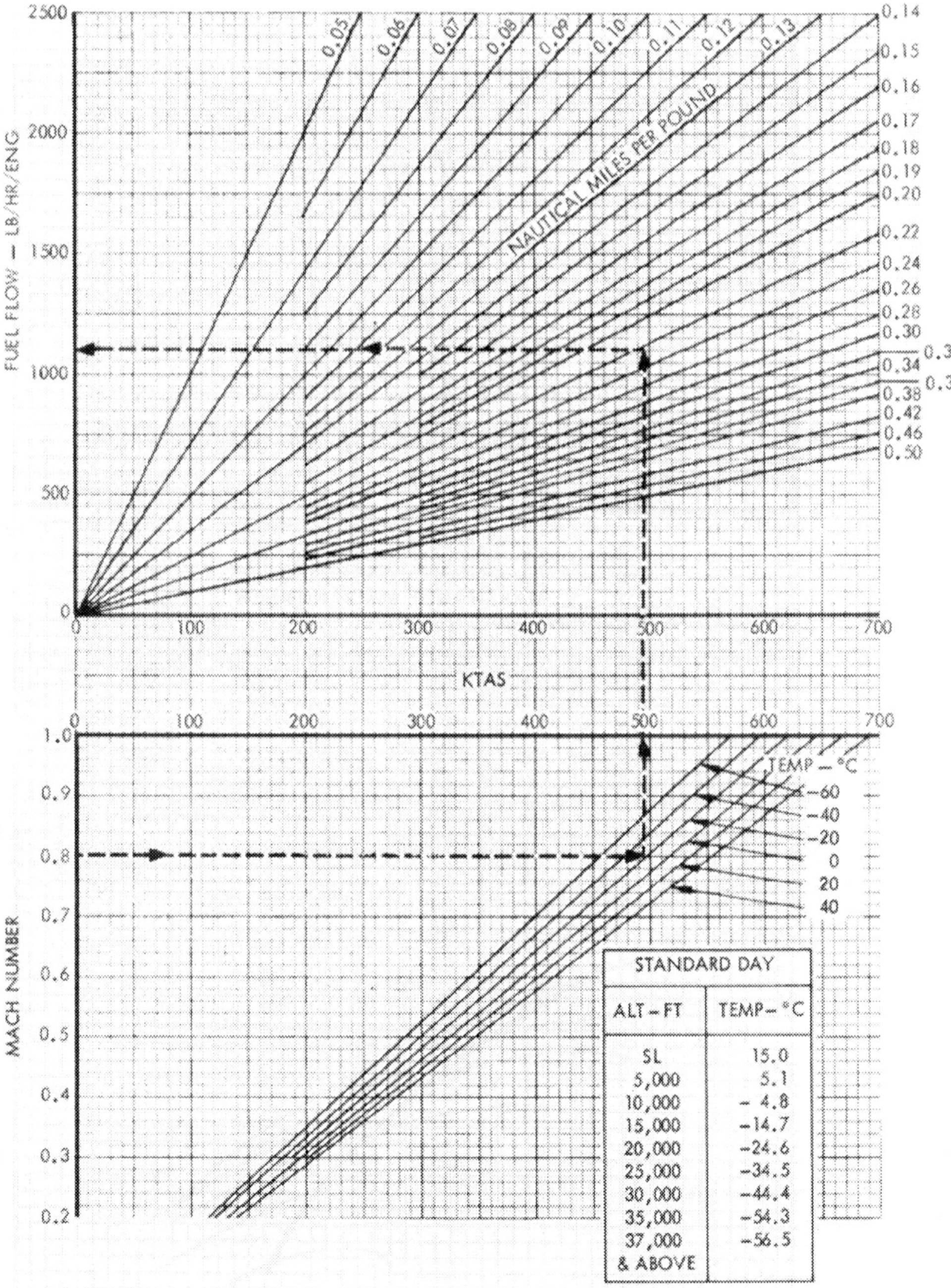


| STANDARD DAY | |
|---|---|
| ALT – FT | TEMP – °C |
| SL | 15.0 |
| 5,000 | 5.1 |
| 10,000 | – 4.8 |
| 15,000 | –14.7 |
| 20,000 | –24.6 |
| 25,000 | –34.5 |
| 30,000 | –44.4 |
| 35,000 | –54.3 |
| 37,000 & ABOVE | –56.5 |

T-38A 1-652(3)C

**FA4-1. (Sheet 4)**

# CRUISE

## MACH NUMBER AND REFERENCE NUMBER

SINGLE ENGINE

MODEL: T-38A
DATE: 1 OCTOBER 1976
DATA BASIS: ESTIMATED DATA

ENGINE: (2) J85-GE-5
FUEL GRADE: JP-4
FUEL DENSITY: 6.5 LB/US G.

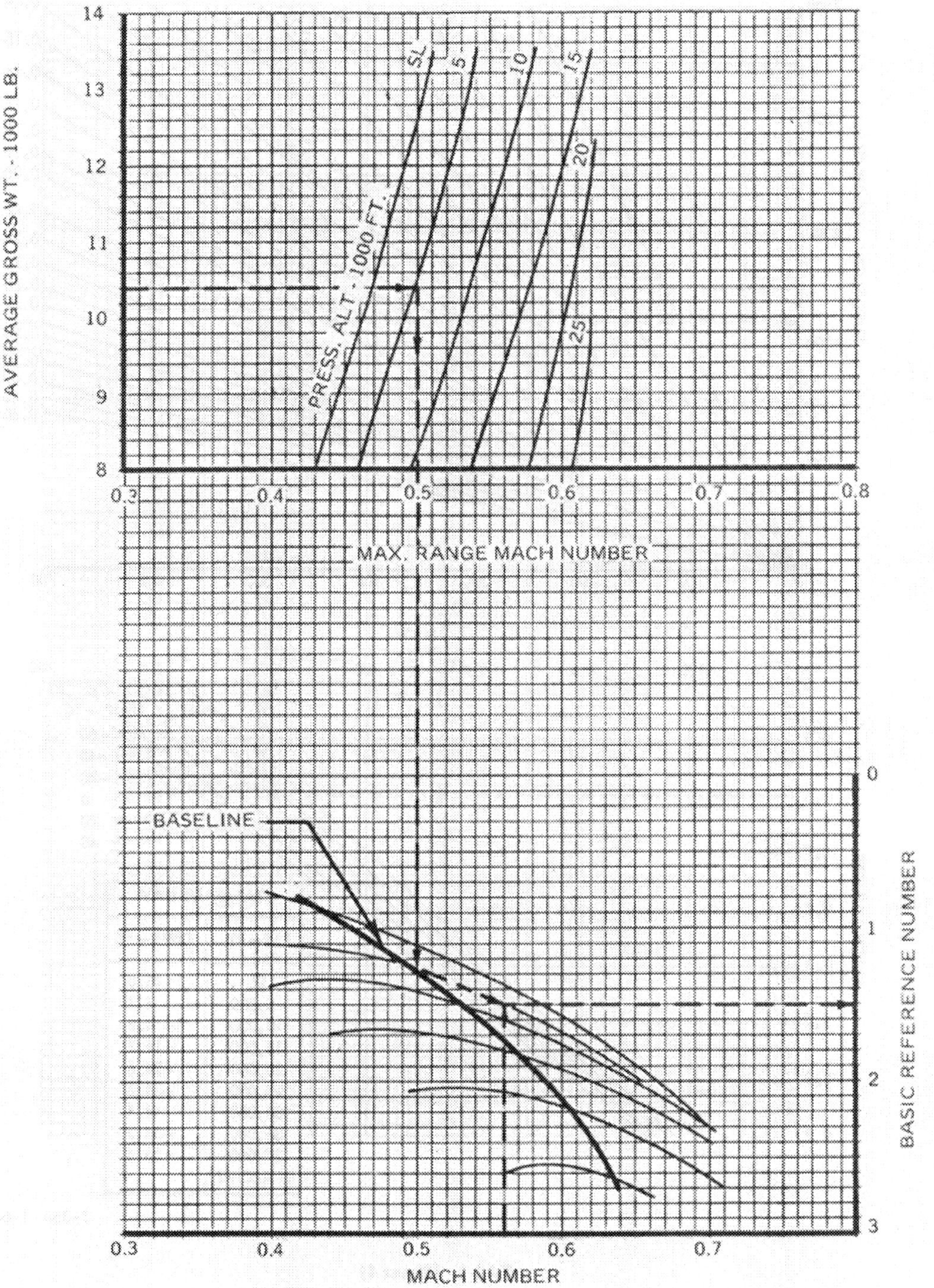


FA4-2. (Sheet 1)

# CRUISE

## CORRECTED REFERENCE NUMBER FOR EXTERNAL STORES

MODEL: T-38A
DATE: 1 OCTOBER 1976
DATA BASIS: FLIGHT TEST

ENGINES: (2) J85-GE-5
FUEL GRADE: JP-4
FUEL DENSITY: 6.5 LB/US GAL

SINGLE ENGINE

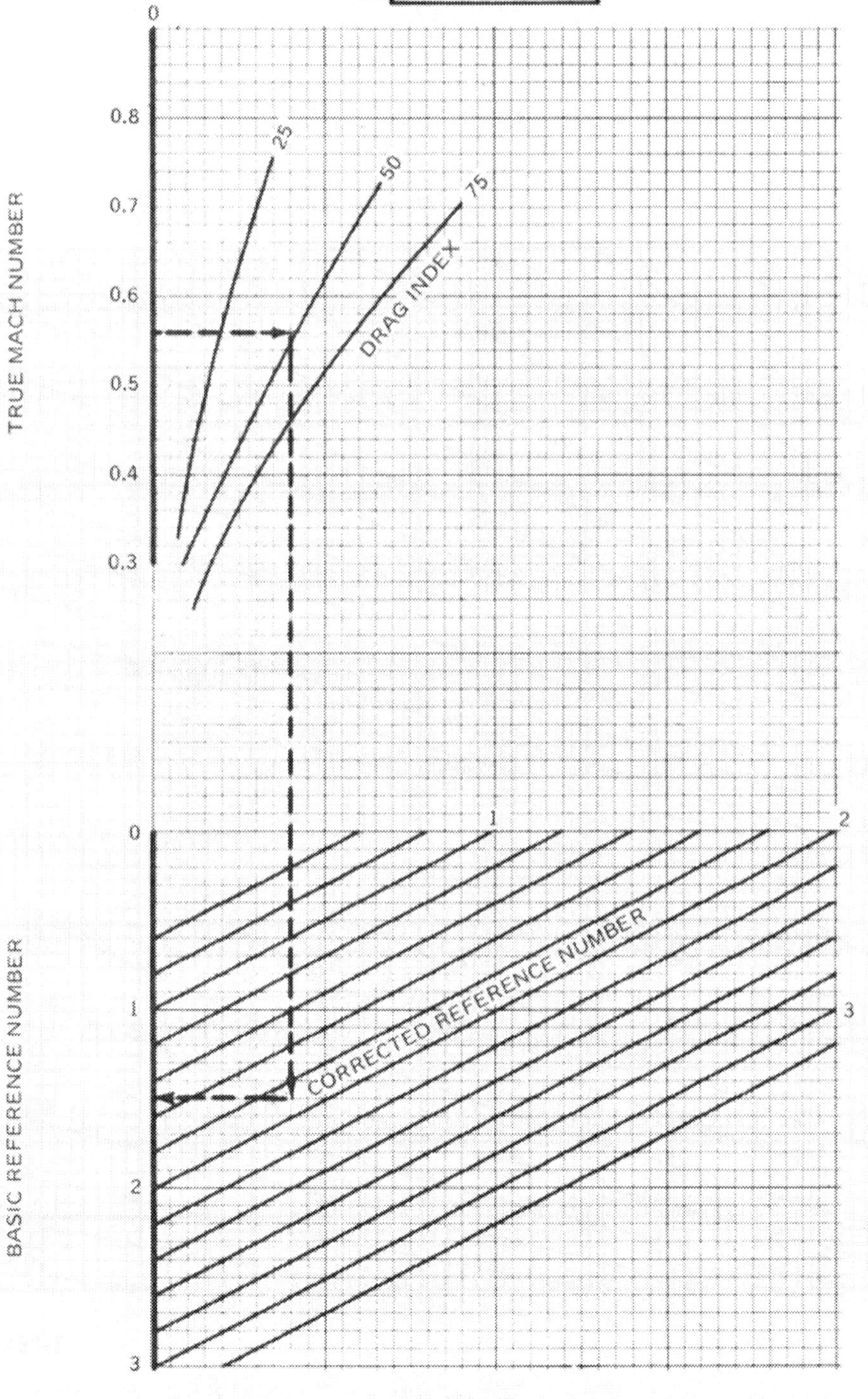


FA4-2. (Sheet 2)

MODEL: T-38A
DATE: 1 OCTOBER 1976
DATA ESTIMATED DATA

# CRUISE

## NAUTICAL MILES PER POUND

ENGINE: (2) J85-GE-5
FUEL GRADE: JP-4
FUEL DENSITY: 6.5 LB/US GAL

SINGLE ENGINE

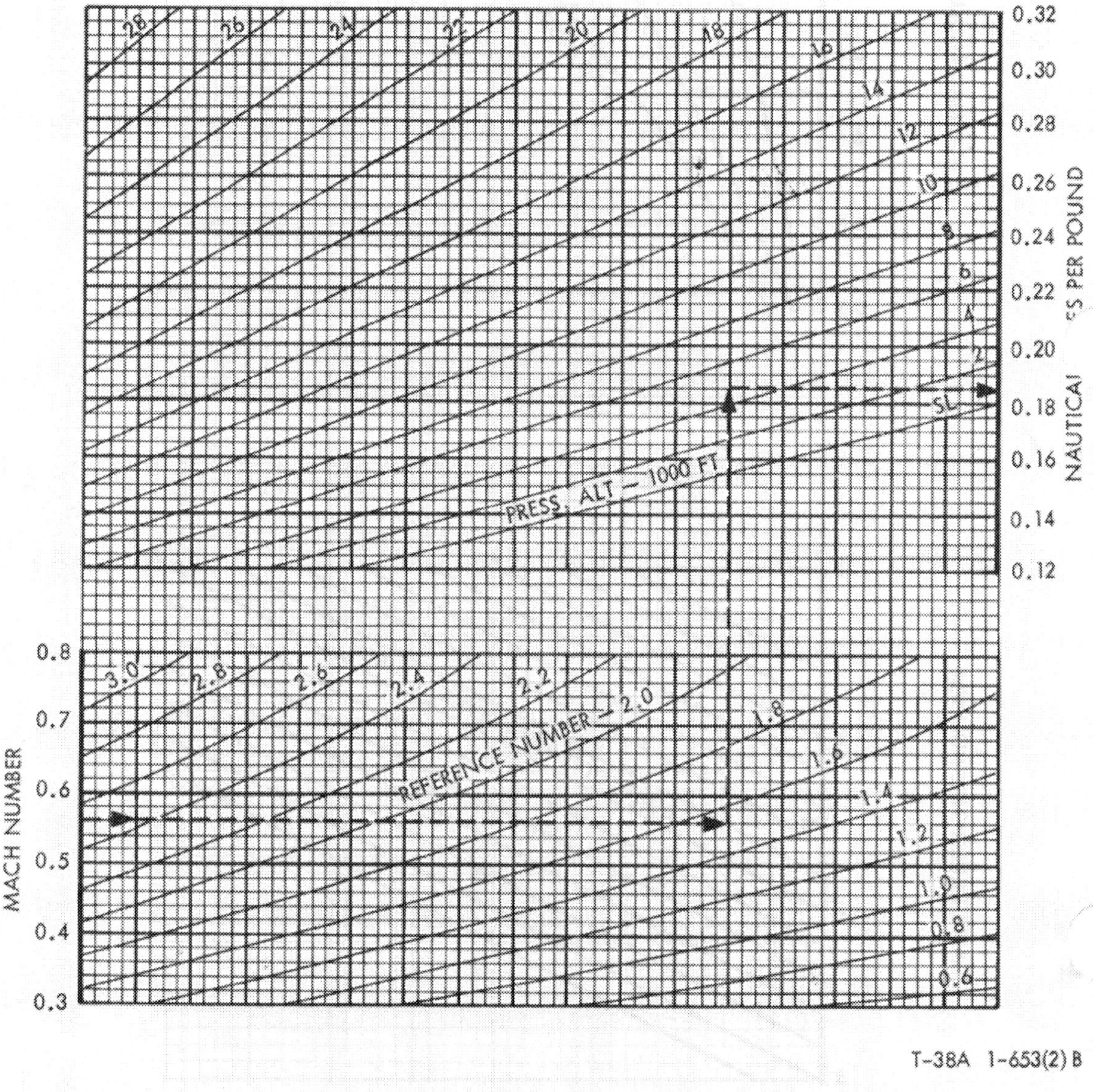


T-38A 1-653(2) B

FA4-2. (Sheet 3)

# CRUISE
## FUEL FLOW AND TRUE AIRSPEED

MODEL: T-38A
DATE: 1 OCTOBER 1976
DATA BASIS: ESTIMATED DATA

SINGLE ENGINE

ENGINES: (2) J85-GE-5
FUEL GRADE: JP-4
FUEL DENSITY: 6.5 LB/US GAL

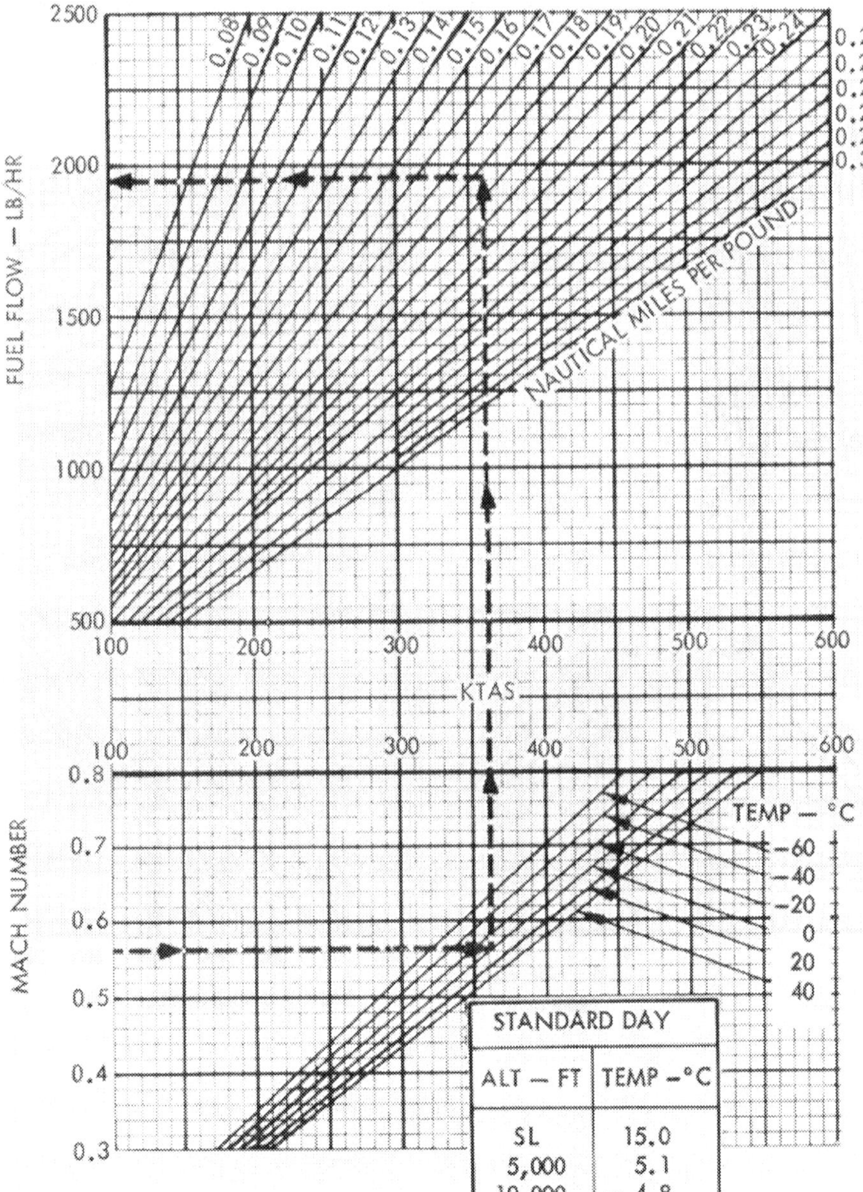


| STANDARD DAY | |
|---|---|
| ALT — FT | TEMP —°C |
| SL | 15.0 |
| 5,000 | 5.1 |
| 10,000 | – 4.8 |
| 15,000 | –14.7 |
| 20,000 | –24.6 |
| 25,000 | –34.5 |
| 30,000 | –44.4 |

T-38A 1-653(3)C

FA4-2. (Sheet 4)

# CONSTANT ALTITUDE CRUISE
## TIME AND AIRSPEED

MODEL: T-38A
DATE: 1 OCTOBER 1976
DATA BASIS: ESTIMATED DATA

ENGINE: (2) J85-GE-5
FUEL GRADE: JP-4
FUEL DENSITY: 6.5 LB/US GAL

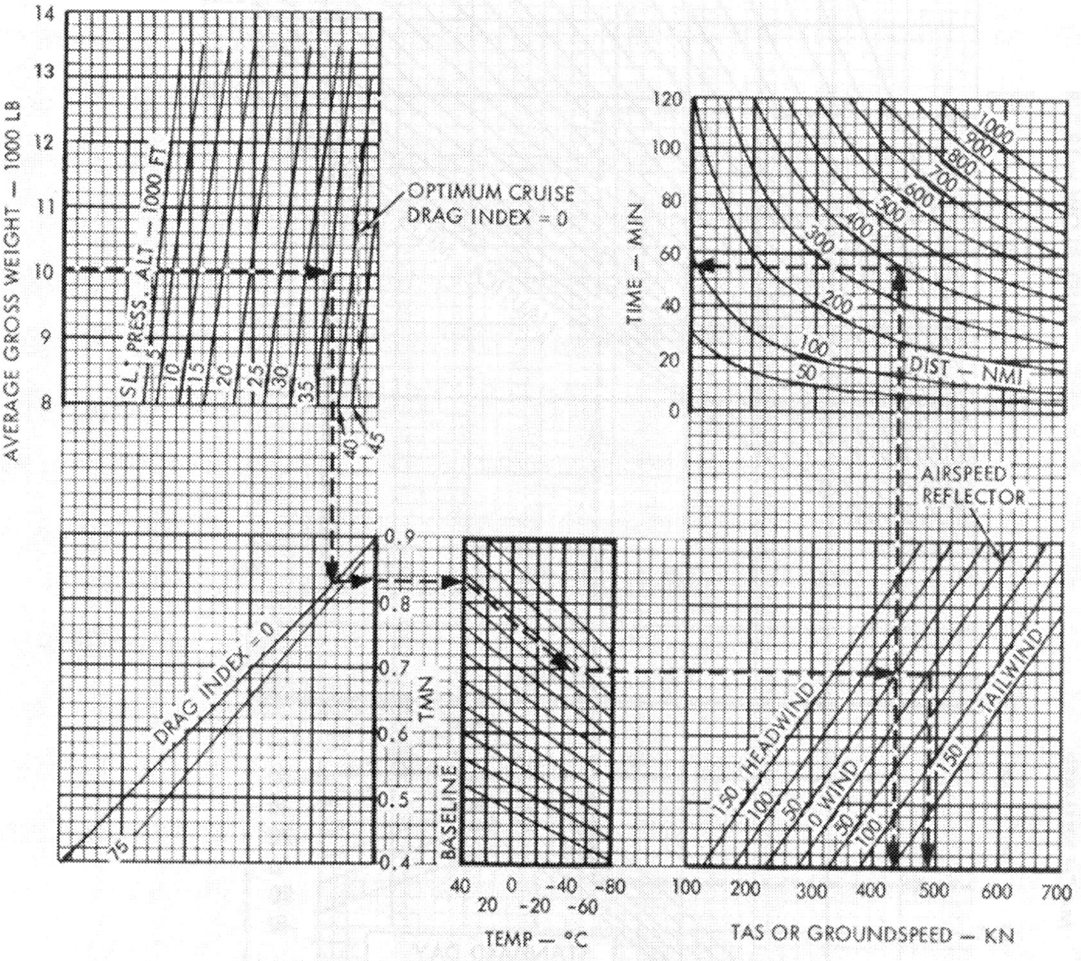


T-38A 1-551(1)C

FA4-3. (Sheet 1)

# CONSTANT ALTITUDE CRUISE
## FUEL FLOW AND FUEL REQUIRED

MODEL: T-38A
DATE: 1 OCTOBER 1976
DATA BASIS: ESTIMATED DATA

ENGINE: (2) J85-GE-5
FUEL GRADE: JP-4
FUEL DENSITY: 6.5 LB/US GAL

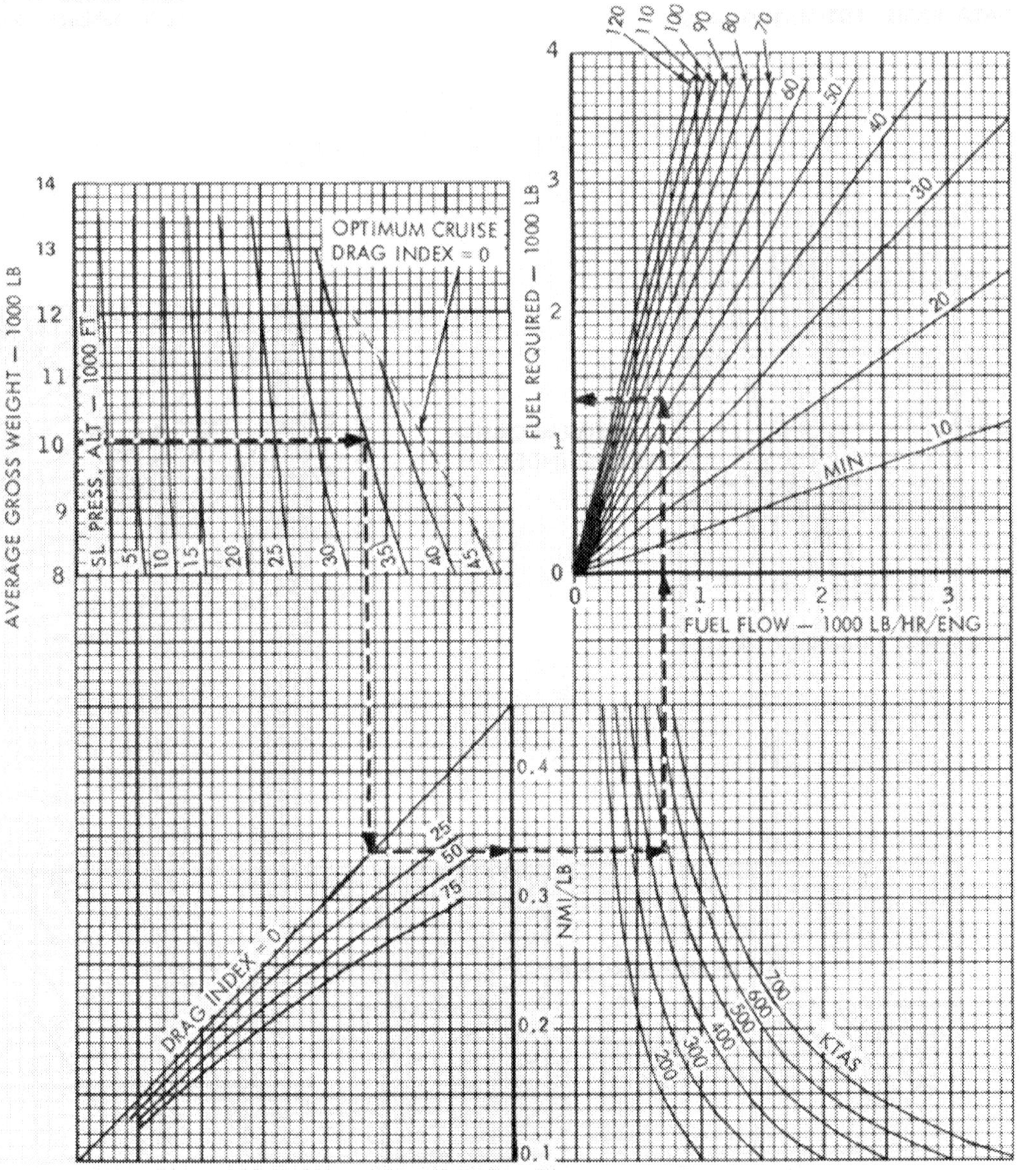


T-38A 1-551(2)C

FA4-3. (Sheet 2)

# CONSTANT ALTITUDE CRUISE
## TIME AND AIRSPEED

SINGLE ENGINE

MODEL: T-38A
DATE: 1 OCTOBER 1976
DATA BASIS: ESTIMATED DATA

ENGINE: (2) J85-GE-5
FUEL GRADE: JP-4
FUEL DENSITY: 6.5 LB/US GAL

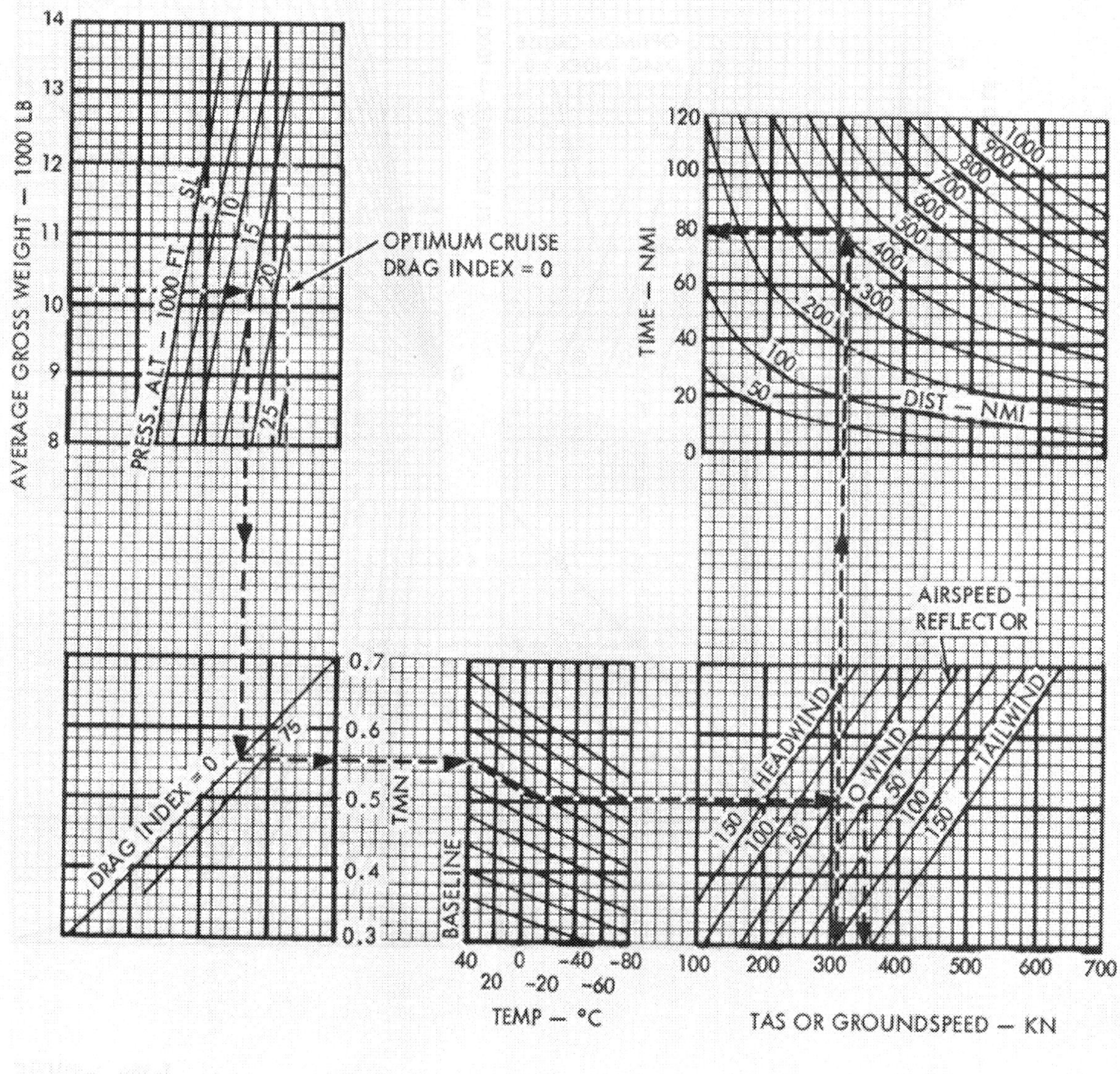


T-38A 1-553(1) B

FA4-4. (Sheet 1)

# CONSTANT ALTITUDE CRUISE
## FUEL FLOW AND FUEL REQUIRED

MODEL: T-38A
DATE: 1 OCTOBER 1976
DATA BASIS: ESTIMATED DATA

SINGLE ENGINE

ENGINE: (2) J85-GE-5
FUEL GRADE: JP-4
FUEL DENSITY: 6.5 LB/US GAL

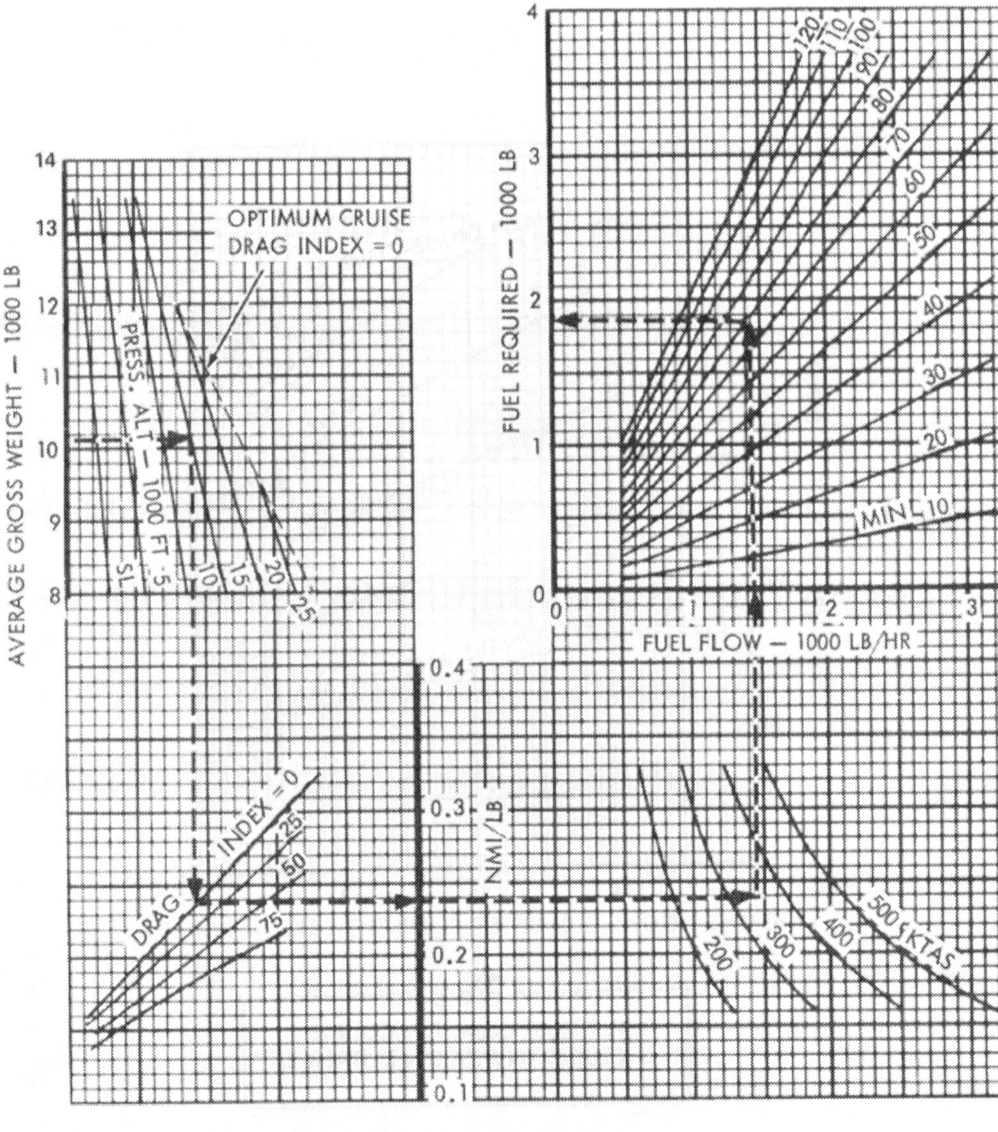


FA4-4. (Sheet 2)

# OPTIMUM CRUISE ALTITUDE FOR SHORT RANGE MISSIONS

CONSTANT ALTITUDE CRUISE
DRAG INDEX = 0

MODEL: T-38A
DATE: 1 AUGUST 1965
DATA BASIS: **FLIGHT TEST**

ENGINE: (2) J85-GE-5
FUEL GRADE: JP-4
FUEL DENSITY: 6.5 LB/US GAL

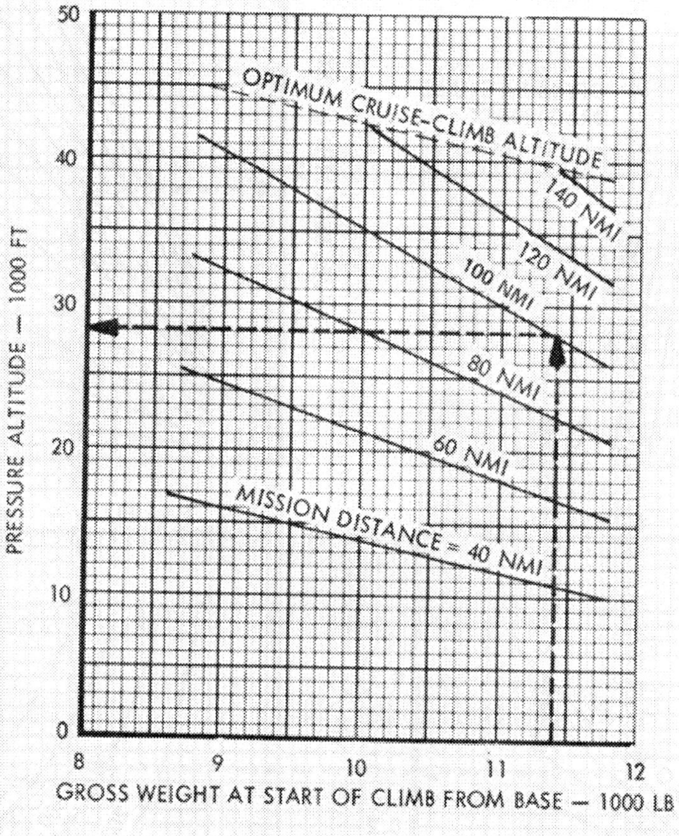


***Note***

MIL THRUST CLIMB ON COURSE INCLUDED IN MISSION DISTANCE.

T-38A 1-550 B

FA4-5.

## DIVERSION RANGE SUMMARY TABLE

CONSTANT ALTITUDE CRUISE
STANDARD DAY ZERO WIND

MODEL: **T-38A**
DATE: **1 JULY 1978**
DATA BASIS: **ESTIMATED DATA**

**TWO ENGINE**

**DRAG INDEX = 0**

ENGINES: (2) J85-GE-5
FUEL GRADE: JP-4
FUEL DENSITY: 6.5 LB/US GAL.

RANGE AND TIME REMAINING WITH 300-LB RESERVES AT SEA LEVEL

| FUEL | 1000 FT. | SL | 5 | 10 | 15 | 20 | 25 | 30 | 35 | 40 | 45 | PROCEDURE |
|---|---|---|---|---|---|---|---|---|---|---|---|---|
| | | INITIAL ALTITUDE | | | | | | | | | | |
| 600 LB | NMI / MIN | 39 / 7 | 46 / 10 | 51 / 12 | 56 / 13 | 60 / 16 | 65 / 17 | 70 / 18 | 73 / 20 | 76 / 21 | 73 / 23 | CRUISE AT INITIAL ALTITUDE TO BASE (1) |
| | 1000 FT | 10/20 | 10/25 | 15/30 | 20/35 | 25/35 | 30/40 | 30/40 | 35/45 | 40/45 | 40/45 | OPTIMUM ALTITUDE (5) |
| | NMI / MIN | 43 / 8 | 48 / 9 | 54 / 10 | 60 / 12 | 66 / 13 | 74 / 14 | 82 / 15 | 87 / 17 | 94 / 18 | 95 / 18 | USE OPTIMUM ALTITUDE UNTIL OVER BASE |
| | NMI / MIN | 64 / 11 | 74 / 12 | 84 / 13 | 95 / 16 | 105 / 17 | 117 / 18 | 128 / 19 | 138 / 21 | 148 / 22 | 154 / 23 | USE OPTIMUM ALTITUDE AND DESCEND ON COURSE (3) |
| 800 LB | NMI / MIN | 65 / 11 | 76 / 15 | 86 / 18 | 96 / 20 | 106 / 22 | 118 / 24 | 130 / 26 | 139 / 29 | 148 / 30 | 147 / 31 | CRUISE AT INITIAL ALTITUDE TO BASE (1) |
| | 1000 FT | 25/35 | 30/40 | 30/40 | 30/45 | 35/45 | 40/45 | 40/45 | 40/45 | 40/45 | 45 | OPTIMUM ALTITUDE |
| | NMI / MIN | 84 / 15 | 95 / 17 | 106 / 18 | 116 / 20 | 126 / 22 | 137 / 23 | 148 / 24 | 157 / 25 | 166 / 20 | 168 / 27 | USE OPTIMUM ALTITUDE UNTIL OVER BASE (2) |
| | NMI / MIN | 124 / 19 | 139 / 20 | 152 / 22 | 166 / 24 | 179 / 26 | 191 / 27 | 203 / 29 | 213 / 30 | 222 / 31 | 228 / 31 | USE OPTIMUM ALTITUDE AND DESCEND ON COURSE (3) |
| 1000 LB | NMI / MIN | 90 / 15 | 106 / 20 | 121 / 23 | 137 / 26 | 152 / 29 | 170 / 31 | 189 / 34 | 204 / 37 | 219 / 39 | 220 / 40 | CRUISE AT INITIAL ALTITUDE TO BASE (1) |
| | 1000 FT | 35/45 | 40/45 | 40/45 | 40/45 | 40/45 | 40/45 | 40/45 | 40/45 | 40/45 | 45 | OPTIMUM ALTITUDE |
| | NMI / MIN | 143 / 24 | 157 / 25 | 171 / 27 | 185 / 29 | 197 / 30 | 208 / 32 | 219 / 33 | 229 / 34 | 237 / 35 | 240 / 35 | USE OPTIMUM ALTITUDE UNTIL OVER BASE (2) |
| | NMI / MIN | 196 / 28 | 212 / 30 | 226 / 31 | 240 / 33 | 252 / 34 | 264 / 36 | 276 / 37 | 286 / 38 | 295 / 39 | 301 / 40 | USE OPTIMUM ALTITUDE AND DESCEND ON COURSE (3) |
| 1400 LB | NMI / MIN | 142 / 24 | 166 / 29 | 192 / 34 | 217 / 38 | 244 / 42 | 273 / 45 | 306 / 50 | 332 / 54 | 360 / 56 | 360 / 56 | CRUISE AT INITIAL ALTITUDE TO BASE (1) |
| | 1000 FT | 40/45 | 40/45 | 40/45 | 40/45 | 40/45 | 40/45 | 40/45 | 40/45 | 40/45 | 45 | OPTIMUM ALTITUDE |
| | NMI / MIN | 282 / 41 | 297 / 43 | 311 / 44 | 324 / 46 | 337 / 47 | 348 / 49 | 359 / 50 | 369 / 51 | 378 / 53 | 381 / 52 | USE OPTIMUM ALTITUDE UNTIL OVER BASE (2) |
| | NMI / MIN | 339 / 45 | 352 / 47 | 369 / 48 | 382 / 50 | 395 / 51 | 406 / 53 | 418 / 54 | 428 / 55 | 437 / 56 | 441 / 56 | USE OPTIMUM ALTITUDE AND DESCEND ON COURSE (3) |
| CRUISE ALT | | SL | 5 | 10 | 15 | 20 | 25 | 30 | 35 | 40 | 45 | *Note* WITH MORE THAN 1400 POUNDS FUEL, CRUISE AT 0.90 MACH, PRESSURE ALTITUDE 43,000 FEET. |
| CRUISE MACH NO. | | 0.54 | 0.56 | 0.59 | 0.64 | 0.68 | 0.73 | 0.77 | 0.81 | 0.85 | 0.89 | |
| APPROX FUEL FLOW LB/HR | | 1325 | 1200 | 1050 | 975 | 900 | 825 | 750 | 700 | 675 | 650 | |
| DESCEND 240 KCAS IDLE (4) | NMI REMAINING | | 8 | 16 | 24 | 32 | 40 | 49 | 59 | 70 | 81 | |
| | MIN REMAINING | | 2 | 4 | 5 | 7 | 8 | 9 | 11 | 12 | 14 | |
| | FUEL REMAINING | | 312 | 328 | 340 | 352 | 363 | 375 | 386 | 397 | 407 | |

1. FUEL AND TIME INCLUDED FOR DESCENT AT DESTINATION WITHOUT DISTANCE CREDIT, SPEED BRAKE CLOSED.
2. TIME AND FUEL INCLUDED FOR CLIMB TO OPTIMUM ALTITUDE AND DESCENT AT DESTINATION; NO DISTANCE CREDIT FOR DESCENT TO SEA LEVEL DESTINATION, SPEED BRAKE OPEN.
3. TIME AND FUEL INCLUDED FOR CLIMB TO OPTIMUM ALTITUDE AND DESCENT AT DESTINATION, RANGE INCLUDES DISTANCE FOR ON-COURSE DESCENT TO SEA LEVEL DESTINATION, SPEED BRAKE CLOSED.
4. DESCENT DATA TABULATED FOR SPEED BRAKE CLOSED; WITH SPEED BRAKE OPEN, USE ONE-HALF OF THE VALUES.
5. CLIMB USING FOLLOWING MIL THRUST CLIMB SPEED SCHEDULE:

| PRESS. ALT (1000 FT.) | SL | 5 | 10 | 15 | 20 | 25 | 30 | 35 | 40 | 45 |
|---|---|---|---|---|---|---|---|---|---|---|
| TRUE MACH | 0.75 | 0.76 | 0.78 | 0.79 | 0.81 | 0.83 | 0.84 | 0.86 | 0.87 | 0.87 |
| KCAS | 496 | 466 | 435 | 406 | 377 | 349 | 322 | 295 | 264 | 236 |

FA4-6.

# DIVERSION RANGE SUMMARY TABLE

CONSTANT ALTITUDE CRUISE
STANDARD DAY ZERO WIND

MODEL: **T-38A**
DATE: **1 JULY 1978**
DATA BASIS: **ESTIMATED DATA**

**TWO ENGINE**

**DRAG INDEX = 25**

ENGINES: (2) J85-GE-5
FUEL GRADE: JP-4
FUEL DENSITY: 6.5 LB/US GAL.

| RANGE AND TIME REMAINING WITH 300-LB RESERVES AT SEA LEVEL | | | | | | | | | | | | PROCEDURE |
|---|---|---|---|---|---|---|---|---|---|---|---|---|
| | | INITIAL ALTITUDE | | | | | | | | | | |
| FUEL | 1000 FT. | SL | 5 | 10 | 15 | 20 | 25 | 30 | 35 | 40 | 45 | |
| 600 LB | NMI<br>MIN | 37<br>7 | 43<br>9 | 48<br>11 | 53<br>13 | 57<br>14 | 62<br>16 | 68<br>18 | 71<br>19 | 75<br>20 | 75<br>21 | CRUISE AT INITIAL ALTITUDE TO BASE ① |
| | 1000 FT | 5/15 | 10/20 | 10/25 | 15/30 | 20/35 | 25/40 | 30/40 | 35/40 | 40/45 | 45 | OPTIMUM ALTITUDE ② |
| | NMI<br>MIN | 39<br>7 | 44<br>8 | 50<br>10 | 56<br>11 | 63<br>12 | 70<br>13 | 77<br>15 | 82<br>16 | 89<br>16 | 89<br>17 | USE OPTIMUM ALTITUDE UNTIL OVER BASE |
| | NMI<br>MIN | 54<br>10 | 66<br>11 | 74<br>12 | 87<br>14 | 98<br>16 | 108<br>17 | 119<br>18 | 128<br>19 | 138<br>20 | 142<br>21 | USE OPTIMUM ALTITUDE AND DESCEND ON COURSE ③ |
| 800 LB | NMI<br>MIN | 61<br>11 | 72<br>14 | 81<br>16 | 91<br>19 | 101<br>21 | 111<br>23 | 124<br>26 | 132<br>27 | 141<br>28 | 141<br>29 | CRUISE AT INITIAL ALTITUDE TO BASE ① |
| | 1000 FT | 20/35 | 25/35 | 30/40 | 30/40 | 35/40 | 40/45 | 40/45 | 40/45 | 40/45 | 45 | OPTIMUM ALTITUDE |
| | NMI<br>MIN | 75<br>14 | 85<br>15 | 96<br>17 | 107<br>19 | 116<br>20 | 126<br>21 | 137<br>22 | 147<br>24 | 155<br>25 | 157<br>25 | USE OPTIMUM ALTITUDE UNTIL OVER BASE ② |
| | NMI<br>MIN | 108<br>17 | 123<br>19 | 137<br>20 | 151<br>22 | 163<br>23 | 175<br>25 | 187<br>26 | 197<br>27 | 207<br>28 | 210<br>29 | USE OPTIMUM ALTITUDE AND DESCEND ON COURSE ③ |
| 1000 LB | NMI<br>MIN | 86<br>16 | 100<br>19 | 114<br>22 | 129<br>25 | 144<br>27 | 160<br>30 | 179<br>33 | 192<br>35 | 207<br>36 | 208<br>36 | CRUISE AT INITIAL ALTITUDE TO BASE ① |
| | 1000 FT | 30/40 | 35/45 | 40/45 | 40/45 | 40/45 | 40/45 | 40/45 | 40/45 | 45 | 45 | OPTIMUM ALTITUDE |
| | NMI<br>MIN | 126<br>21 | 140<br>23 | 154<br>25 | 168<br>26 | 180<br>28 | 191<br>29 | 203<br>30 | 212<br>32 | 222<br>32 | 224<br>32 | USE OPTIMUM ALTITUDE UNTIL OVER BASE ② |
| | NMI<br>MIN | 171<br>25 | 188<br>27 | 204<br>28 | 218<br>30 | 231<br>31 | 243<br>33 | 255<br>34 | 265<br>35 | 275<br>36 | 277<br>36 | USE OPTIMUM ALTITUDE AND DESCEND ON COURSE ③ |
| 1400 LB | NMI<br>MIN | 134<br>24 | 156<br>29 | 180<br>32 | 204<br>36 | 230<br>40 | 257<br>43 | 288<br>48 | 310<br>50 | 336<br>52 | —<br>— | CRUISE AT INITIAL ALTITUDE TO BASE ① |
| | 1000 FT | 40/45 | 45 | 45 | 45 | 45 | 45 | 45 | 45 | 45 | 45 | OPTIMUM ALTITUDE |
| | NMI<br>MIN | 251<br>36 | 268<br>38 | 284<br>40 | 298<br>42 | 312<br>43 | 325<br>45 | 339<br>46 | 341<br>48 | 350<br>49 | 355<br>49 | USE OPTIMUM ALTITUDE UNTIL OVER BASE ② |
| | NMI<br>MIN | 304<br>40 | 321<br>42 | 337<br>44 | 352<br>46 | 365<br>47 | 378<br>49 | 392<br>51 | 395<br>51 | 398<br>52 | 400<br>53 | USE OPTIMUM ALTITUDE AND DESCEND ON COURSE ③ |
| CRUISE ALT | | SL | 5 | 10 | 15 | 20 | 25 | 30 | 35 | 40 | 45 | |
| CRUISE MACH NO. | | 0.50 | 0.54 | 0.58 | 0.63 | 0.67 | 0.71 | 0.76 | 0.80 | 0.85 | 0.89 | |
| APPROX FUEL FLOW LB/HR | | 1350 | 1225 | 1100 | 1025 | 950 | 875 | 800 | 750 | 725 | 725 | |
| DESCEND 240 KCAS IDLE ④ | NMI REMAINING | | 7 | 15 | 22 | 30 | 38 | 45 | 53 | 62 | 69 | |
| | MIN REMAINING | | 2 | 3 | 5 | 6 | 7 | 9 | 10 | 11 | 12 | |
| | FUEL REMAINING | | 312 | 225 | 337 | 347 | 356 | 365 | 373 | 382 | 390 | |

***Note***

WITH MORE THAN 1400 POUNDS FUEL, CRUISE AT 0.88 MACH, PRESSURE ALTITUDE 42,000 FEET.

① FUEL AND TIME INCLUDED FOR DESCENT AT DESTINATION WITHOUT DISTANCE CREDIT, SPEED BRAKE CLOSED.
② TIME AND FUEL INCLUDED FOR CLIMB TO OPTIMUM ALTITUDE AND DESCENT AT DESTINATION; NO DISTANCE CREDIT FOR DESCENT TO SEA LEVEL DESTINATION, SPEED BRAKE OPEN.
③ TIME AND FUEL INCLUDED FOR CLIMB TO OPTIMUM ALTITUDE AND DESCENT AT DESTINATION, RANGE INCLUDES DISTANCE FOR ON-COURSE DESCENT TO SEA LEVEL DESTINATION, SPEED BRAKE CLOSED.
④ DESCENT DATA TABULATED FOR SPEED BRAKE CLOSED; WITH SPEED BRAKE OPEN, USE ONE-HALF OF THE VALUES.
5. CLIMB USING FOLLOWING MIL THRUST CLIMB SPEED SCHEDULE:

| PRESS. ALT (1000 FT.) | SL | 5 | 10 | 15 | 20 | 25 | 30 | 35 | 40 | 45 |
|---|---|---|---|---|---|---|---|---|---|---|
| TRUE MACH | 0.75 | 0.76 | 0.78 | 0.79 | 0.81 | 0.83 | 0.84 | 0.86 | 0.87 | 0.87 |
| KCAS | 496 | 466 | 435 | 406 | 377 | 349 | 322 | 295 | 264 | 236 |

FA4-7.

# DIVERSION RANGE SUMMARY TABLE

CONSTANT ALTITUDE CRUISE
STANDARD DAY ZERO WIND

MODEL: **T-38A**
DATE: **1 JULY 1978**
DATA BASIS: **ESTIMATED DATA**

TWO ENGINE

**DRAG INDEX = 50**

ENGINES: (2) J85-GE-5
FUEL GRADE: JP-4
FUEL DENSITY: 6.5 LB/US GAL.

RANGE AND TIME REMAINING WITH 300-LB RESERVES AT SEA LEVEL

| FUEL | 1000 FT. | SL | 5 | 10 | 15 | 20 | 25 | 30 | 35 | 40 | 45 | PROCEDURE |
|---|---|---|---|---|---|---|---|---|---|---|---|---|
| | | INITIAL ALTITUDE | | | | | | | | | | |
| 600 LB | NMI<br>MIN | 35<br>7 | 48<br>8 | 45<br>11 | 50<br>12 | 55<br>14 | 59<br>16 | 65<br>17 | 68<br>18 | 71<br>19 | —<br>— | CRUISE AT INITIAL ALTITUDE TO BASE (1) |
| | 1000 FT | 5/15 | 5/20 | 10/25 | 15/30 | 20/30 | 25/35 | 30/40 | 35/40 | 40 | 40 | OPTIMUM ALTITUDE (5) |
| | NMI<br>MIN | 36<br>6 | 42<br>8 | 47<br>10 | 53<br>10 | 59<br>12 | 66<br>13 | 73<br>14 | 77<br>15 | 82<br>15 | 83<br>15 | USE OPTIMUM ALTITUDE UNTIL OVER BASE |
| | NMI<br>MIN | 48<br>8 | 59<br>10 | 69<br>12 | 80<br>13 | 89<br>14 | 99<br>15 | 110<br>17 | 120<br>18 | 128<br>19 | 131<br>19 | USE OPTIMUM ALTITUDE AND DESCEND ON COURSE (3) |
| 800 LB | NMI<br>MIN | 58<br>11 | 68<br>13 | 77<br>16 | 86<br>18 | 96<br>21 | 106<br>22 | 117<br>24 | 124<br>25 | 132<br>26 | —<br>— | CRUISE AT INITIAL ALTITUDE TO BASE (1) |
| | 1000 FT | 15/30 | 20/30 | 25/35 | 30/40 | 30/40 | 35/40 | 40 | 40 | 40 | 40 | OPTIMUM ALTITUDE |
| | NMI<br>MIN | 67<br>12 | 77<br>14 | 87<br>15 | 98<br>17 | 107<br>18 | 116<br>20 | 126<br>21 | 135<br>22 | 143<br>23 | 144<br>23 | USE OPTIMUM ALTITUDE UNTIL OVER BASE (2) |
| | NMI<br>MIN | 94<br>15 | 108<br>17 | 122<br>18 | 136<br>20 | 148<br>21 | 160<br>23 | 171<br>24 | 181<br>25 | 189<br>26 | 192<br>27 | USE OPTIMUM ALTITUDE AND DESCEND ON COURSE (3) |
| 1000 LB | NMI<br>MIN | 81<br>16 | 95<br>18 | 108<br>21 | 122<br>23 | 137<br>27 | 151<br>29 | 169<br>31 | 180<br>33 | 192<br>34 | —<br>— | CRUISE AT INITIAL ALTITUDE TO BASE (1) |
| | 1000 FT | 30/40 | 30/40 | 35/40 | 40 | 40 | 40 | 40 | 40 | 40 | 40 | OPTIMUM ALTITUDE |
| | NMI<br>MIN | 110<br>19 | 125<br>21 | 138<br>22 | 151<br>24 | 163<br>25 | 175<br>27 | 186<br>28 | 196<br>29 | 204<br>30 | 204<br>31 | USE OPTIMUM ALTITUDE UNTIL OVER BASE (2) |
| | NMI<br>MIN | 148<br>21 | 166<br>24 | 182<br>25 | 196<br>27 | 208<br>29 | 220<br>30 | 232<br>32 | 241<br>33 | 249<br>34 | 252<br>35 | USE OPTIMUM ALTITUDE AND DESCEND ON COURSE (3) |
| 1400 LB | NMI<br>MIN | 127<br>24 | 148<br>27 | 170<br>32 | 193<br>34 | 218<br>39 | 242<br>41 | 271<br>45 | 289<br>47 | 311<br>49 | —<br>— | CRUISE AT INITIAL ALTITUDE TO BASE (1) |
| | 1000 FT | 40 | 40 | 40 | 40 | 40 | 40 | 40 | 40 | 40 | 40 | OPTIMUM ALTITUDE |
| | NMI<br>MIN | 221<br>32 | 239<br>35 | 255<br>37 | 269<br>38 | 282<br>40 | 294<br>41 | 305<br>43 | 314<br>44 | 322<br>45 | 323<br>45 | USE OPTIMUM ALTITUDE UNTIL OVER BASE (2) |
| | NMI<br>MIN | 226<br>36 | 284<br>38 | 300<br>40 | 314<br>42 | 327<br>44 | 339<br>45 | 350<br>46 | 360<br>48 | 368<br>49 | 370<br>49 | USE OPTIMUM ALTITUDE AND DESCEND ON COURSE (3) |
| CRUISE ALT | | SL | 5 | 10 | 15 | 20 | 25 | 30 | 35 | 40 | 45 | |
| CRUISE MACH NO. | | 0.47 | 0.52 | 0.56 | 0.61 | 0.65 | 0.70 | 0.75 | 0.79 | 0.84 | DESCEND TO 40,000 FT. | |
| APPROX FUEL FLOW LB/HR | | 1350 | 1250 | 1125 | 1075 | 1000 | 900 | 850 | 775 | 775 | | |
| DESCEND 240 KCAS IDLE (4) | NMI REMAINING | | 7 | 14 | 20 | 27 | 34 | 42 | 49 | 57 | 64 | |
| | MIN REMAINING | | 1 | 3 | 4 | 6 | 7 | 8 | 9 | 10 | 11 | |
| | FUEL REMAINING | | 310 | 323 | 333 | 342 | 352 | 360 | 368 | 375 | 382 | |

*Note*

WITH MORE THAN 1400 POUNDS FUEL, CRUISE AT 0.86 MACH, PRESSURE ALTITUDE 41,000 FEET.

(1) FUEL AND TIME INCLUDED FOR DESCENT AT DESTINATION WITHOUT DISTANCE CREDIT, SPEED BRAKE CLOSED.
(2) TIME AND FUEL INCLUDED FOR CLIMB TO OPTIMUM ALTITUDE AND DESCENT AT DESTINATION; NO DISTANCE CREDIT FOR DESCENT TO SEA LEVEL DESTINATION, SPEED BRAKE OPEN.
(3) TIME AND FUEL INCLUDED FOR CLIMB TO OPTIMUM ALTITUDE AND DESCENT AT DESTINATION, RANGE INCLUDES DISTANCE FOR ON-COURSE DESCENT TO SEA LEVEL DESTINATION, SPEED BRAKE CLOSED.
(4) DESCENT DATA TABULATED FOR SPEED BRAKE CLOSED; WITH SPEED BRAKE OPEN, USE ONE-HALF OF THE VALUES.
5. CLIMB USING FOLLOWING MIL THRUST CLIMB SPEED SCHEDULE:

| PRESS. ALT (1000 FT.) | SL | 5 | 10 | 15 | 20 | 25 | 30 | 35 | 40 | 45 |
|---|---|---|---|---|---|---|---|---|---|---|
| TRUE MACH | 0.75 | 0.76 | 0.78 | 0.79 | 0.81 | 0.83 | 0.84 | 0.86 | 0.87 | 0.87 |
| KCAS | 496 | 466 | 435 | 406 | 377 | 349 | 322 | 295 | 264 | 236 |

FA4-8.

# DIVERSION RANGE SUMMARY TABLE

CONSTANT ALTITUDE CRUISE
STANDARD DAY ZERO WIND

MODEL: T-38A
DATE: 1 JULY 1978
DATA BASIS: ESTIMATED DATA

TWO ENGINE

DRAG INDEX = 75

ENGINES: (2) J85-GE-5
FUEL GRADE: JP-4
FUEL DENSITY: 6.5 LB/US GAL.

| RANGE AND TIME REMAINING WITH 300-LB RESERVES AT SEA LEVEL | | | | | | | | | | | | PROCEDURE |
|---|---|---|---|---|---|---|---|---|---|---|---|---|
| | | INITIAL ALTITUDE | | | | | | | | | | |
| FUEL | 1000 FT. | SL | 5 | 10 | 15 | 20 | 25 | 30 | 35 | 40 | 45 | |
| 600 LB | NMI<br>MIN | 33<br>6 | 39<br>8 | 44<br>10 | 48<br>12 | 52<br>13 | 57<br>14 | 62<br>16 | 64<br>17 | 67<br>17 | —<br>— | CRUISE AT INITIAL ALTITUDE TO BASE ❶ |
| | 1000 FT | 0/10 | 5/15 | 10/20 | 15/25 | 20/30 | 25/30 | 30/35 | 35/40 | 40 | 40 | OPTIMUM ALTITUDE ❷ |
| | NMI<br>MIN | 33<br>6 | 39<br>8 | 45<br>9 | 50<br>10 | 56<br>11 | 62<br>12 | 69<br>13 | 73<br>14 | 77<br>14 | 77<br>14 | USE OPTIMUM ALTITUDE UNTIL OVER BASE |
| | NMI<br>MIN | 42<br>7 | 52<br>9 | 62<br>10 | 72<br>11 | 83<br>13 | 92<br>14 | 101<br>15 | 112<br>16 | 120<br>17 | 123<br>18 | USE OPTIMUM ALTITUDE AND DESCEND ON COURSE ❸ |
| 800 LB | NMI<br>MIN | 55<br>11 | 64<br>13 | 73<br>15 | 82<br>17 | 91<br>19 | 100<br>20 | 110<br>22 | 117<br>24 | 123<br>25 | —<br>— | CRUISE AT INITIAL ALTITUDE TO BASE ❶ |
| | 1000 FT | 10/25 | 15/30 | 20/30 | 25/35 | 30/40 | 30/40 | 30/40 | 35/40 | 40 | 40 | OPTIMUM ALTITUDE |
| | NMI<br>MIN | 60<br>11 | 69<br>13 | 79<br>14 | 88<br>15 | 99<br>17 | 108<br>18 | 117<br>20 | 125<br>21 | 133<br>21 | 134<br>22 | USE OPTIMUM ALTITUDE UNTIL OVER BASE ❷ |
| | NMI<br>MIN | 80<br>12 | 95<br>14 | 109<br>16 | 121<br>18 | 135<br>19 | 147<br>21 | 159<br>22 | 168<br>23 | 176<br>25 | 179<br>26 | USE OPTIMUM ALTITUDE AND DESCEND ON COURSE ❸ |
| 1000 LB | NMI<br>MIN | 77<br>15 | 90<br>17 | 103<br>20 | 116<br>23 | 129<br>24 | 143<br>27 | 159<br>29 | 169<br>31 | 179<br>32 | —<br>— | CRUISE AT INITIAL ALTITUDE TO BASE ❶ |
| | 1000 FT | 25/30 | 30/35 | 30/40 | 30/40 | 35/40 | 40 | 40 | 40 | 40 | 40 | OPTIMUM ALTITUDE |
| | NMI<br>MIN | 96<br>16 | 110<br>18 | 125<br>20 | 137<br>22 | 149<br>23 | 160<br>24 | 172<br>26 | 181<br>27 | 189<br>28 | 189<br>29 | USE OPTIMUM ALTITUDE UNTIL OVER BASE ❷ |
| | NMI<br>MIN | 126<br>18 | 144<br>21 | 162<br>22 | 177<br>24 | 191<br>26 | 203<br>28 | 215<br>29 | 224<br>30 | 232<br>32 | 234<br>32 | USE OPTIMUM ALTITUDE AND DESCEND ON COURSE ❸ |
| 1400 LB | NMI<br>MIN | 121<br>23 | 141<br>27 | 162<br>30 | 183<br>33 | 205<br>36 | 228<br>39 | 254<br>42 | 272<br>44 | 289<br>45 | —<br>— | CRUISE AT INITIAL ALTITUDE TO BASE ❶ |
| | 1000 FT | 40 | 40 | 40 | 40 | 40 | 40 | 40 | 40 | 40 | 40 | OPTIMUM ALTITUDE |
| | NMI<br>MIN | 191<br>28 | 211<br>31 | 229<br>33 | 244<br>35 | 258<br>37 | 270<br>38 | 282<br>40 | 291<br>41 | 299<br>42 | 301<br>43 | USE OPTIMUM ALTITUDE UNTIL OVER BASE ❷ |
| | NMI<br>MIN | 234<br>31 | 254<br>34 | 272<br>36 | 287<br>38 | 301<br>40 | 313<br>41 | 325<br>43 | 334<br>44 | 342<br>45 | 346<br>46 | USE OPTIMUM ALTITUDE AND DESCEND ON COURSE ❸ |

| CRUISE ALT | | SL | 5 | 10 | 15 | 20 | 25 | 30 | 35 | 40 | |
|---|---|---|---|---|---|---|---|---|---|---|---|
| CRUISE MACH NO. | | 0.47 | 0.52 | 0.56 | 0.61 | 0.65 | 0.70 | 0.75 | 0.79 | 0.84 | DESCEND TO 40,000 FT. |
| APPROX FUEL FLOW LB/HR | | 1400 | 1300 | 1200 | 1125 | 1050 | 975 | 900 | 850 | 825 | |
| DECEND 240 KCAS IDLE ❹ | NMI REMAINING | 6 | 13 | 19 | 25 | 32 | 39 | 45 | 53 | | |
| | MIN REMAINING | 1 | 3 | 4 | 5 | 6 | 7 | 8 | 9 | | |
| | FUEL REMAINING | 310 | 320 | 330 | 339 | 348 | 356 | 363 | 370 | | |

*Note*

WITH MORE THAN 1400 POUNDS FUEL, CRUISE AT 0.86 MACH, PRESSURE ALTITUDE 39,000 FEET.

❶ FUEL AND TIME INCLUDED FOR DESCENT AT DESTINATION WITHOUT DISTANCE CREDIT, SPEED BRAKE CLOSED.
❷ TIME AND FUEL INCLUDED FOR CLIMB TO OPTIMUM ALTITUDE AND DESCENT AT DESTINATION; NO DISTANCE CREDIT FOR DESCENT TO SEA LEVEL DESTINATION, SPEED BRAKE OPEN.
❸ TIME AND FUEL INCLUDED FOR CLIMB TO OPTIMUM ALTITUDE AND DESCENT AT DESTINATION, RANGE INCLUDES DISTANCE FOR ON-COURSE DESCENT TO SEA LEVEL DESTINATION, SPEED BRAKE CLOSED.
❹ DESCENT DATA TABULATED FOR SPEED BRAKE CLOSED; WITH SPEED BRAKE OPEN, USE ONE-HALF OF THE VALUES.
5. CLIMB USING FOLLOWING MIL THRUST CLIMB SPEED SCHEDULE:

| PRESS. ALT (1000 FT.) | SL | 5 | 10 | 15 | 20 | 25 | 30 | 35 | 40 | 45 |
|---|---|---|---|---|---|---|---|---|---|---|
| TRUE MACH | 0.75 | 0.76 | 0.78 | 0.79 | 0.81 | 0.83 | 0.84 | 0.86 | 0.87 | 0.87 |
| KCAS | 496 | 466 | 435 | 406 | 377 | 349 | 322 | 295 | 264 | 236 |

FA4-9.

# DIVERSION RANGE SUMMARY TABLE

CONSTANT ALTITUDE CRUISE
STANDARD DAY ZERO WIND

MODEL: T-38A
DATE: 1 JULY 1978
DATA BASIS: ESTIMATED DATA

SINGLE ENGINE

DRAG INDEX = 0

ENGINES: (2) J85-GE-5
FUEL GRADE: JP-4
FUEL DENSITY: 6.5 LB/US GAL.

RANGE AND TIME REMAINING WITH 300-LB RESERVES AT SEA LEVEL

| FUEL | 1000 FT. | INITIAL ALTITUDE SL | 5 | 10 | 15 | 20 | 25 | 30 | 35 | 40 | 45 | PROCEDURE |
|---|---|---|---|---|---|---|---|---|---|---|---|---|
| 600 LB | NMI<br>MIN | 53<br>11 | 60<br>14 | 65<br>16 | 71<br>18 | 76<br>20 | 78<br>21 | —<br>— | —<br>— | —<br>— | —<br>— | CRUISE AT INITIAL ALTITUDE TO BASE (1) |
| | 1000 FT | 5/15 | 10/20 | 10/20 | 15/20 | 20/25 | 25 | 25 | 25 | 25 | 25 | OPTIMUM ALTITUDE (2) |
| | NMI<br>MIN | 55<br>12 | 60<br>14 | 67<br>15 | 73<br>16 | 80<br>17 | 83<br>17 | 88<br>18 | 97<br>20 | 106<br>20 | 116<br>22 | USE OPTIMUM ALTITUDE UNTIL OVER BASE |
| | NMI<br>MIN | 70<br>14 | 80<br>16 | 91<br>17 | 100<br>19 | 110<br>20 | 118<br>21 | 124<br>21 | 132<br>23 | 142<br>24 | 151<br>25 | USE OPTIMUM ALTITUDE AND DESCEND ON COURSE (3) |
| 800 LB | NMI<br>MIN | 87<br>18 | 99<br>22 | 110<br>25 | 119<br>26 | 131<br>29 | 134<br>30 | —<br>— | —<br>— | —<br>— | —<br>— | CRUISE AT INITIAL ALTITUDE TO BASE (1) |
| | 1000 FT | 10/20 | 15/25 | 20/25 | 20/25 | 20/25 | 25 | 20/25 | 20/25 | 20/25 | 20/25 | OPTIMUM ALTITUDE |
| | NMI<br>MIN | 97<br>21 | 106<br>22 | 116<br>24 | 126<br>25 | 134<br>26 | 139<br>26 | 146<br>27 | 155<br>29 | 164<br>30 | 174<br>32 | USE OPTIMUM ALTITUDE UNTIL OVER BASE (2) |
| | NMI<br>MIN | 123<br>23 | 134<br>25 | 146<br>26 | 156<br>28 | 166<br>29 | 174<br>30 | 180<br>30 | 188<br>32 | 198<br>33 | 208<br>34 | USE OPTIMUM ALTITUDE AND DESCEND ON COURSE (3) |
| 1000 LB | NMI<br>MIN | 122<br>25 | 138<br>29 | 154<br>33 | 168<br>35 | 185<br>38 | 189<br>38 | —<br>— | —<br>— | —<br>— | —<br>— | CRUISE AT INITIAL ALTITUDE TO BASE (1) |
| | 1000 FT | 20/25 | 20/25 | 20/25 | 20/25 | 20/25 | 25 | 20/25 | 20/25 | 20/25 | 20/25 | OPTIMUM ALTITUDE |
| | NMI<br>MIN | 148<br>29 | 159<br>31 | 169<br>33 | 179<br>34 | 188<br>35 | 194<br>35 | 200<br>36 | 209<br>38 | 218<br>39 | 228<br>41 | USE OPTIMUM ALTITUDE UNTIL OVER BASE (2) |
| | NMI<br>MIN | 178<br>32 | 190<br>34 | 201<br>35 | 211<br>36 | 221<br>38 | 229<br>38 | 235<br>39 | 244<br>41 | 253<br>42 | 263<br>43 | USE OPTIMUM ALTITUDE AND DESCEND ON COURSE (3) |
| 1400 LB | NMI<br>MIN | 189<br>39 | 214<br>44 | 239<br>49 | 261<br>51 | 288<br>55 | 294<br>55 | —<br>— | —<br>— | —<br>— | —<br>— | CRUISE AT INITIAL ALTITUDE TO BASE (1) |
| | 1000 FT | 20/25 | 20/25 | 20/25 | 20/25 | 20/25 | 25 | 25 | 25 | 25 | 25 | OPTIMUM ALTITUDE |
| | NMI<br>MIN | 250<br>47 | 261<br>48 | 272<br>50 | 282<br>51 | 291<br>53 | 299<br>52 | 304<br>53 | 313<br>54 | 322<br>55 | 332<br>57 | USE OPTIMUM ALTITUDE UNTIL OVER BASE (2) |
| | NMI<br>MIN | 282<br>49 | 294<br>51 | 305<br>52 | 316<br>53 | 325<br>55 | 334<br>55 | 339<br>56 | 348<br>58 | 358<br>58 | 367<br>60 | USE OPTIMUM ALTITUDE AND DESCEND ON COURSE (3) |

| | | | | | | | |
|---|---|---|---|---|---|---|---|
| CRUISE ALT | | SL | 5 | 10 | 15 | 20 | 25 |
| CRUISE MACH NO. | | 0.44 | 0.47 | 0.50 | 0.54 | 0.58 | 0.62 |
| APPROX FUEL FLOW LB/HR | | 1650 | 1550 | 1400 | 1350 | 1300 | 1275 |
| DECEND 240 KCAS IDLE (4) | NMI REMAINING | | 8 | 16 | 24 | 32 | 40 |
| | MIN REMAINING | | 2 | 4 | 5 | 7 | 8 |
| | FUEL REMAINING | | 306 | 314 | 320 | 326 | 332 |

DESCEND TO 20,000 OR 25,000 FT.

USE IDLE THRUST AND 240 KCAS WITH SPEED BRAKE CLOSED.

*Note*

WITH MORE THAN 1400 POUNDS FUEL, CRUISE AT 0.62 MACH, PRESSURE ALTITUDE 23,000 FEET.

1. FUEL AND TIME INCLUDED FOR DESCENT AT DESTINATION WITHOUT DISTANCE CREDIT, SPEED BRAKE CLOSED.
2. TIME AND FUEL INCLUDED FOR CLIMB TO OPTIMUM ALTITUDE AND DESCENT AT DESTINATION; NO DISTANCE CREDIT FOR DESCENT TO SEA LEVEL DESTINATION, SPEED BRAKE OPEN.
3. TIME AND FUEL INCLUDED FOR CLIMB TO OPTIMUM ALTITUDE AND DESCENT AT DESTINATION, RANGE INCLUDES DISTANCE FOR ON-COURSE DESCENT TO SEA LEVEL DESTINATION, SPEED BRAKE CLOSED.
4. DESCENT DATA TABULATED FOR SPEED BRAKE CLOSED; WITH SPEED BRAKE OPEN, USE ONE-HALF OF THE VALUES.
5. CLIMB USING FOLLOWING MIL THRUST CLIMB SPEED SCHEDULE:

| PRESS. ALT (1000 FT.) | SL | 5 | 10 | 15 | 20 | 25 |
|---|---|---|---|---|---|---|
| TRUE MACH | 0.43 | 0.46 | 0.49 | 0.52 | 0.56 | 0.59 |
| KCAS | 281 | 278 | 271 | 264 | 256 | 246 |

FA4-10.

# DIVERSION RANGE SUMMARY TABLE

CONSTANT ALTITUDE CRUISE
STANDARD DAY ZERO WIND

MODEL: T-38A
DATE: 1 JULY 1978
DATA BASIS: ESTIMATED DATA

SINGLE ENGINE

DRAG INDEX = 25

ENGINES: (2) J85-GE-5
FUEL GRADE: JP-4
FUEL DENSITY: 6.5 LB/US GAL.

RANGE AND TIME REMAINING WITH 300-LB RESERVES AT SEA LEVEL

| FUEL | 1000 FT. | INITIAL ALTITUDE SL | 5 | 10 | 15 | 20 | 25 | 30 | 35 | 40 | 45 | PROCEDURE |
|---|---|---|---|---|---|---|---|---|---|---|---|---|
| 600 LB | NMI<br>MIN | 49<br>11 | 55<br>13 | 61<br>15 | 65<br>17 | 71<br>18 | —<br>— | —<br>— | —<br>— | —<br>— | —<br>— | CRUISE AT INITIAL ALTITUDE TO BASE (1) |
| | 1000 FT | 5/15 | 5/15 | 10/20 | 15/20 | 20 | 20 | 20 | 20 | 20 | 20 | OPTIMUM ALTITUDE (2) |
| | NMI<br>MIN | 50<br>11 | 56<br>12 | 62<br>13 | 67<br>15 | 74<br>16 | 76<br>16 | 85<br>18 | 92<br>19 | 100<br>20 | 106<br>21 | USE OPTIMUM ALTITUDE UNTIL OVER BASE |
| | NMI<br>MIN | 63<br>13 | 72<br>14 | 84<br>16 | 93<br>17 | 101<br>18 | 107<br>19 | 112<br>21 | 119<br>21 | 127<br>22 | 133<br>23 | USE OPTIMUM ALTITUDE AND DESCEND ON COURSE (3) |
| 800 LB | NMI<br>MIN | 81<br>18 | 92<br>21 | 102<br>23 | 110<br>25 | 120<br>27 | —<br>— | —<br>— | —<br>— | —<br>— | —<br>— | CRUISE AT INITIAL ALTITUDE TO BASE (1) |
| | 1000 FT | 10/20 | 15/20 | 20 | 20 | 20/25 | 20 | 20 | 20 | 20 | 20 | OPTIMUM ALTITUDE |
| | NMI<br>MIN | 89<br>19 | 97<br>20 | 106<br>22 | 116<br>23 | 123<br>24 | 126<br>25 | 135<br>27 | 142<br>27 | 150<br>28 | 156<br>29 | USE OPTIMUM ALTITUDE UNTIL OVER BASE (2) |
| | NMI<br>MIN | 111<br>21 | 123<br>23 | 133<br>24 | 143<br>26 | 153<br>26 | 156<br>27 | 162<br>29 | 169<br>30 | 177<br>31 | 183<br>32 | USE OPTIMUM ALTITUDE AND DESCEND ON COURSE (3) |
| 1000 LB | NMI<br>MIN | 113<br>25 | 128<br>28 | 142<br>30 | 154<br>33 | 169<br>35 | —<br>— | —<br>— | —<br>— | —<br>— | —<br>— | CRUISE AT INITIAL ALTITUDE TO BASE (1) |
| | 1000 FT | 20 | 20/25 | 20/25 | 20/25 | 20 | 20 | 20 | 20 | 20 | 20 | OPTIMUM ALTITUDE |
| | NMI<br>MIN | 133<br>27 | 144<br>29 | 155<br>30 | 165<br>31 | 172<br>33 | 174<br>33 | 183<br>35 | 190<br>36 | 198<br>37 | 204<br>37 | USE OPTIMUM ALTITUDE UNTIL OVER BASE (2) |
| | NMI<br>MIN | 160<br>29 | 171<br>31 | 185<br>32 | 194<br>33 | 199<br>35 | 205<br>36 | 210<br>37 | 217<br>38 | 225<br>39 | 231<br>40 | USE OPTIMUM ALTITUDE AND DESCEND ON COURSE (3) |
| 1400 LB | NMI<br>MIN | 176<br>39 | 198<br>43 | 221<br>45 | 240<br>48 | 262<br>51 | —<br>— | —<br>— | —<br>— | —<br>— | —<br>— | CRUISE AT INITIAL ALTITUDE TO BASE (1) |
| | 1000 FT | 20 | 20 | 20 | 20 | 20 | 20 | 20 | 20 | 20 | 20 | OPTIMUM ALTITUDE |
| | NMI<br>MIN | 226<br>43 | 237<br>45 | 248<br>46 | 257<br>47 | 265<br>48 | 267<br>49 | 276<br>51 | 283<br>52 | 291<br>53 | 297<br>53 | USE OPTIMUM ALTITUDE UNTIL OVER BASE (2) |
| | NMI<br>MIN | 253<br>45 | 264<br>47 | 275<br>49 | 284<br>50 | 292<br>51 | 297<br>52 | 303<br>53 | 310<br>54 | 318<br>55 | 324<br>56 | USE OPTIMUM ALTITUDE AND DESCEND ON COURSE (3) |

| | | | | | | | |
|---|---|---|---|---|---|---|---|
| CRUISE ALT | | SL | 5 | 10 | 15 | 20 | 25 | DESCEND TO 20,000 FT. |
| CRUISE MACH NO. | | 0.41 | 0.45 | 0.49 | 0.53 | 0.57 | 0.62 | |
| APPROX FUEL FLOW LB/HR | | 1650 | 1575 | 1500 | 1450 | 1375 | 1400 | USE IDLE THRUST AND 240 KCAS WITH SPEED BRAKE CLOSED. |
| DECEND 240 KCAS IDLE (4) | NMI REMAINING | | 7 | 15 | 22 | 30 | 28 | |
| | MIN REMAINING | | 2 | 3 | 5 | 6 | 7 | |
| | FUEL REMAINING | | 306 | 313 | 319 | 325 | 330 | |

*Note*

WITH MORE THAN 1400 POUNDS FUEL, CRUISE AT 0.60 MACH, PRESSURE ALTITUDE 20,000 FEET.

(1) FUEL AND TIME INCLUDED FOR DESCENT AT DESTINATION WITHOUT DISTANCE CREDIT, SPEED BRAKE CLOSED.
(2) TIME AND FUEL INCLUDED FOR CLIMB TO OPTIMUM ALTITUDE AND DESCENT AT DESTINATION; NO DISTANCE CREDIT FOR DESCENT TO SEA LEVEL DESTINATION, SPEED BRAKE OPEN.
(3) TIME AND FUEL INCLUDED FOR CLIMB TO OPTIMUM ALTITUDE AND DESCENT AT DESTINATION, RANGE INCLUDES DISTANCE FOR ON-COURSE DESCENT TO SEA LEVEL DESTINATION, SPEED BRAKE CLOSED.
(4) DESCENT DATA TABULATED FOR SPEED BRAKE CLOSED; WITH SPEED BRAKE OPEN, USE ONE-HALF OF THE VALUES.
5. CLIMB USING FOLLOWING MIL THRUST CLIMB SPEED SCHEDULE:

| PRESS. ALT (1000 FT.) | SL | 5 | 10 | 15 | 20 | 25 |
|---|---|---|---|---|---|---|
| TRUE MACH | 0.43 | 0.46 | 0.49 | 0.52 | 0.56 | 0.59 |
| KCAS | 281 | 278 | 271 | 264 | 256 | 246 |

FA4-11.

# DIVERSION RANGE SUMMARY TABLE

CONSTANT ALTITUDE CRUISE
STANDARD DAY ZERO WIND

MODEL: **T-38A**
DATE: **1 JULY 1978**
DATA BASIS: **ESTIMATED DATA**

SINGLE ENGINE

**DRAG INDEX = 50**

ENGINES: (2) J85-GE-5
FUEL GRADE: JP-4
FUEL DENSITY: 6.5 LB/US GAL.

| RANGE AND TIME REMAINING WITH 300-LB RESERVES AT SEA LEVEL | | | | | | | | | | | | |
|---|---|---|---|---|---|---|---|---|---|---|---|---|
| FUEL | 1000 FT. | INITIAL ALTITUDE SL | 5 | 10 | 15 | 20 | 25 | 30 | 35 | 40 | 45 | PROCEDURE |
| 600 LB | NMI<br>MIN | 46<br>10 | 52<br>12 | 56<br>14 | 60<br>15 | 64<br>17 | — | — | — | — | — | CRUISE AT INITIAL ALTITUDE TO BASE (1) |
| | 1000 FT | 5/10 | 5/15 | 10/15 | 15/20 | 20 | 20 | 20 | 20 | 20 | 20 | OPTIMUM ALTITUDE (5) |
| | NMI<br>MIN | 47<br>10 | 52<br>11 | 58<br>13 | 62<br>14 | 67<br>15 | 69<br>15 | 79<br>16 | 85<br>17 | 92<br>18 | 98<br>19 | USE OPTIMUM ALTITUDE UNTIL OVER BASE |
| | NMI<br>MIN | 57<br>12 | 66<br>13 | 74<br>14 | 85<br>16 | 91<br>17 | 96<br>18 | 103<br>19 | 109<br>20 | 117<br>21 | 123<br>22 | USE OPTIMUM ALTITUDE AND DESCEND ON COURSE (3) |
| 800 LB | NMI<br>MIN | 77<br>17 | 86<br>19 | 95<br>22 | 102<br>23 | 109<br>25 | — | — | — | — | — | CRUISE AT INITIAL ALTITUDE TO BASE (1) |
| | 1000 FT | 10/15 | 10/20 | 15/20 | 20 | 20 | 20 | 20 | 20 | 20 | 20 | OPTIMUM ALTITUDE |
| | NMI<br>MIN | 82<br>18 | 89<br>19 | 97<br>20 | 105<br>22 | 111<br>23 | 113<br>23 | 123<br>24 | 130<br>25 | 137<br>26 | 143<br>27 | USE OPTIMUM ALTITUDE UNTIL OVER BASE (2) |
| | NMI<br>MIN | 98<br>19 | 111<br>21 | 121<br>23 | 130<br>24 | 136<br>25 | 141<br>26 | 148<br>27 | 154<br>28 | 161<br>29 | 168<br>29 | USE OPTIMUM ALTITUDE AND DESCEND ON COURSE (3) |
| 1000 LB | NMI<br>MIN | 107<br>24 | 120<br>26 | 133<br>29 | 143<br>31 | 153<br>33 | — | — | — | — | — | CRUISE AT INITIAL ALTITUDE TO BASE (1) |
| | 1000 FT | 15/20 | 20 | 20 | 20 | 20 | 20 | 20 | 20 | 20 | 20 | OPTIMUM ALTITUDE |
| | NMI<br>MIN | 121<br>25 | 130<br>27 | 140<br>28 | 149<br>30 | 155<br>31 | 157<br>31 | 167<br>32 | 174<br>33 | 181<br>34 | 187<br>35 | USE OPTIMUM ALTITUDE UNTIL OVER BASE (2) |
| | NMI<br>MIN | 145<br>28 | 155<br>29 | 165<br>31 | 174<br>32 | 180<br>33 | 185<br>34 | 192<br>35 | 198<br>36 | 205<br>36 | 212<br>37 | USE OPTIMUM ALTITUDE AND DESCEND ON COURSE (3) |
| 1400 LB | NMI<br>MIN | 165<br>37 | 185<br>40 | 206<br>43 | 221<br>46 | 236<br>48 | — | — | — | — | — | CRUISE AT INITIAL ALTITUDE TO BASE (1) |
| | 1000 FT | 20 | 20 | 20 | 20 | 20 | 20 | 20 | 20 | 20 | 20 | OPTIMUM ALTITUDE |
| | NMI<br>MIN | 204<br>40 | 215<br>42 | 225<br>43 | 233<br>44 | 239<br>45 | 241<br>46 | 251<br>47 | 257<br>48 | 265<br>49 | 271<br>50 | USE OPTIMUM ALTITUDE UNTIL OVER BASE (2) |
| | NMI<br>MIN | 229<br>43 | 239<br>44 | 249<br>46 | 258<br>47 | 263<br>48 | 269<br>49 | 276<br>50 | 282<br>50 | 289<br>51 | 295<br>52 | USE OPTIMUM ALTITUDE AND DESCEND ON COURSE (3) |

| CRUISE ALT | | SL | 5 | 10 | 15 | 20 | 25 | DESCEND TO 20,000 FT |
|---|---|---|---|---|---|---|---|---|
| CRUISE MACH NO. | | 0.41 | 0.45 | 0.49 | 0.53 | 0.57 | | USE IDLE THRUST AND 240 KCAS WITH SPEED BRAKE CLOSED. |
| APPROX FUEL FLOW LB/HR | | 1850 | 1700 | 1625 | 1575 | 1525 | | |
| DESCEND 240 KCAS IDLE (4) | NMI REMAINING | | 7 | 14 | 20 | 27 | | |
| | MIN REMAINING | | 1 | 3 | 4 | 6 | | |
| | FUEL REMAINING | | 305 | 312 | 317 | 321 | | |

*Note*

WITH MORE THAN 1400 POUNDS FUEL, CRUISE AT 0.58 MACH, PRESSURE ALTITUDE 20,000 FEET.

(1) FUEL AND TIME INCLUDED FOR DESCENT AT DESTINATION WITHOUT DISTANCE CREDIT, SPEED BRAKE CLOSED.
(2) TIME AND FUEL INCLUDED FOR CLIMB TO OPTIMUM ALTITUDE AND DESCENT AT DESTINATION; NO DISTANCE CREDIT FOR DESCENT TO SEA LEVEL DESTINATION, SPEED BRAKE OPEN.
(3) TIME AND FUEL INCLUDED FOR CLIMB TO OPTIMUM ALTITUDE AND DESCENT AT DESTINATION, RANGE INCLUDES DISTANCE FOR ON-COURSE DESCENT TO SEA LEVEL DESTINATION, SPEED BRAKE CLOSED.
(4) DESCENT DATA TABULATED FOR SPEED BRAKE CLOSED; WITH SPEED BRAKE OPEN, USE ONE-HALF OF THE VALUES.
5. CLIMB USING FOLLOWING MIL THRUST CLIMB SPEED SCHEDULE:

| PRESS. ALT (1000 FT.) | SL | 5 | 10 | 15 | 20 | 25 |
|---|---|---|---|---|---|---|
| TRUE MACH | 0.43 | 0.46 | 0.49 | 0.52 | 0.56 | 0.59 |
| KCAS | 281 | 278 | 271 | 264 | 256 | 246 |

FA4-12.

# DIVERSION RANGE SUMMARY TABLE

CONSTANT ALTITUDE CRUISE
STANDARD DAY ZERO WIND

MODEL: **T-38A**
DATE: **1 JULY 1978**
DATA BASIS: **ESTIMATED DATA**

**SINGLE ENGINE**

**DRAG INDEX = 75**

ENGINES: (2) J85-GE-5
FUEL GRADE: JP-4
FUEL DENSITY: 6.5 LB/US GAL.

RANGE AND TIME REMAINING WITH 300-LB RESERVES AT SEA LEVEL

| FUEL | 1000 FT. | INITIAL ALTITUDE SL | 5 | 10 | 15 | 20 | 25 | 30 | 35 | 40 | 45 | PROCEDURE |
|---|---|---|---|---|---|---|---|---|---|---|---|---|
| 600 LB | NMI<br>MIN | 44<br>10 | 48<br>11 | 53<br>13 | 56<br>15 | 59<br>16 | —<br>— | —<br>— | —<br>— | —<br>— | —<br>— | CRUISE AT INITIAL ALTITUDE TO BASE (1) |
| | 1000 FT | 5/10 | 5/10 | 10/15 | 15 | 20 | 20 | 20 | 20 | 20 | 20 | OPTIMUM ALTITUDE (5) |
| | NMI<br>MIN | 44<br>10 | 49<br>11 | 54<br>12 | 58<br>13 | 61<br>13 | 63<br>14 | 73<br>15 | 78<br>16 | 85<br>17 | 91<br>18 | USE OPTIMUM ALTITUDE UNTIL OVER BASE |
| | NMI<br>MIN | 53<br>11 | 60<br>12 | 69<br>13 | 75<br>15 | 84<br>16 | 89<br>16 | 96<br>17 | 101<br>18 | 108<br>19 | 113<br>20 | USE OPTIMUM ALTITUDE AND DESCEND ON COURSE (3) |
| 800 LB | NMI<br>MIN | 72<br>16 | 81<br>18 | 89<br>20 | 95<br>22 | 100<br>23 | —<br>— | —<br>— | —<br>— | —<br>— | —<br>— | CRUISE AT INITIAL ALTITUDE TO BASE (1) |
| | 1000 FT | 10/15 | 10/15 | 15 | 20 | 20 | 20 | 20 | 20 | 20 | 20 | OPTIMUM ALTITUDE |
| | NMI<br>MIN | 76<br>17 | 83<br>18 | 90<br>19 | 97<br>20 | 102<br>21 | 104<br>21 | 114<br>22 | 119<br>23 | 126<br>24 | 132<br>25 | USE OPTIMUM ALTITUDE UNTIL OVER BASE (2) |
| | NMI<br>MIN | 91<br>18 | 100<br>19 | 108<br>21 | 120<br>22 | 125<br>23 | 130<br>24 | 136<br>24 | 142<br>25 | 149<br>26 | 154<br>27 | USE OPTIMUM ALTITUDE AND DESCEND ON COURSE (3) |
| 1000 LB | NMI<br>MIN | 101<br>23 | 113<br>25 | 124<br>27 | 133<br>29 | 131<br>28 | —<br>— | —<br>— | —<br>— | —<br>— | —<br>— | CRUISE AT INITIAL ALTITUDE TO BASE (1) |
| | 1000 FT | 10/15 | 15/20 | 20 | 20 | 20 | 20 | 20 | 20 | 20 | 20 | OPTIMUM ALTITUDE |
| | NMI<br>MIN | 111<br>23 | 120<br>25 | 129<br>26 | 138<br>27 | 140<br>27 | 144<br>28 | 154<br>30 | 159<br>30 | 166<br>31 | 172<br>32 | USE OPTIMUM ALTITUDE UNTIL OVER BASE (2) |
| | NMI<br>MIN | 129<br>25 | 142<br>27 | 152<br>28 | 161<br>29 | 156<br>28 | 170<br>31 | 177<br>32 | 182<br>33 | 189<br>33 | 195<br>34 | USE OPTIMUM ALTITUDE AND DESCEND ON COURSE (3) |
| 1400 LB | NMI<br>MIN | 156<br>35 | 174<br>38 | 192<br>41 | 206<br>43 | 210<br>44 | —<br>— | —<br>— | —<br>— | —<br>— | —<br>— | CRUISE AT INITIAL ALTITUDE TO BASE (1) |
| | 1000 FT | 20 | 20 | 20 | 20 | 20 | 20 | 20 | 20 | 20 | 20 | OPTIMUM ALTITUDE |
| | NMI<br>MIN | 187<br>37 | 197<br>38 | 208<br>40 | 217<br>41 | 220<br>41 | 224<br>42 | 229<br>43 | 234<br>44 | 237<br>45 | 242<br>46 | USE OPTIMUM ALTITUDE UNTIL OVER BASE (2) |
| | NMI<br>MIN | 210<br>39 | 220<br>40 | 231<br>42 | 236<br>43 | 240<br>43 | 244<br>44 | 250<br>45 | 254<br>46 | 261<br>47 | 265<br>48 | USE OPTIMUM ALTITUDE AND DESCEND ON COURSE (3) |

| CRUISE ALT | SL | 5 | 10 | 15 | 20 | 25 | DESCEND TO 20,000 FT |
|---|---|---|---|---|---|---|---|
| CRUISE MACH NO. | 0.40 | 0.44 | 0.48 | 0.51 | 0.55 | | USE IDLE THRUST AND 240 KCAS WITH SPEED BRAKE CLOSED. |
| APPROX FUEL FLOW LB/HR | 1825 | 1750 | 1700 | 1625 | 1600 | | |

| DECEND 240 KCAS IDLE (4) | | | | | |
|---|---|---|---|---|---|
| NMI REMAINING | 6 | 13 | 19 | 25 | |
| MIN REMAINING | 1 | 3 | 4 | 5 | |
| FUEL REMAINING | 305 | 310 | 315 | 320 | |

***Note***

WITH MORE THAN 1400 POUNDS FUEL, CRUISE AT 0.56 MACH, PRESSURE ALTITUDE 18,000 FEET.

1. FUEL AND TIME INCLUDED FOR DESCENT AT DESTINATION WITHOUT DISTANCE CREDIT, SPEED BRAKE CLOSED.
2. TIME AND FUEL INCLUDED FOR CLIMB TO OPTIMUM ALTITUDE AND DESCENT AT DESTINATION; NO DISTANCE CREDIT FOR DESCENT TO SEA LEVEL DESTINATION, SPEED BRAKE OPEN.
3. TIME AND FUEL INCLUDED FOR CLIMB TO OPTIMUM ALTITUDE AND DESCENT AT DESTINATION, RANGE INCLUDES DISTANCE FOR ON-COURSE DESCENT TO SEA LEVEL DESTINATION, SPEED BRAKE CLOSED.
4. DESCENT DATA TABULATED FOR SPEED BRAKE CLOSED; WITH SPEED BRAKE OPEN, USE ONE-HALF OF THE VALUES.
5. CLIMB USING FOLLOWING MIL THRUST CLIMB SPEED SCHEDULE:

| PRESS. ALT (1000 FT.) | SL | 5 | 10 | 15 | 20 | 25 |
|---|---|---|---|---|---|---|
| TRUE MACH | 0.43 | 0.46 | 0.49 | 0.52 | 0.56 | 0.59 |
| KCAS | 281 | 278 | 271 | 264 | 256 | 246 |

FA4-13.

# PART 5 ENDURANCE

## TABLE OF CONTENTS



## PURPOSE OF CHARTS.

Endurance charts determine the optimum mach number and fuel required to loiter at a given altitude for a specific period of time. A correction grid to gross weight for bank angle and a temperature correction grid (hotter-than-standard conditions) to fuel flow are provided for optional use.

NOTE

> The effects of temperature for colder-than-standard day conditions are considered negligible. Use standard day (baseline) for temperatures below standard day.

The altitude for maximum loiter time is defined in the charts by the drag index curves titled "optimum maximum endurance altitude" contained in the gross weight grid. The endurance chart for two-engine operation provides data for drag indices of 0 thru 75. The single-engine endurance chart provides data for drag indices of 0 thru 75.

## USE.

Enter the appropriate two-engine or single-engine chart (FA5-1 or FA5-2) with gross weight. If the loiter period requires turning flight, gross weight should be corrected for bank angle. To use the bank angle correction grid, enter with gross weight and contour the nearest guideline to the right while simultaneously entering the bank angle scale with desired degree of bank angle and projecting up. At the point of intersection of the two projections, proceed left and read gross weight corrected for bank angle.

Gross weight (corrected for bank angle, if required) is then projected right from the gross weight scale of the chart to the pressure altitude. If maximum loiter time is desired, stop momentarily at the optimum maximum endurance altitude drag index curve (interpolate, if necessary). Mark this position location on the chart for further use.

From the point of intersection with pressure altitude, proceed up to the configuration drag index in the upper left grid of the chart, then left to read the indicated mach number for loiter. Return to the plotted point intersection of the gross weight and pressure altitude and proceed down to the drag index at the lower left portion of the chart, then right to the gross weight curve. From this point proceed up to the baseline of the temperature correction grid (standard day). For hotter-than-standard day condition, contour the guidelines to the temperature increase. (If no increase is required, proceed directly thru.) Fuel flow can be read while proceeding up to the desired loiter time. Project right to read fuel required for loiter.

If loiter is already known, project left from the fuel required scale and simultaneously intersect the vertical plot projected from the temperature grid to read loiter time.

The chase-thru lines on FA5-1 represent an average gross weight of 10,500 pounds, bank angle of 20 degrees, drag index of 50, loiter time of 30 minutes, and a temperature deviation of 10°C hotter than standard. These lines show corrected gross weight of 11,200, an optimum maximum endurance altitude for a drag index of 50 of 32,000 feet, a loiter speed of 0.71 mach, a fuel flow of 1800 lbs/hr, and 900 pounds of fuel required. For loiter times of long duration greater accuracy requires use of average gross weight during loiter to calculate the fuel required. To obtain average loiter weight, the fuel required to loiter must first be determined based on gross weight at start or end of loiter and then is recalculated based on start or end gross weight, decreased or increased, respectively, by half the calculated loiter fuel.

MODEL: T-38A
DATE: 1 JULY 1978
DATA BASIS: ESTIMATED DATA

# MAXIMUM ENDURANCE
## FLAPS UP
FUEL REQUIRED AND
MACH NUMBER
STANDARD DAY
DRAG INDEX= 0 TO 75

ENGINE: (2) J85-GE-5
FUEL GRADE: JP-4
FUEL DENSITY: 6.5 LB/US GAL

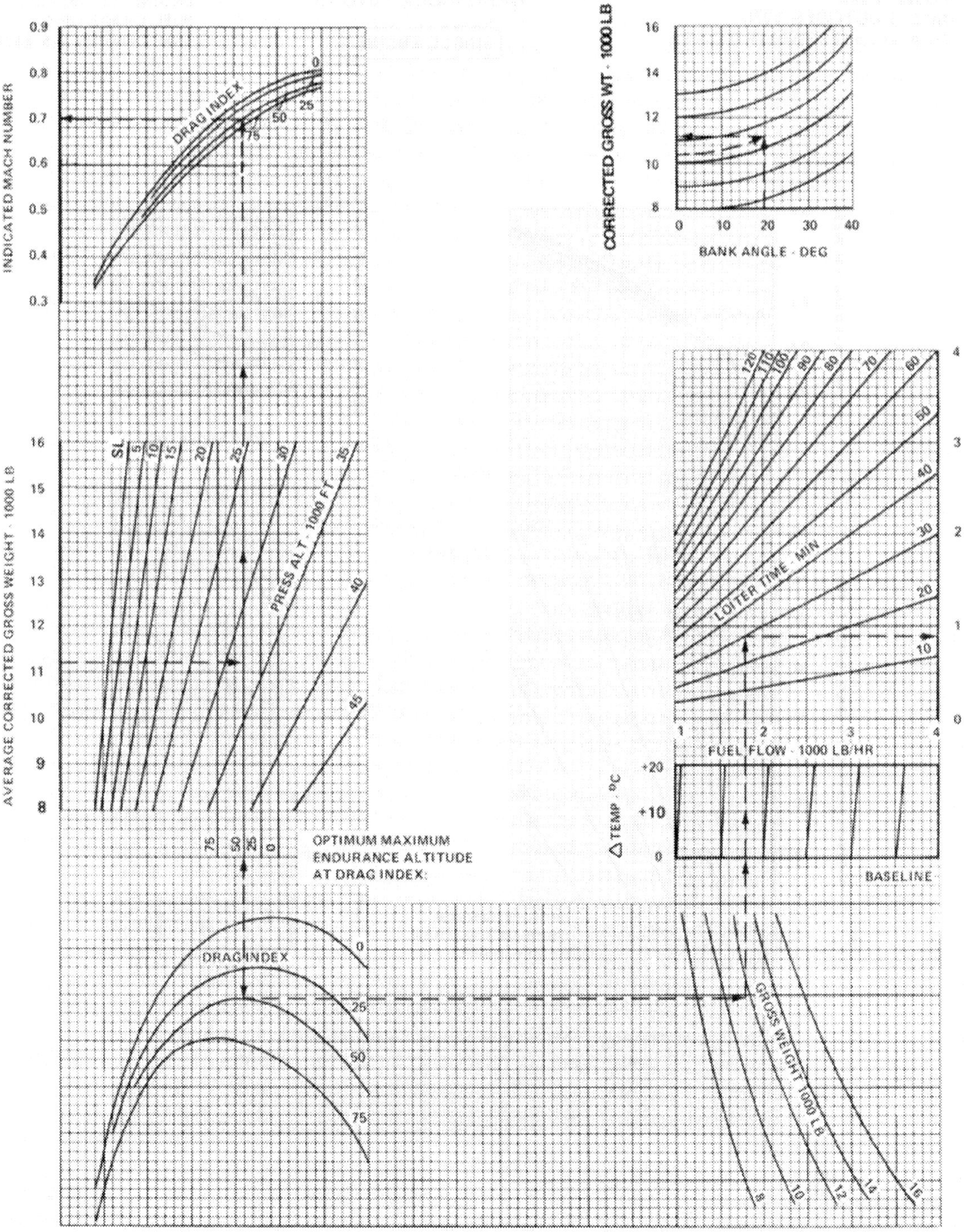


FA5-1.

# MAXIMUM ENDURANCE
## FLAPS UP
### FUEL REQUIRED AND MACH NUMBER

STANDARD DAY
DRAG INDEX = 0TO 75

SINGLE ENGINE

MODEL: T-38A
DATE: 1 OCTOBER 1976
DATA BASIS: **ESTIMATED DATA**

ENGINE: (2) J85-GE-5
FUEL GRADE: JP-4
FUEL DENSITY: 6.5 LB/US GA

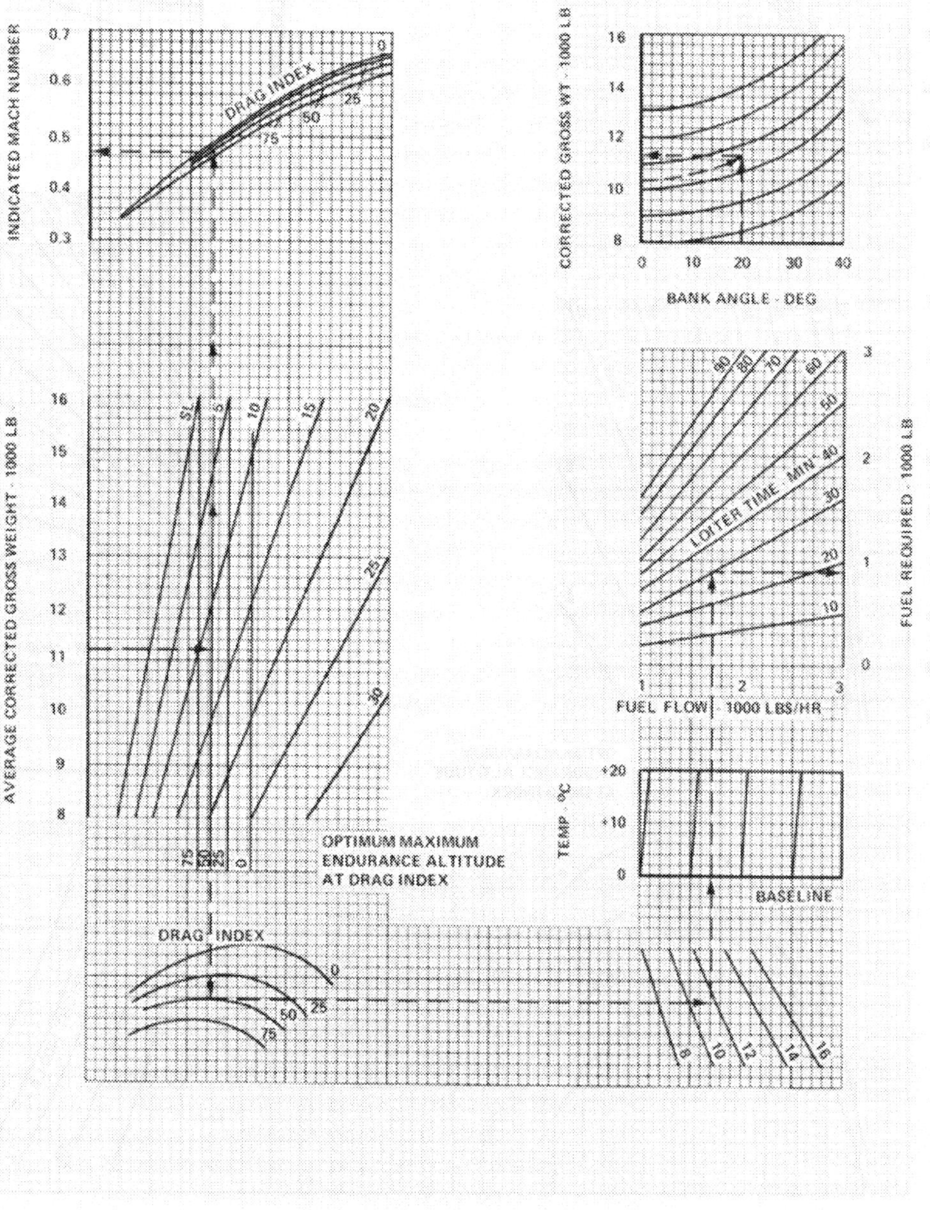


FA5-2.

# PART 6 DESCENT

TABLE OF CONTENTS



## PURPOSE OF CHARTS.

The descent charts provide a means of determining the fuel, time, and distance required to descend from altitude with speed brake closed or open.

## DESCENT CHARTS.

The maximum range descent chart (FA6-1) shows the performance for maximum range. This range is obtained by using idle thrust and maintaining an airspeed of 240 KCAS. FA6-2 and FA6-3 give the performance for penetration descent. This chart requires 80% RPM and an airspeed of 280 KCAS. The descent charts may be used for descending from one altitude to another by reading the incremental values between the initial and final altitudes.

## USE OF DESCENT CHARTS.

Assume that maximum range descent is desired from a pressure altitude of 40,000 feet to 10,000 feet. From FA6-1, the chase-thru lines show the fuel to descend is 70 pounds (98 − 28), the time is 9 minutes (12.5 − 3.5), and the distance is 54 nautical miles (70 − 16).

# MAXIMUM RANGE DESCENT
## IDLE THRUST 240 KCAS

MODEL: T-38A
DATE: 1 OCTOBER 1976
DATA BASIS: ESTIMATED DATA

STANDARD DAY
DRAG INDEX 0 TO 75
ALL GROSS WEIGHTS
SPEED BRAKE IN

ENGINE: (2) J85-GE-5
FUEL GRADE: JP-4
FUEL DENSITY: 6.5 LB/US GAL

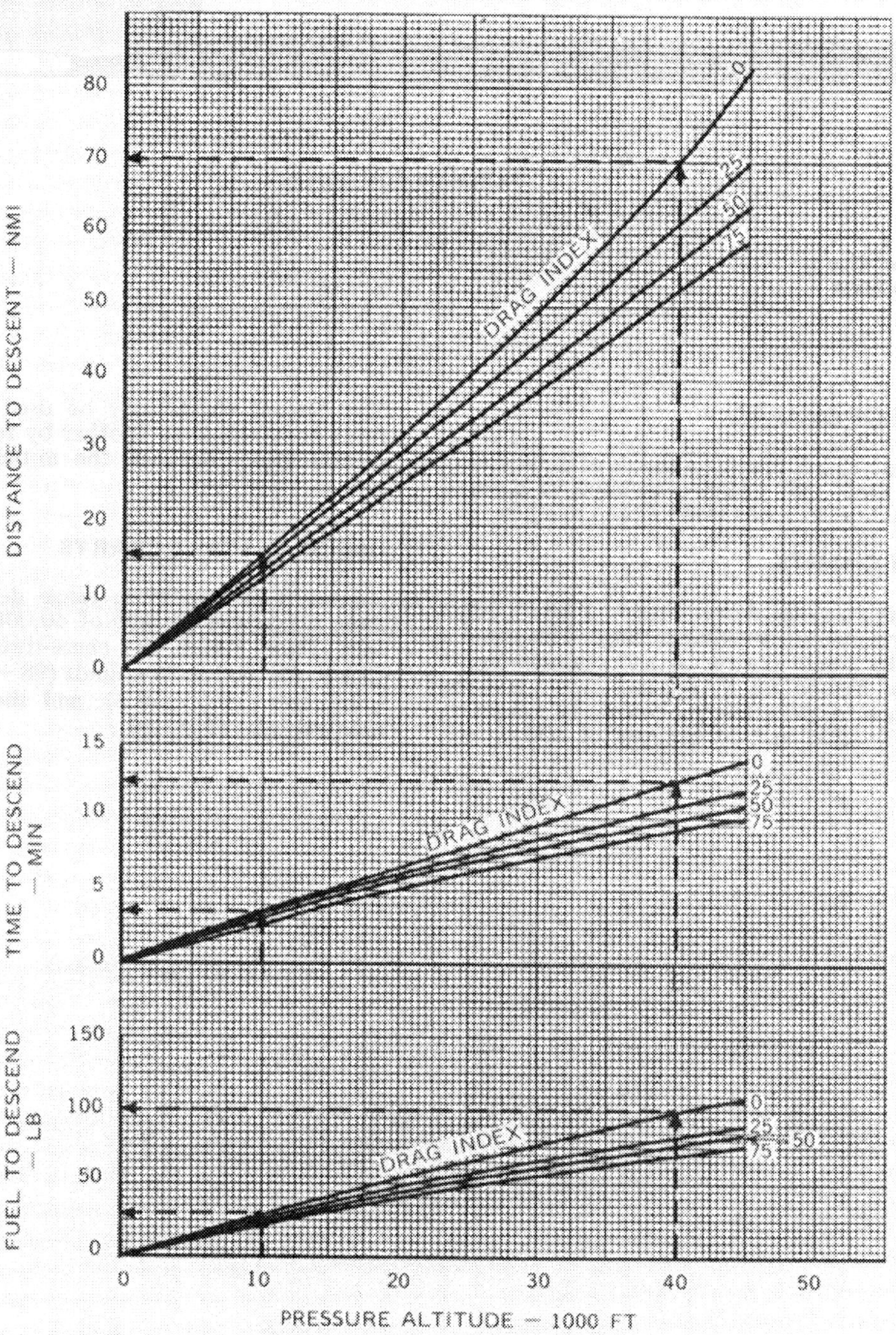


FA6-1

# PENETRATION DESCENT

80% RPM 280 KCAS
STANDARD DAY

MODEL: T-38A
DATE: 1 OCTOBER 1976
DATA BASIS: ESTIMATED DATA

DRAG INDEX 0 TO 75
SPEED BRAKE—CLOSED

ENGINE: (2) J85-GE-5
FUEL GRADE: JP-4
FUEL DENSITY: 6.5 LB/US GAL

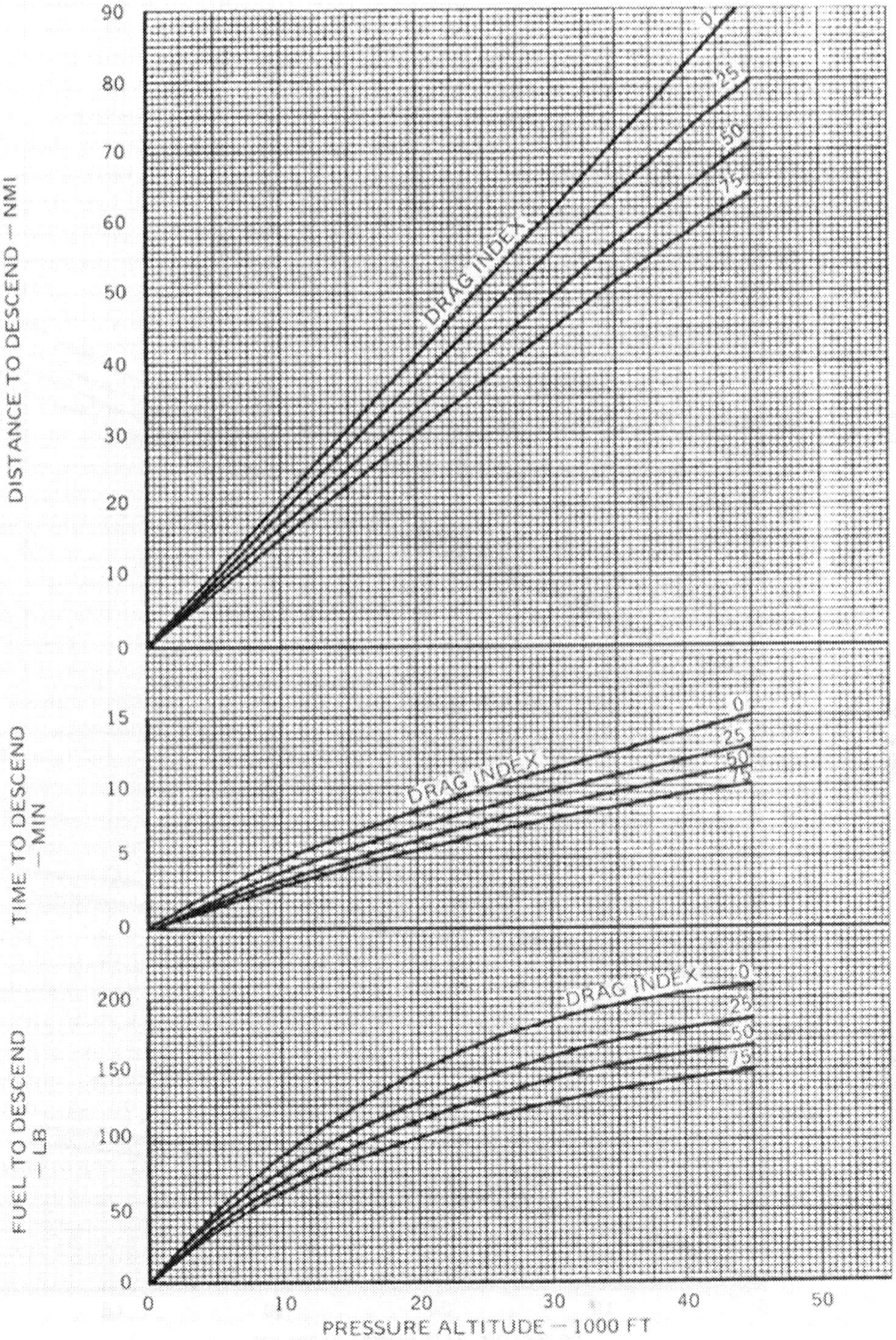


FA6-2.

## PENETRATION DESCENT

80% RPM — 280 KCAS — STANDARD DAY
SPEED BRAKE — OUT
DRAG INDEX 0 TO 75

MODEL: T-38A
DATE: 1 OCTOBER 1976
DATA BASIS: ESTIMATED DATA

ENGINE: (2) J85-GE-5
FUEL GRADE: JP-4
FUEL DENSITY: 6.5 LB/US GAL

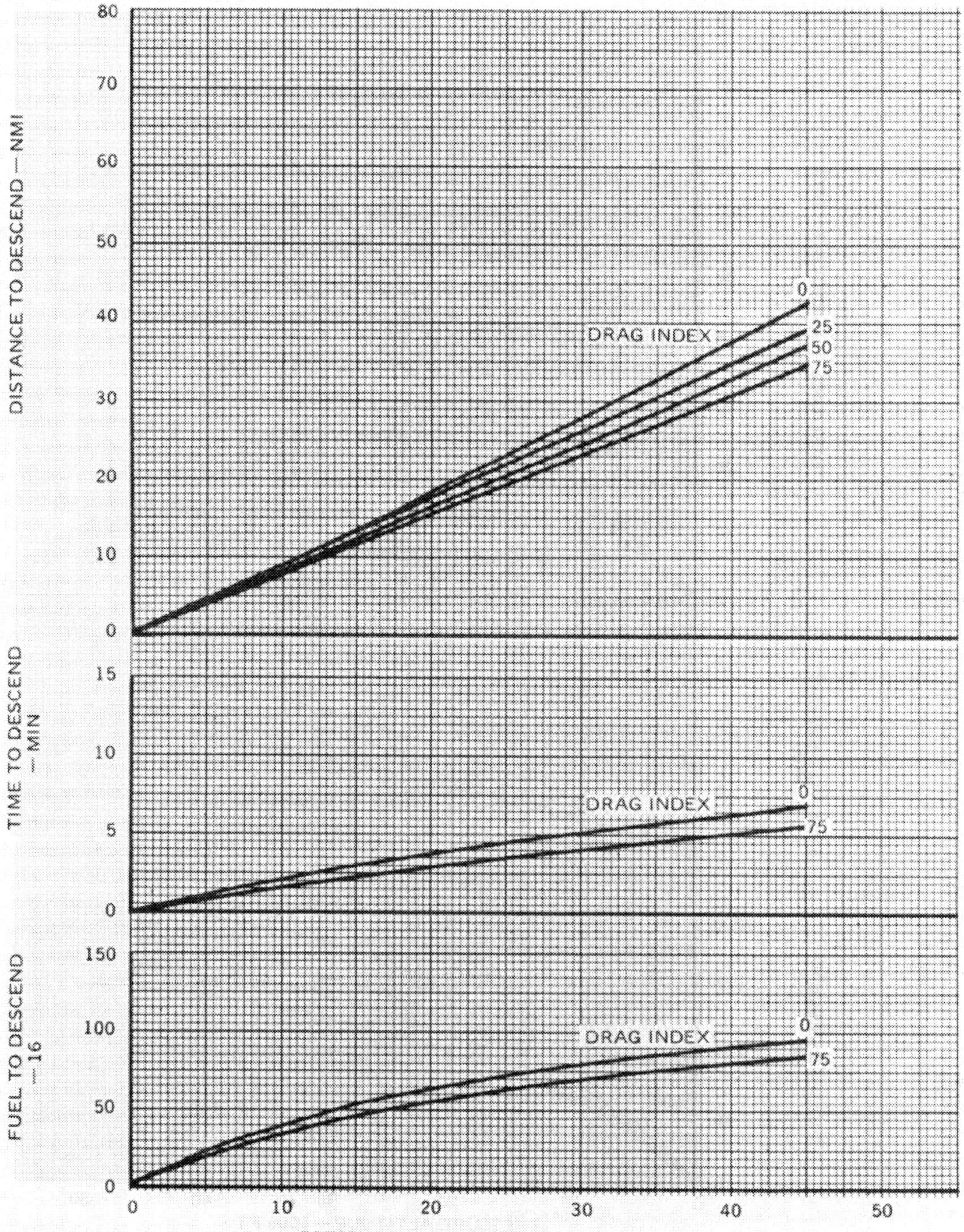


FA6-3.

TABLE OF CONTENTS



## LANDING DISTANCE.

The landing distance chart (FA7-1) shows ground distance and associated landing speeds. The ground roll distance is based on full flaps. The chart shows data for landing at the appropriate chart landing speeds, maintaining a 12-degree nose high attitude until just prior to loss of elevator authority, then lowering the nosewheel to the runway, and applying optimum braking. If the landing technique differs, landing distances will vary from those given in the charts. A 5-percent variation in touchdown speed causes approximately a 10-percent variation in landing distance. Insufficient aerodynamic or wheel braking could further increase the ground roll distance by as much as 50 percent.

### LANDING SPEED.

The landing speed chart (FA7-1) shows the normal landing final approach speed, minimum roll landing final approach speed, and touchdown speed. The landing speeds in the normal landing distance chart (FA7-1) are compatible with the normal landing pattern speed rule in section II, which indicates that final turn, final approach, and touchdown speeds be increased 1 knot for each 100 pounds of fuel above 1000 pounds of fuel remaining.

### USE OF LANDING DISTANCE CHART.

The chase-thru lines in the landing distance chart (FA7-1) show a landing with two engines operating, with a runway air temperature of 15°C at 2000 feet pressure altitude, a gross weight of 9000 pounds, and a 20-knot headwind.

## EFFECT OF RUNWAY CONDITION (RCR) ON GROUND ROLL DISTANCE.

FA7-2 provides the means of correcting the landing ground roll distance for the effect of various runway surface conditions. The corrections are shown as a function of Runway Condition Readings (RCR), which is a number indicating the degree of braking effectiveness available during the ground roll. RCR values vary from 23 to 5 for dry to icy runways. RCR of 12 is provided for a wet runway but for conditions of heavy rain or standing water lower RCR values should be selected to determine the approximate stopping distance. When wet conditions prevail the runway will be reported as wet (no RCR provided).

**CAUTION**

RCR values provide an approximation of the required stopping distance. RCR is only valid for dry or icy runways. Selection of an RCR for a wet runway does not insure a safe landing and stopping distance. If hydroplaning occurs, it is not possible to predict the actual stopping distance.

### USE OF THE CORRECTION CHART FOR RUNWAY SURFACE CONDITIONS.

Using the ground roll distance of 2700 feet for a dry, hard surfaced runway and an RCR of 12, the chase-thru lines in FA7-2 show 3600 feet required for this runway condition.

## SINGLE-ENGINE THRUST REQUIRED AND AVAILABLE.

The single-engine thrust required and available charts (FA7-3 thru FA7-5) show thrust required and available versus airspeed for go-around configuration with 0%, 60%, and 100% flaps with gear down. These charts are for several weights and temperatures from sea level to 6,000 feet and include both single-engine MAX and MIL thrusts.

### USE OF SINGLE-ENGINE THRUST REQUIRED AND AVAILABLE CHART.

Assume an airspeed of 160 KIAS, a weight of 11,000 pounds, and an ambient temperature of 30°C with MAX thrust. The chase-thru lines in FA7-3 show the thrust required is 2230 pounds for all altitudes, and the thrust available is 2810 pounds at sea level. When the pressure altitude is 2000 feet, the thrust available is 2625 pounds.

## EFFECT OF BANK ANGLE ON VERTICAL VELOCITY.

The effect of bank angle on vertical velocity charts show the climb capability of the aircraft as a function of ambient temperature, gross weight (fuel remaining), bank angle, and thrust setting. FA7-6 shows two and single-engine performance for MIL thrust settings and FA7-7 shows two and single-engine performance for MAX thrust settings. All of the charts are for landing gear extended and 60% flaps, which is the recommended flap setting for single-engine approaches. The two engine charts are for comparison purposes and are based on a 60% flap setting. The rate-of-climb determined from the charts is valid only for the recommended approach turn speed, which may be computed from the curve in the upper left corner of each chart.

### USE OF EFFECT OF BANK ANGLE ON VERTICAL VELOCITY CHARTS.

Assume a pattern altitude at 2000 feet, ambient temperature of 25°C and 1000 pounds of fuel remaining. Entering the MIL thrust, single-engine chart, FA7-6 (sheet 2), the chase-thru lines show an approach turn speed of 175 KIAS. Reentering the chart at an ambient temperature of 25°C the chase-thru lines show a climb capability of 300 fpm with a 0° bank angle. If a 30° bank angle were used in turn, the chase-thru lines show a negative climb capability of −300 fpm in the gray area. In the MAX thrust, single-engine chart FA7-7 (sheet 2), for the same conditions, the chase-thru lines show a 2300 fpm climb capability at 0° bank angle and 1700 fpm climb capability at a 30° bank angle.

## LANDING DISTANCE

DRY, HARD SURFACED RUNWAY
FULL FLAPS

MODEL: T-38A
DATE: 1 JULY 1978
DATA BASIS: ESTIMATED DATA

ENGINES: (2)J85-GE-5
FUEL GRADE: JP-4
FUEL DENSITY: 6.5 LB/US GAL

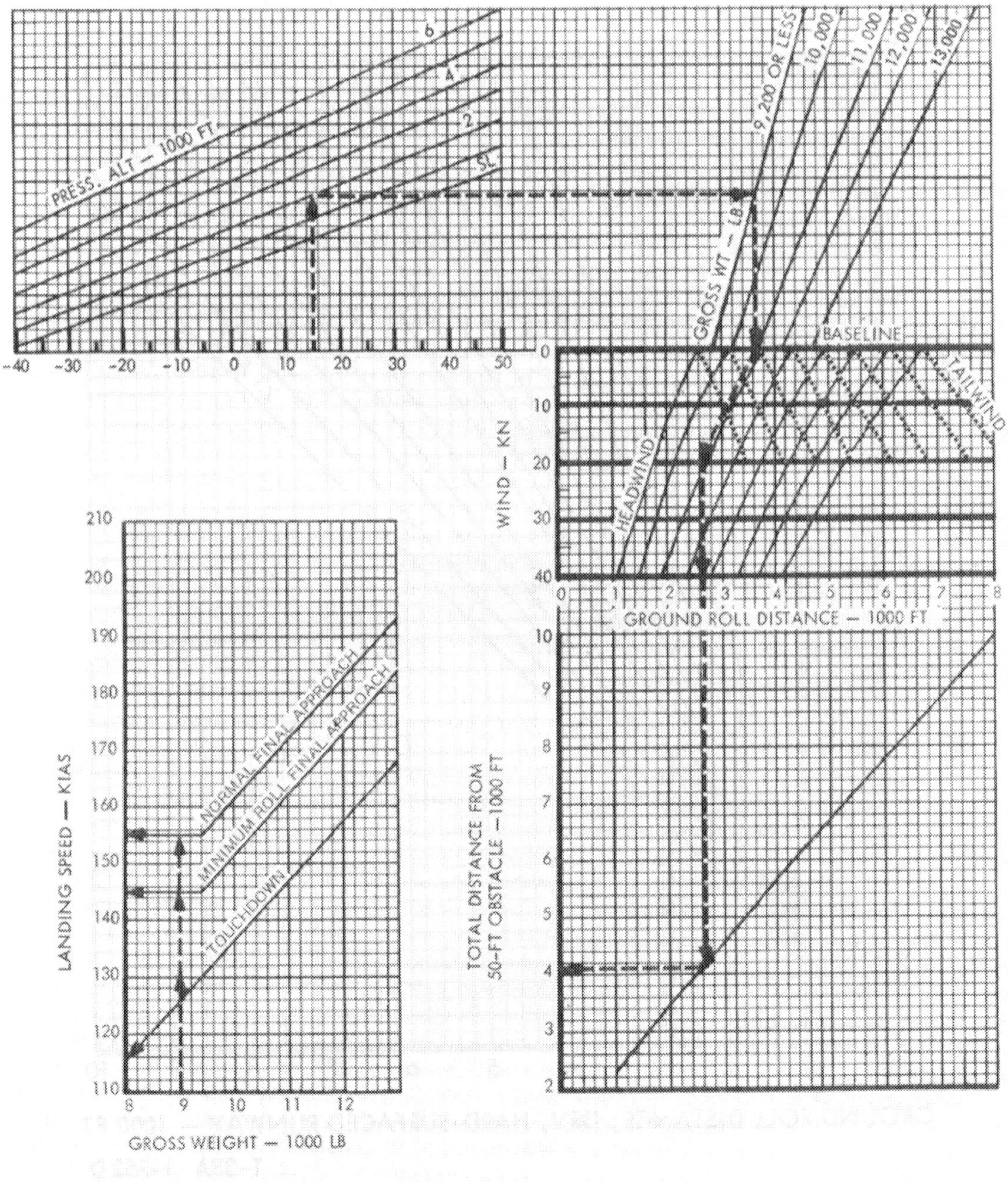


FA7-1.

T-38A 1-756D

# EFFECT OF RUNWAY CONDITION (RCR)

## ON GROUND ROLL DISTANCE
## FULL FLAPS

MODEL: T-38A
DATE: 1 AUGUST 1965
DATA BASIS: **FLIGHT TEST**

ENGINE: (2) J85-GE-5
FUEL GRADE: JP-4
FUEL DENSITY: 6.5 LB/US GAL

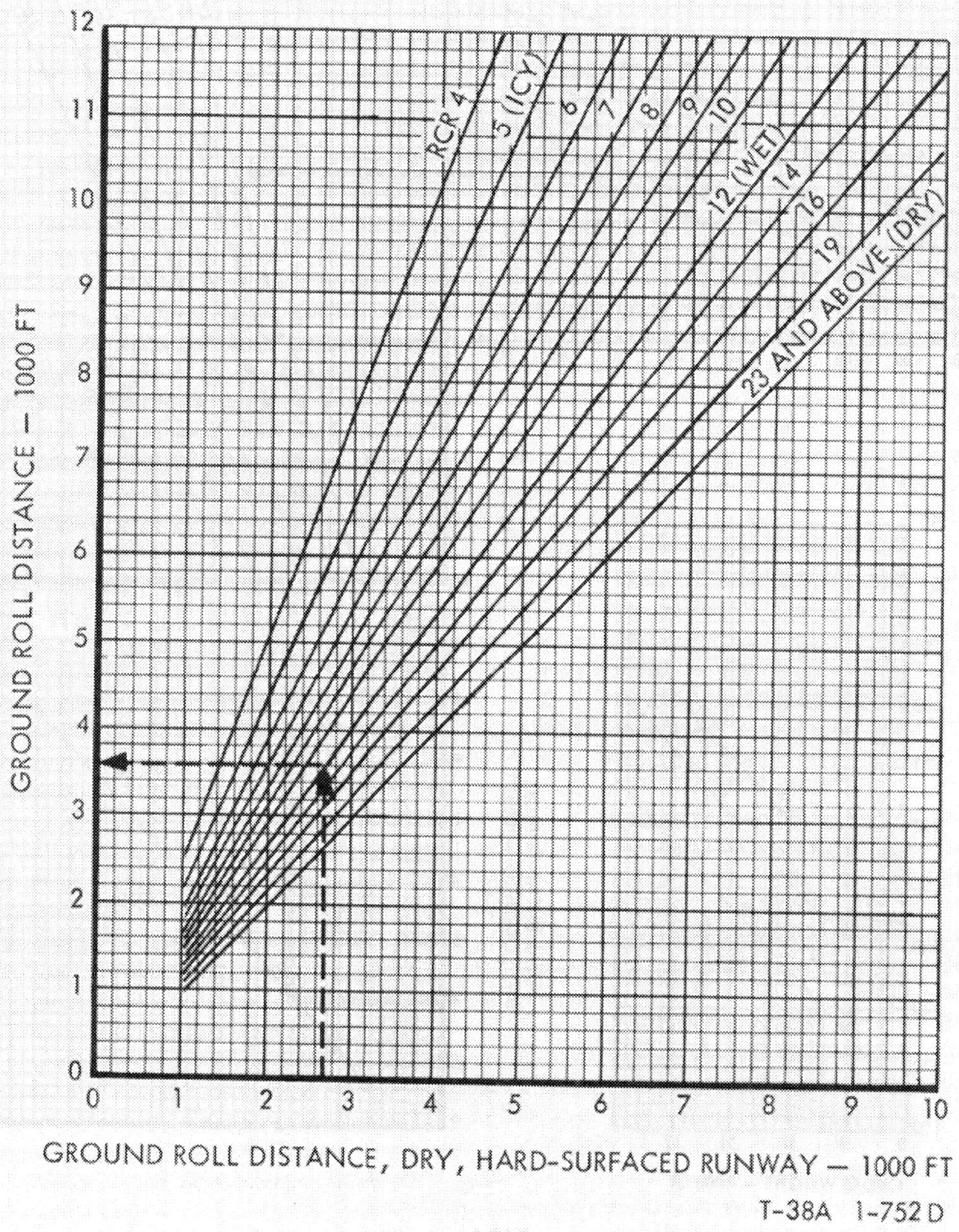


T-38A 1-752 D



FA7-2.

# SINGLE-ENGINE THRUST REQUIRED AND AVAILABLE

WITH 60% FLAPS AND GEAR DOWN
SEA LEVEL TO 6000 FEET

MODEL: T-38A
DATE: 1 JULY 1978
DATA BASIS: ESTIMATED DATE

ENGINE: (2) J85-GE-5
FUEL GRADE: JP-4
FUEL DENSITY: 6.5 LB/US GAL

*Note*

DECREASE SEA LEVEL THRUST AVAILABLE BY 3.3% FOR EACH 1000 FEET INCREASE IN PRESSURE ALTITUDE.

———— THRUST REQUIRED FOR CONSTANT SPEED LEVEL FLIGHT, ALL ALTITUDES & TEMPERATURES
— — — MAXIMUM THRUST AVAILABLE AT SEA LEVEL
- - - — MILITARY THRUST AVAILABLE AT SEA LEVEL

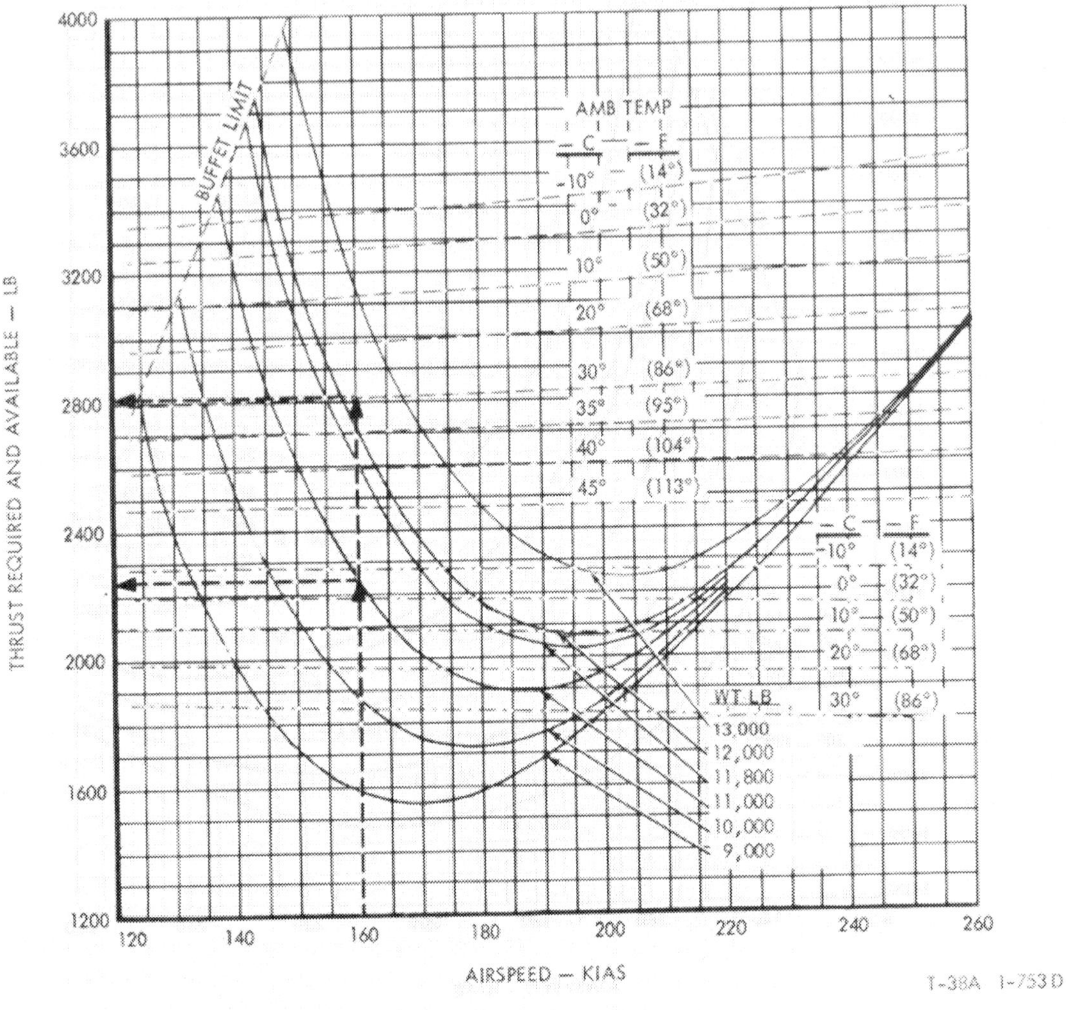


FA7-3.

## SINGLE-ENGINE THRUST REQUIRED AND AVAILABLE

**0% FLAPS, GEAR DOWN**

SEA LEVEL TO 6000 FEET

MODEL: **T-38A**
DATE: **1 JULY 1978**
DATA BASIS: **ESTIMATED DATA**

ENGINE: (2) J85-GE-5
FUEL GRADE: JP-4
FUEL DENSITY: 6.5 LB/US GAL.

*Note*

DECREASE SEA LEVEL THRUST AVAILABLE BY 3.3% FOR EACH 1000 FEET INCREASE IN PRESSURE ALTITUDE.

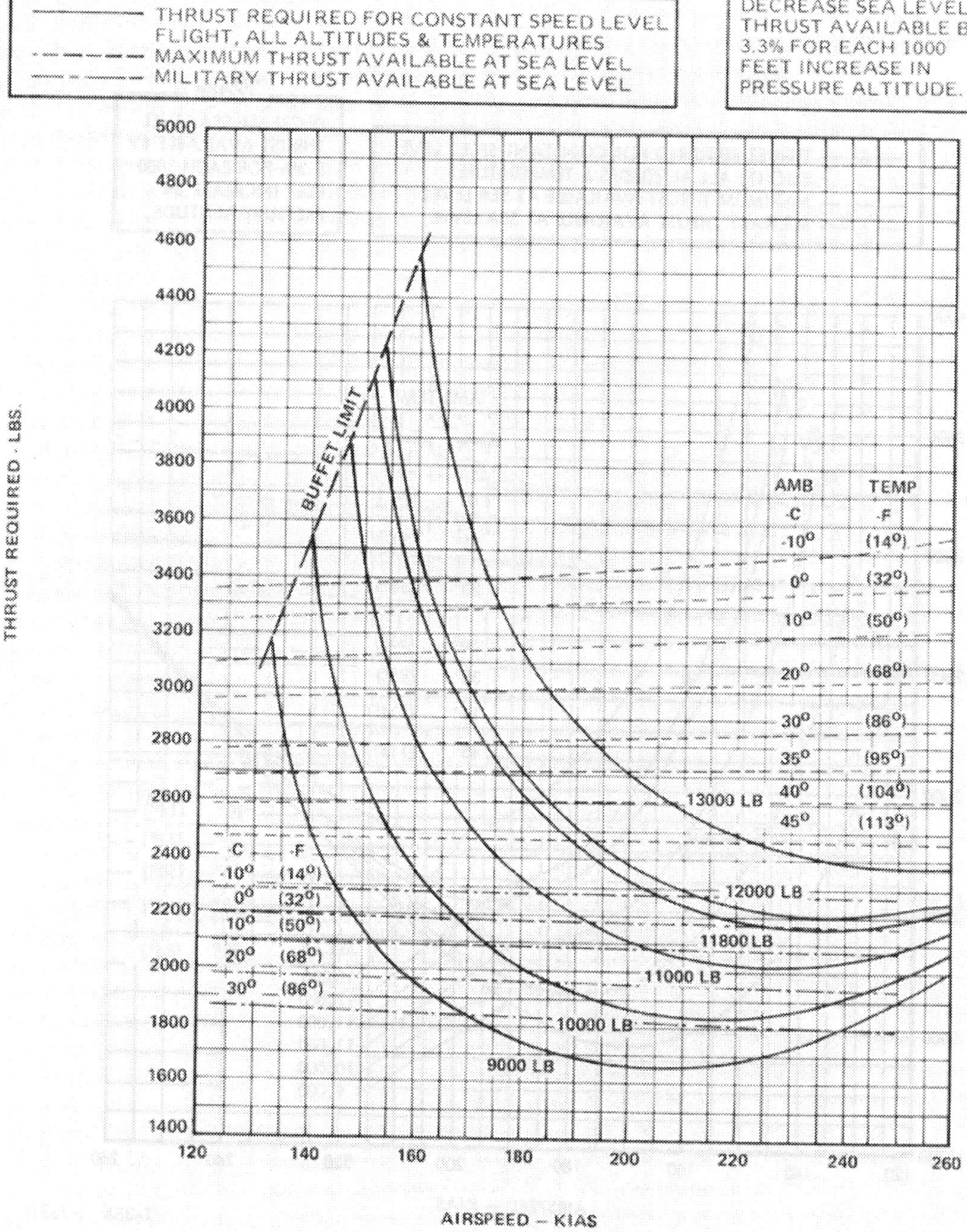


FA7-4.

SINGLE-ENGINE THRUST REQUIRED AND AVAILABLE

**100% FLAPS, GEAR DOWN**

SEA LEVEL TO 6000 FEET

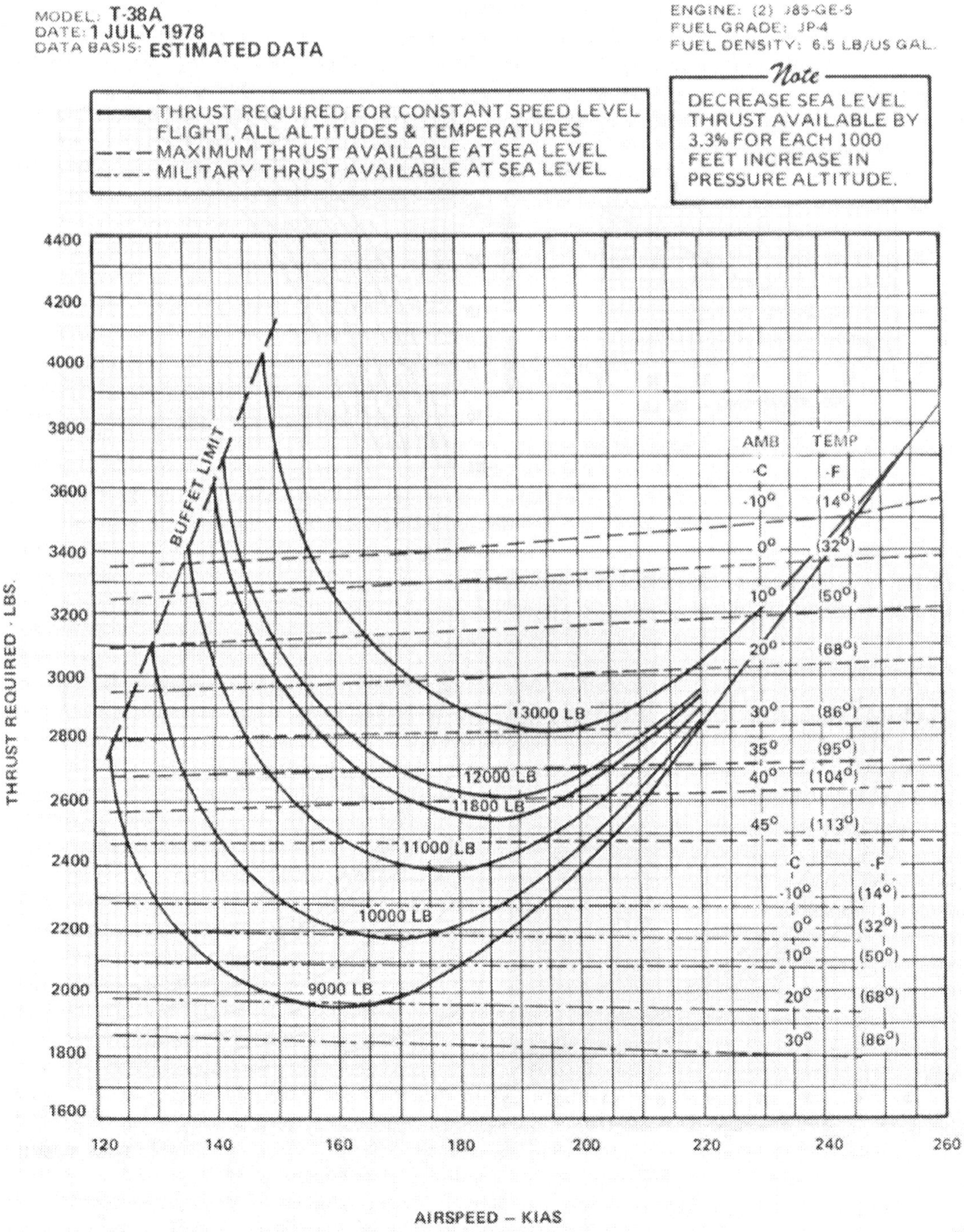


FA7-5.

## EFFECT OF BANK ANGLE ON VERTICAL VELOCITY

MIL THRUST
WITH 60% FLAPS AND GEAR DOWN

MODEL: T-38A
DATE: 1 APRIL 1969
DATA BASIS: **FLIGHT TEST**

ENGINES: (2) J85-GE-5
FUEL GRADE: JP-4
FUEL DENSITY: 6.5 LB/US GAL

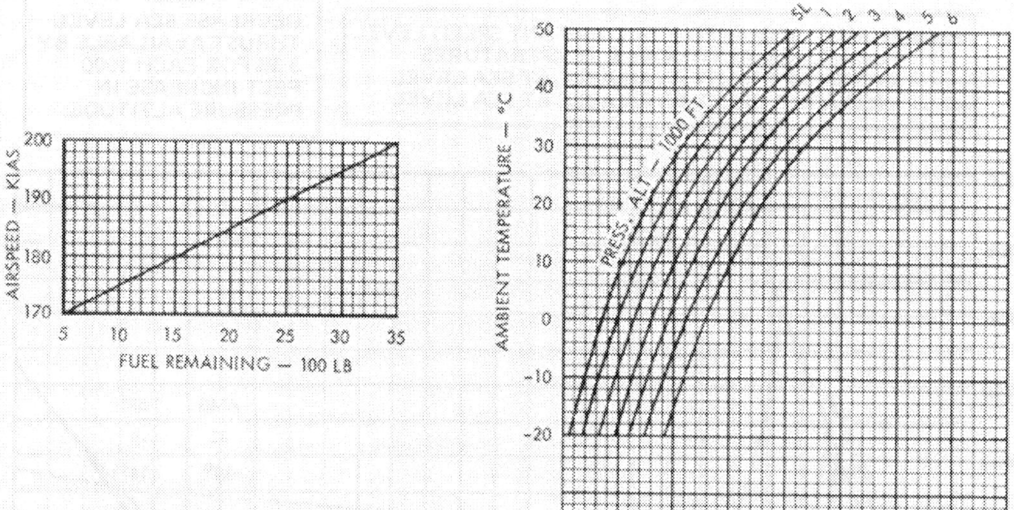


T-38A 1-754(1)

FA7-6. (Sheet 1)

## EFFECT OF BANK ANGLE ON VERTICAL VELOCITY

MIL THRUST
WITH 60% FLAPS AND GEAR DOWN

SINGLE ENGINE

MODEL: T-38A
DATE: 1 APRIL 1969
DATA BASIS: **FLIGHT TEST**

ENGINES: (2) J85-GE-5
FUEL GRADE: JP-4
FUEL DENSITY: 6.5 LB/US GAL

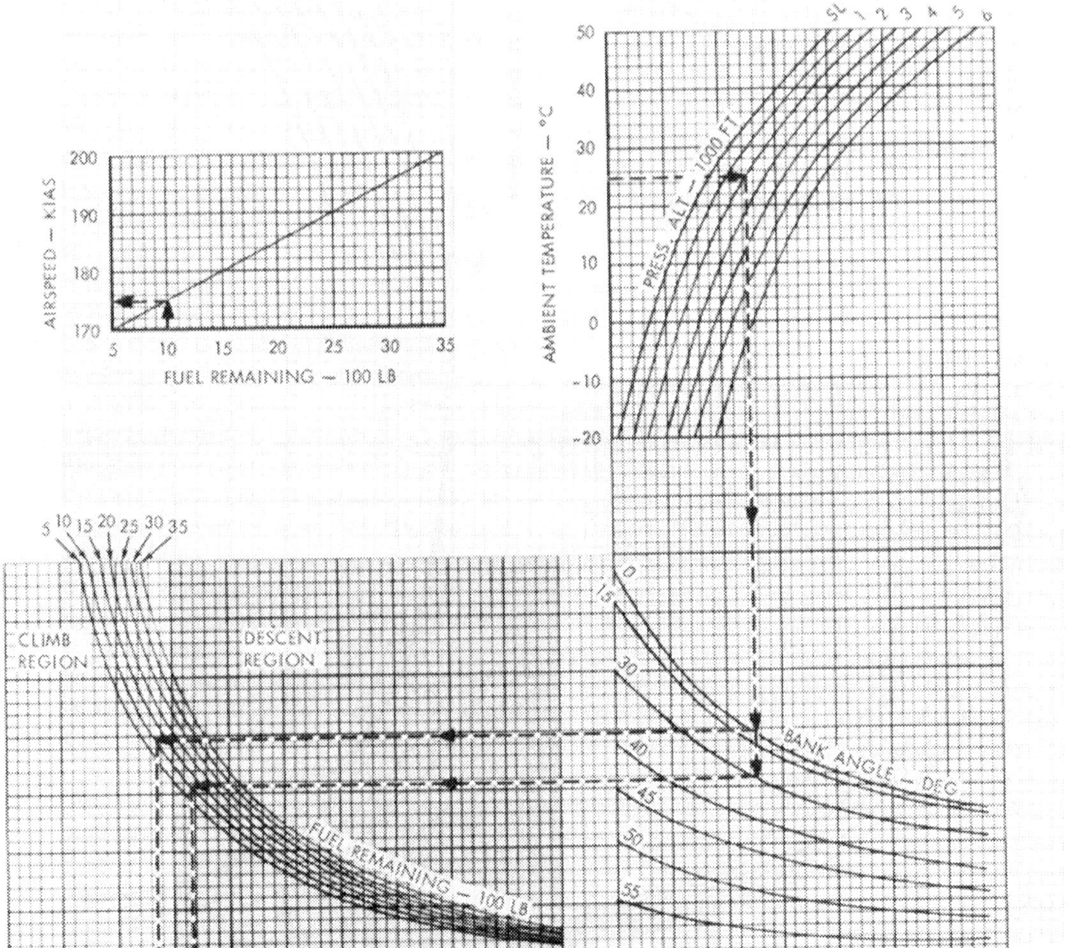


FA7-6. (Sheet 2)

## EFFECT OF BANK ANGLE ON VERTICAL VELOCITY

MAX THRUST
WITH 60% FLAPS AND GEAR DOWN

MODEL: T-38A
DATE: 1 APRIL 1969
DATA BASIS: **FLIGHT TEST**

ENGINES: (2) J85-GE-5
FUEL GRADE: JP-4
FUEL DENSITY: 6.5 LB/US GAL

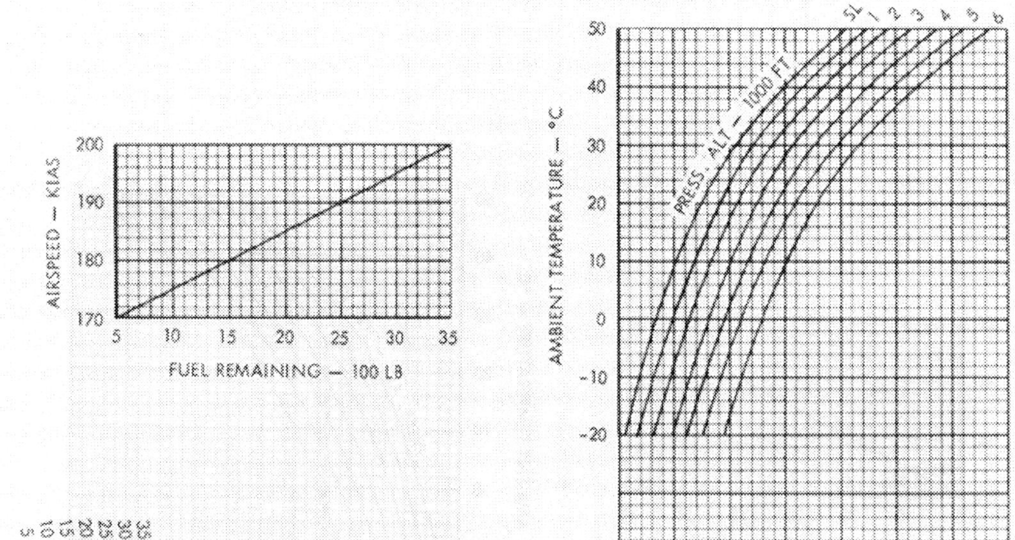


FA7-7. (Sheet 1)

# EFFECT OF BANK ANGLE ON VERTICAL VELOCITY

MAX THRUST
WITH 60% FLAPS AND GEAR DOWN

MODEL: T-38A
DATE: 1 APRIL 1969
DATA BASIS: **FLIGHT TEST**

SINGLE ENGINE

ENGINES: (2) J85-GE-5
FUEL GRADE: JP-4
FUEL DENSITY: 6.5 LB/US GAL

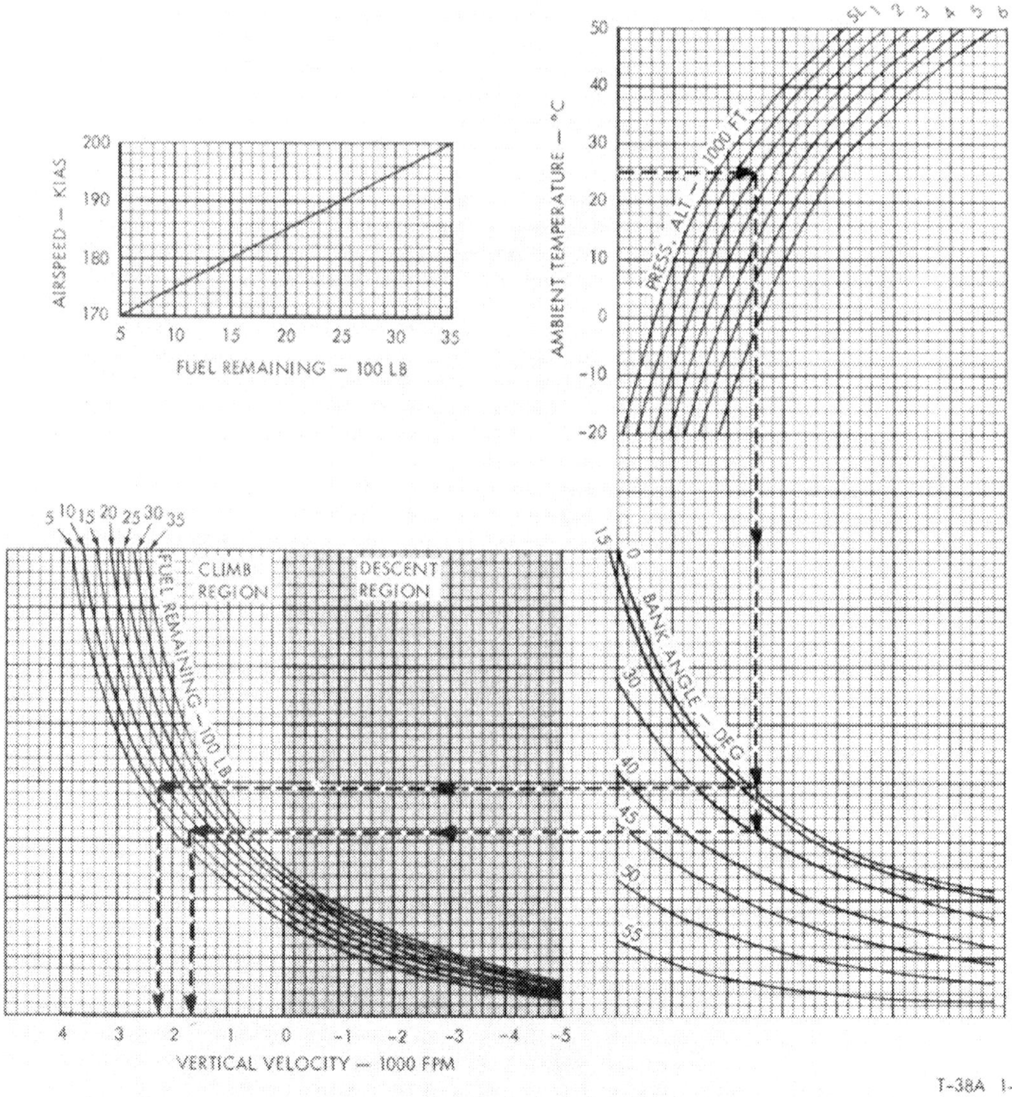


T-38A 1-755(2)

FA7-7. (Sheet 2)

# PART 8 MISSION PLANNING

TABLE OF CONTENTS



## PURPOSE OF MISSION PLANNING.

Mission planning can be termed preflight planning. The purpose of preflight planning is to obtain optimum performance from the aircraft for any specific mission. Optimum performance will vary, for example, from maximum time on station to maximum radius with no time on station. Exact requirements will vary, depending upon the types of missions to be flown.

## MISSION PLANNING SAMPLE PROBLEM.

The following problem is an exercise in the use of the performance charts. It is not intended to reflect actual or proposed missions employing this aircraft on a typical cross country flight.

### FLIGHT PLAN DATA.

A mission profile is to be flown, assuming the following conditions:

1. Takeoff data.

   a. Takeoff weight (solo) . . . . 11,800 lb.
   b. Wind . . . . . . . 10-knot headwind.
   c. Runway temperature . . . . . . 15°C.
   d. Pressure altitude . . . . . . . 4000 ft.
   e. Runway . . . . . . . . . . 7000 ft.
   f. RCR . . . . . . . . . . . . . 12.

2. Climb data to 35,000 ft.

   a. Temperature deviation from standard +10°C.
   b. Wind. . . . . . . . . 15-knot headwind.

3. 35,000 ft. cruise data.

   a. Temperature. . . . . . . . . . −46°C.
   b. Wind . . . . . . . . 50-knot headwind.
   c. Speed . . . . . . . . . . . Optimum.

4. Descent data to 3000 ft.

   a. Temperature deviation from standard . . . . . . . . . . . Zero.
   b. Wind . . . . . . . . 15-knot tailwind.

5. Enter pattern 1000 ft above terrain with 1000 lb. fuel reserve.

6. Landing data.

   a. Landing weight. . . . . . . . 9000 lb.
   b. Wind . . . . . . . 20-knot headwind.
   c. Temperature . . . . . . . . . . 15°C.
   d. Pressure altitude . . . . . . . 2000 ft.
   e. Runway length . . . . . . . 7000 ft.
   f. RCR . . . . . . . . . . . . . . 12.

### TAKEOFF.

1. MAX thrust takeoff factor (FA2-2) . . . 3.45.
2. Takeoff speed (FA2-3) . . . . 154 KIAS.
3. Takeoff distance (FA2-4) . . . . 3050 ft.
4. Critical field length (FA2-6).
   a. RCR = 23 . . . . . . . . . . 5800 ft.
   b. RCR = 12 . . . . . . . . . . 6500 ft.

5. Critical engine failure speed (FA2-7).

   a. RCR = 23 . . . . . . . . . 132 KIAS.
   b. RCR = 12 . . . . . . . . . 112 KIAS.

6. Acceleration check speed at 1500 feet from brake release.

   a. Normal (FA2-10) . . . . . . . 110 KIAS.
   b. Minimum (7000 — 6500) = 500,

   $110 - (\frac{500}{1000} \times 3)$. . . . . . . 108 KIAS.

7. Single-engine takeoff speed (FA2-3). (154 + 8) . . . . . . . . . . . 162 KIAS.

8. Refusal speed (FA2-7).

   a. RCR = 23 . . . . . . . . . . 143 KIAS.
   b. RCR = 12 . . . . . . . . . . 116 KIAS.

## INITIAL CLIMB FROM 4000 FT TO FLIGHT LEVEL 350.

(Using MIL Thrust Climb Chart FA3-1)

1. Aircraft weight at start of climb is 11,800 lb minus the allowance for taxi, takeoff, and acceleration to climb speed (11,800 — 300). . . . . . . . . . . . 11,500 lb.

2. Obtain time to climb, fuel to climb, and climb range.
   Time (8.2 — 0.6) . . . . . . 7.6 min.
   Fuel (565 — 55) . . . . . . 510 lb.
   Range (67 — 3) . . . . . . 64 nmi.

3. Compute distance lost due to headwind (15 X 7.6/60) . . . . . . . . . . . 2 nmi.

4. Adjusted climb range (64 — 2) . . . 62 nmi.

5. Weight at level-off (11,500 — 510) . . . . . . . . 10,990 lb.

## PENETRATION DESCENT TO 3000 FT, SPEED BRAKE OPENED (280 KCAS, 80% RPM).

1. Obtain time, fuel, and no wind range from FA6-3.
   Fuel (85 — 15) . . . . . . . 70 lb.
   Time (5.6 — 0.6) . . . . . 5.0 min.
   Range (32 — 2) . . . . . 30 nmi.

2. Compute distance gained due to tailwind (15X 5.0/60) . . . . . . . . . . 1 nmi.

3. Compute ground range (30 + 1) . . 31 nmi.

4. Weight at end of descent (8010 + 1000) . . . . . . . . . . . 9010 lb.

## AVERAGE GROSS WEIGHT.

1. Weight at beginning of cruise . . 10,990 lb.

2. Weight at end of cruise (9010 + 70) . . . . . . . . . . . 9080 lb.

3. Compute fuel for cruise (10,990 — 9080) . . . . . . . . . . 1910 lb.

4. Average weight (10,990 — 1/2 X 1910) . . . . 10,035 lb.

## CRUISE AT FLIGHT LEVEL 350.

(Using Cruise chart FA4-1).

1. Maximum range mach number . . . . 0.83

2. Basic reference number . . . . . . . . . 4

3. Nautical miles per pound . . . . . . . 0.338

4. True airspeed . . . . . . . . . . 485 kn.

5. Fuel flow lb/hr/eng. . . . . . . . . . . 715

6. Groundspeed (485 — 50) . . . . . 435 kn.

7. Time $(\frac{1980 \times 60}{715 \times 2})$ . . . . . . . . 83 min.

8. Ground distance $(83 \times \frac{435}{60})$ . . . 602 nmi.

## CRUISE AT FLIGHT LEVEL 350.

(Using Constant Altitude Cruise chart FA4-3).

1. True mach number . . . . . . . . . 0.83

2. True airspeed . . . . . . . . . . 485 kn.

3. True groundspeed . . . . . . . . 435 kn.

4. Fuel flow lb/hr/eng . . . . . . . . . 715

5. Time $(\frac{1980 \times 60}{715 \times 2})$ . . . . . . . . 83 min.

6. Ground distance $(83 \times \frac{435}{60})$ . . . 602 nmi.

## LANDING.

(Using Normal Landing Distance chart FA7-1).

1. Final turn speed . . . . . . . 175 KIAS.

2. Normal landing final approach speed . . . . . . . 155 KIAS.

3. Minimum roll landing final approach speed . . . . . . . 145 KIAS.

4. Touchdown speed . . . . . . 125 KIAS.

5. Ground roll distance
   a. RCR = 23 . . . . . . . . . . . 2700 ft.
   b. RCR = 12 (FA7-2) . . . . . . 3600 ft.

## MISSION SUMMARY.

1. Total time . . . . . . . . . . . 95.6 min.
2. Total range . . . . . . . . . . . 695 nmi.

## TAKEOFF AND LANDING DATA CARD.

The takeoff and landing data card is included in the Flight Crew Checklist normal procedures. The takeoff and landing data was computed during mission planning from Part 2 and Part 7 respectively. The landing weight for immediately after takeoff is the takeoff gross weight less an average fuel allowance of 300 lb for takeoff and go-around. Landing immediately after takeoff for the conditions stated in the mission planning takeoff data is computed as follows:

## LANDING (Immediately After Takeoff).

1. Landing gross weight . . . . . . 11,500 lb.
2. Final turn speed . . . . . . . . 198 KIAS.
3. Normal landing final approach speed (FA7-1) . . . . 178 KIAS.
4. Touchdown speed (FA7-1) . . . 153 KIAS.
5. Ground roll distance.
   a. RCR = 23 (FA7-1) . . . . . . . 5100 ft.
   b. RCR = 12 (FA7-2) . . . . . . . 6800 ft.

The takeoff and landing information for mission planning is entered on the data card as a ready reference for review prior to takeoff and landing as shown in figure FA8-1.

# TAKEOFF & LANDING DATA CARD

## CONDITIONS

| | |
|---|---|
| RUNWAY LENGTH | 7000 FT |
| WIND COMPONENT | 10 HW KN |
| RUNWAY TEMPERATURE | 15 °C |
| PRESSURE ALTITUDE | 4000 FT |

## TAKEOFF

| | | |
|---|---|---|
| ACCELERATION CHECK | 108 KIAS | 1500 FT |
| CRITICAL ENGINE FAILURE SPEED | | 112 KIAS |
| REFUSAL SPEED | | 116 KIAS |
| SINGLE-ENGINE TAKEOFF SPEED | | 162 KIAS |
| TAKEOFF DISTANCE | | 3050 FT |

## LANDING

| | IMMEDIATELY AFTER TAKEOFF | FINAL |
|---|---|---|
| FINAL TURN SPEED | 198 KIAS | 175 KIAS |
| FINAL APPROACH SPEED | 178 KIAS | 155 KIAS |
| TOUCHDOWN SPEED | 153 KIAS | 125 KIAS |
| GROUND ROLL: DRY | 5100 FT | 2700 FT |
| GROUND ROLL: WET | 6800 FT | 3600 FT |

T-38A 1-751 D

FA8-1.

# ALPHABETICAL INDEX

T-38A 1-105

## A



(Boldface Type Denotes Illustration)

# B



# C



**(Boldface Type Denotes Illustration)**

(Boldface Type Denotes Illustration)

(Boldface Type Denotes Illustration)

**(Boldface Type Denotes Illustration)**

(Boldface Type Denotes Illustration)

(Boldface Type Denotes Illustration)

## U



## V



## W



(Boldface Type Denotes Illustration)

HISTORIC SUBMARINE, DEEP SEA DIVING AND
RESEARCH DVDs NOW AVAILABLE FROM
WWW.PERISCOPEFILM.COM
Historic Submarine and Documentary and Training Films DVD
The Silent Service in World War II
Historic Submarine and Documentary and Training Films DVD
The Nuclear Pioneers
Historic Submarine Documentary and Training Films DVD
The Cutting Edge
Historic Submarine Documentary and Training Films DVD
The Cold War Below
FEATURING
HISTORIC FILMS
RESCUED FROM
U.S. NAVY
ARCHIVES!
Historic Submarine Documentary and Training Films DVD
The Deep Horizon
HISTORIC SUBMARINE DOCUMENTARY AND TRAINING FILMS DVD
SUBMARINES, FROGMEN AND DIVERS
DVD

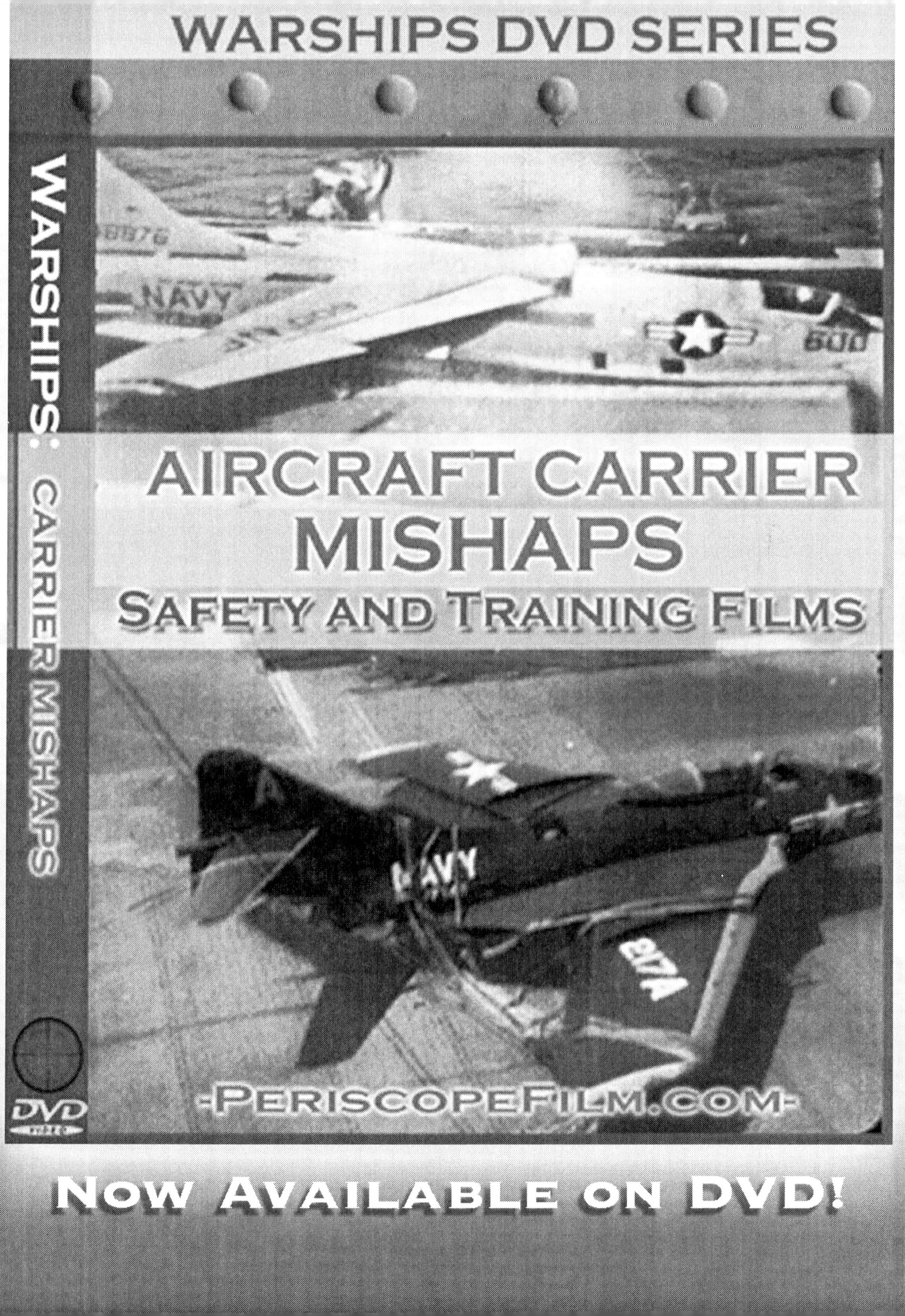
WARSHIPS DVD SERIES
WARSHIPS: CARRIER MISHAPS
NAVY
AIRCRAFT CARRIER
MISHAPS
SAFETY AND TRAINING FILMS
NAVY
-PERISCOPEFILM.COM-
DVD
NOW AVAILABLE ON DVD!

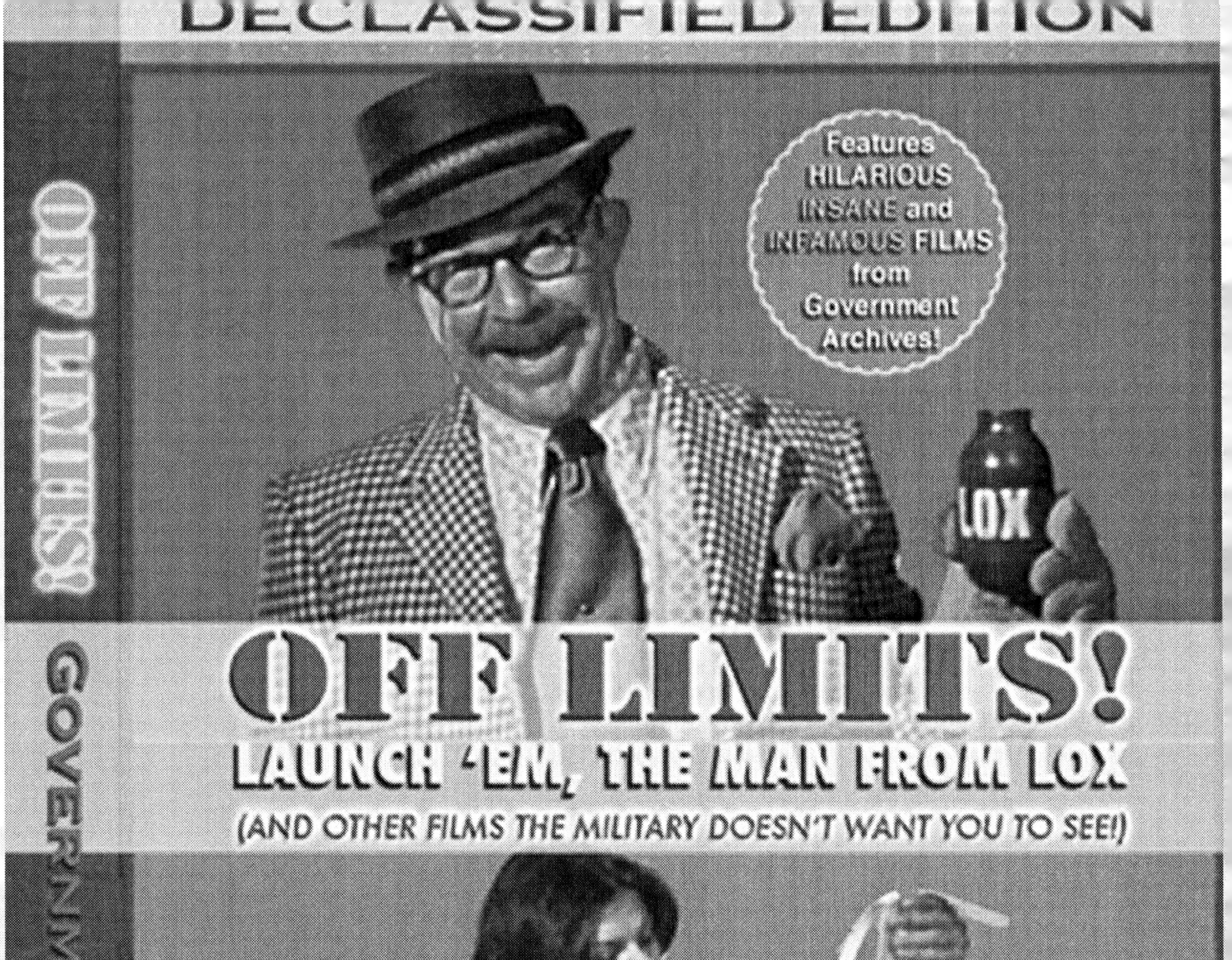
DECLASSIFIED EDITION
Features HILARIOUS INSANE and INFAMOUS FILMS from Government Archives!
LOX
OFF LIMITS!
LAUNCH 'EM, THE MAN FROM LOX
(AND OTHER FILMS THE MILITARY DOESN'T WANT YOU TO SEE!)
OFF LIMITS!
GOVERNMENT FILMS
DVD

EPIC BATTLES
OF WWII
WWII BATTLE DVD SERIES
WWII BATTLES: OKINAWA AND IWO JIMA
FEATURES RARE ORIGINAL WWII COLOR FILMS!
OKINAWA AND IWO JIMA
WWII PACIFIC CAMPAIGN • 1945
WWW.PERISCOPEFILM.COM
EPIC BATTLE DVD SERIES
EPIC BATTLES: D-DAY TO V-E DAY 1944-45
FEATURES RARE WWII FILMS FROM ARMY ARCHIVES!
THE TRUE GLORY
D-DAY TO V-E DAY
WWW.PERISCOPEFILM.COM
EPIC BATTLE DVD SERIES
EPIC BATTLES: ITALIAN CAMPAIGN 1943-45
FEATURES RARE WWII FILMS FROM ARMY ARCHIVES!
THE ROAD TO ROME
ITALIAN CAMPAIGN 1943-45
ANZIO
CASSINO
WWW.PERISCOPEFILM.COM
NOW AVAILABLE ON DVD!

WARSHIPS DVD SERIES
WARSHIPS DVD SERIES
WARSHIPS: LEYTE GULF TO TOKYO BAY
LEYTE GULF TO TOKYO BAY
THE AIRCRAFT CARRIER WAR
1944-1945
WARSHIPS DVD VOL. 6
WARSHIPS DVD SERIES
WARSHIPS: THE FIGHTING LADIES
FEATURES RARE ORIGINAL WWII COLOR FILMS!
THE FIGHTING LADIES:
YORKTOWN AND HORNET
THE AIRCRAFT CARRIER WAR
1942-1944
WARSHIPS DVD VOL. 5
WARSHIPS DVD SERIES
WARSHIPS: KRIEGSMARINE 1914-45
KRIEGSMARINE
GERMAN NAVY 1914-45
WARSHIPS DVD VOL. 3
NOW AVAILABLE!

ISBN# 978-1-935327-64-6 1-935327-64-X

www.PeriscopeFilm.com